Science and Technology for Rural Development

Science and Technology for Rural Development

— *Editors* —
Seetha I. Wickremasinghe
Ma. Josefina P. Abilay
Jayasamara Gunaratne

CENTRE FOR SCIENCE & TECHNOLOGY OF THE
NON-ALIGNED AND OTHER DEVELOPING COUNTRIES
(NAM S&T CENTRE)

2012
DAYA PUBLISHING HOUSE®
New Delhi - 110 002

ISBN 9789351241713

Centre for Science and Technology of the Non-Aligned and Other Developing Countries (NAM S&T Centre)
Core-6A, 2nd Floor, India Habitat Centre, Lodhi Road,
New Delhi-110 003 (India)
Phone: +91-11-24644974, 24645134, Fax: +91-11-24644973
E-mail: namstct@gmail.com
Website: www.namstct.org

Published by : **Daya Publishing House®**
A Division of
Astral International Pvt. Ltd.
– ISO 9001:2008 Certified Company –
4760-61/23, Ansari Road, Darya Ganj,
New Delhi - 110 002
Phone: 23245578, 23244987
Fax: (011) 23260116
e-mail : dayabooks@vsnl.com website : www.dayabooks.com

Laser Typesetting : **Classic Computer Services**
Delhi - 110 035

Printed at : **Chawla Offset Printers**
Delhi - 110 052

PRINTED IN INDIA

Republic of the Philippines
DEPTT. OF SCIENCE & TECH.,
OFFICE OF THE SECRETARY

2/F DOST Main Building, General Santos Avenue
Bicutan, Taguig City
Philippines

Tel. Nos. (63-2) 837-20-71 to 82
Fax. no. (63-2) 837-29-37
URL: http://www.dost.gov.ph

Foreword

In the present day rural economy of the developing countries, rural micro-enterprises have come to play a significant role in alleviating socio-economic problems through applications of science and technology with increased production and generation of employment in the villages. As majority of the population in the developing countries live in rural areas, this has become all the more crucial to counter the national economic and social policies which are often urban centric and do not necessarily help in alleviating the problems of rural poverty. The rural household incomes are strongly dependent on the vagaries of agriculture, but these may be complemented by diversifying and creating additional on- and off-farm incomes through small enterprises with accelerated applications of science and technology to upgrade their skills and innovative product development.

Most of the rural areas in developing countries already have a favourable culture and tradition for rural enterprise development, marketing and technical skills upgrading and formation of collective action groups among households and communities, which can motivate and empower village people to make their lives more secure and prosperous. However, most of the rural enterprises, largely still follow the traditional and primitive technologies and the application of modern technologies are yet to be harnessed for efficient and productive use of the material, financial and skill resources of rural people. This, therefore, has become important to identify the relevant technologies which have the potential for increased employment opportunities and livelihood for the rural people. Developing institutional mechanisms to encourage the growth of skill and knowledge-based jobs through the creation of strong linkages between the academe, industry and vocational training institutions is an ideal solution to the problems of unemployment and poverty among the villagers.

I am happy to note that in order to address the above problems of rural areas in the developing countries and to promote South–South cooperation among the developing countries in this direction, the Centre for Science and Technology of the Non-Aligned and Other Developing Countries (NAM S&T Centre) organised an international workshop at Hanoi, Vietnam on 'Application of Science & Technology for Occupational Villages Development' on 2–5 August 2010 jointly with the International Cooperation Department of the Ministry of Science & Technology (MOST), Government of Vietnam and the Centre for Regional Research and Development (CRD) of Vietnam. Based on the deliberations of this workshop, the NAM S&T Centre has brought out this book that includes 17 papers written by various scholars and scientist from the developing countries. These articles primarily dwell upon the applications of science and technology for rural industrialization, employment generation and income enhancement in villages in the developing countries. I am happy that this publication is yet another step in promoting the South–South Cooperation.

Hon. Mario G. Montejo

Secretary,

Department of Science and Technology (DOST)

Preface

The decision makers in developing countries need to pay attention mostly to the rural development programmes since majority of the population of their countries consists of the rural community. In the era of globalization, many developing countries aimed at replicating the industrial development success that was shown by the so-called newly industrialised economies in Asia during the early 1980s. However, with the rules and regulations imposed by the World Trade Organization (WTO), the developing countries faced difficulties in terms of its contribution to the industrial production, exports, employment and creation of an entrepreneurial base in the country. As a result, many developing countries initiated societal programmes that were aimed at deriving optimum benefits by careful management of resources, S&T experience and skills of the rural people in their countries. Through these programmes, it has been possible to bring in specific intervention packages for the benefits of the rural areas, particularly, the village community. Such attempts of various governments have provided unique opportunity of pursuing action oriented research and development work by interested scientists who are committed to work in the rural development. It has also helped in local networking and coordination amongst actors in the government, semi-government, private sector and the Non-Governmental Organizations (NGOs). Recognizing the importance of rural development and increasing income of the village community, the Centre for Science and Technology of the Non-Aligned and Other Developing Countries (NAM S&T Centre) in New Delhi, India organized a 3 day International Workshop on " Application of Science and Technology for occupational villages development" at Hanoi, Vietnam during 2-5 August 2010 jointly with the International Cooperation Department of the Ministry of Science and Technology (MOST), Government of Vietnam and the Centre for Regional Research and Development (CRRD) of Vietnam. Several eminent personnel such as scientists, technologists, policy makers, scholars and experts in the rural development field attended the Workshop. This book is an extended and upgraded version of compilation of the papers presented at the Workshop.

The purpose of the book is to share perspectives and experience of developing countries that use science and technology for rural development and its sustainability. The subjects discussed in the book are divided into three sections. The *Section I* includes six papers that highlight various approaches and perspectives of science and technology that can be used for rural industrialization. The *Section II* has eight contributions that discuss case studies of varied approaches towards rural industrialization. Further, three more papers that review the government policies and initiatives of different developing countries in promoting rural industrialization are categorized under the *Section III* of the book.

In Section I, the first paper by Ma. Josefina P. Abilay highlights the attempts of science and technology (S&T) interventions that are transferred mostly to the community organizations and the agro-based micro, small and medium enterprises (MSMEs), with the aim of promoting rural industrialization in Mimaropa region in the Philippines. The roles of DOST-MIMAROPA, an ISO 9001:2008-certified agency and the newest of 16 administrative regional offices (apart from one autonomous region) in the country are also discussed in detail. M. Bandyopadhyay next elaborates policies, strategies and goals of India for rural industrialization and employment generation through science and technology. However, he concludes that even though considerable work has been done by the field level institutions, efforts are confined largely to relatively urbanized centres and hence, the rate of diffusion of new technologies to rural enterprises has been slow. While discussing the role of science and technology in the rural, small and medium scale enterprises, H. Hajihoseini emphasizes on the medium and small business plants and factories in Iran that have not been able to launch R&D centres or even a laboratory to test their products and raw materials towards technology development. He had highlighted some policy insights in the said direction. According to Syed Nawazish Ali Shah's paper it is noted that even though bestowed with abundant natural resources of all kinds, Pakistan is still an energy deficient country and therefore, the people as well as the industries face severe energy shortage problems. As mentioned, the Aik Hunar Aik Nagar (AHAN) programme based on the "One Village One Product" concept aims to alleviate poverty in the rural and peri-urban areas of Pakistan by supporting rural, micro and small enterprises. Dilip Subba mentions in his paper about the strategies of Nepal towards the use of science and technology to develop village occupation and further states that there is a great opportunity to develop the village occupations, since there are many resources, namely agriculture, forest, natural beauty, rivers and rivulettes. However, he emphasizes on the fact that Nepal belongs to the group of countries that are scientifically lagging behind. The last paper in the Section I by Aye Aye Khai and Mya Mya Oo discusses the current status of R&D activities conducted in Myanmar with the aim of rural community development and highlights that various approaches such as biogas plants, bio- fertilizer, bamboo vinegar, chitosan, natural liquid fertilizer etc., are being tried out. In addition, it draws examples from various on-going projects such as the home-made fermented fish sausage, fruit juice, anti-malarial drug using medicinal plant extracts and the cultivation of *Musa* spp. through plant tissue culture.

The papers in Section I indicate the diversity of experiences from developing countries in the use of science and technology interventions with special reference to micro, small and medium scale industrialization. While some countries have achieved a certain degree of success others are still in the initial stages of application of science and technology towards rural development. Sustainability of occupational villages' development approach seems to have considerable barriers that are of importance and need careful attention.

In Section II, the first paper by V. Madhava Rao, highlights a complementary approach towards rural development has achieved by designing and development of a village Geographical Information System (GIS) for the Rajasthan State. A modern technology such as the (GIS) proves to be a decision supporting tool at the local level for judicious and rational decisions, by integrating all knowledge bases at the village level. The next paper on renewable energy technologies and rural development in Tanzania shows the impacts of renewable energy technologies to rural development by presenting two case studies. It shows how the poverty status of a rural community can be changed with the application of appropriate technology. The author of this paper, Omar Jjumanne Bakari states that the energy technology transfer appears to be a major challenge due to the price factor in relation to the purchase power in the rural community. It further recommends the need of government intervention on creating awareness, capacity building and involvement of rural communities in planning and implementation of different energy technology development projects in the rural areas.

As highlighted by Anjela C. Nwanya *et al.,* Nigeria has continued to lead in developing new ways of using S and T to address the rural industrialization drive of the government while the impact of science and technology in rural based economic enterprises is progressively improving. The Co-operative Societies of Nigeria, equivalent to the occupational village development concept, have been the most effective channel of disseminating innovations in S&T to rural dwellers. Jayasamara Gunaratne elaborates a Sri Lankan attempt towards rural development by taking technology to the village through Vidatha Resource Centre Concept, which has enormous opportunities to uplift the lives of the rural people. This project has become a major development programme at the rural level, providing job opportunities for the unemployed and an extra income for poor families, not only in increasing their monthly income, but also in earning foreign currency, when products are launched in the international markets. Highlighting the issue of the environment protection for sustainable development of occupational villages, Nguyen Van Mien states that the Vietnamese Occupational Village Association has performed a bridging role by transferring the research results and environment pollution treatment technologies developed by researchers to the villages for application in large scale, along with other functional authorities of the research and implement projects. Through such approaches, the occupational villages, in future, will develop in a sustainable manner and promote their economic and cultural values, while contributing to the development of the rural area as well as the whole country. The approach of Indian Institute of Science (IISc), Bangalore, towards rainwater harvesting technology to generate sustainable solutions for drinking water while creating local employment is another attempt to ensure rural development as pointed out by A.R. Shivakumar. According to him, a chain of practically educating elders through young children has created a social change in the villages under consideration while maintaining a sustainable method of roof top rainwater harvesting in each individual house too. An interesting paper on "Spreading the green aura of medicinal and aromatic plants for improving rural economy" by Anil Kumar Singh in which he states that India has clearly established the medicinal and aromatic plants based know-how and technologies that could empower the poor farmers and unemployed youth including women towards self reliance, with an additional income and improved livelihood. In the last paper in the Section II, a recent strategy in Kenya towards socio-economic empowerment of rural communities is being discussed by S.M. Guchu and G. Gichini.

The initiative of establishing the digital villages or popularly known as Parsha Centres, is one of the projects recently rolled out by the Government, towards application of Science, technology and innovation (STI) to empower the wider rural population in Kenya.

In Section III, the government policies and initiatives in promoting rural industrialization are being discussed. The first paper on the situation analysis of Uganda by Balu Tabaaro highlights that the government is carrying out three approaches to enhance science and technology usage for the occupational village development *viz.*, a) enhance educational levels, b) establishment of ICT infrastructure, and c) technologies for value addition of the Agricultural produce. He highlights a need for more interaction among Non Aligned Movement (NAM), scientists and technologists to enhance the diffusion of science and technology for occupational village development. The policies and practical measures adopted by the government of Ethiopia for promoting occupational villages are next discussed by Yared Awgichew and according to him, Ethiopia's economic base moves away from agriculture into manufacturing and services while links between rural and urban areas get strengthened to balance the regional economic growth and poverty reduction. The last paper by Hoeung Kimsay in Section III discusses the country status of Cambodia including economy, livelihood, and development efforts of the government through various policies, national plans, strategies, goals and road maps etc. The National Green Growth Roadmap has focused on accessing clean water and sanitation, energy, information and knowledge, better mobility, finance and investments, food security (agriculture) and non-chemical products. The concept of green jobs is one of the main programmes of the Cambodian government for rural development.

In conclusion, it must be emphasized that the papers of 14 developing countries compiled in this book present a broad scenario with diversity of approaches with regard to the use of science, technology and innovation for uplifting the lives of the rural community of those countries. The readers may find it very useful as reference material to learn about the status and culture, and more interestingly, the feasibility of various strategies implemented by other developing countries in achieving similar goals. It also draws attention of the policy makers and other relevant authorities in developing countries to a more practical concept of establishing occupational villages towards socio-economic development. In fact, the efforts of the Centre for Science and Technology of the Non-Aligned and Other Developing Countries (NAM S&T Centre), New Delhi to bring out this publication are timely and much appreciated.

Seetha I. Wickremasinghe

Ma. Josefina P. Abilay

Jayasamara Gunaratne

Introduction

Agriculture is the primary source of income for most of the working population in the rural areas of the developing countries. However, the rural household income is strongly dependent on the vagaries of nature. To counter this, there is an urgent need to diversify and create additional on and off-farm incomes. However, the national economic and social policies are often urban centric and do not necessarily help in alleviating the problems of rural poverty. There is a clear need for enterprise development, marketing skills and means of coordinating households/communities into collective action groups. In this direction, rural micro-enterprises play a significant role for their development, increased production and generation of employment in the villages. At the same time, there is also a vast network of R&D and academic institutions spread throughout the countries, which can be harnessed to benefit the rural enterprise.

Several programmes for the development and introduction of technology for application in the rural areas and promotion of rural enterprises have been initiated in many of the developing countries to benefit the small and micro enterprises in villages. But they need concerted efforts to generate, transfer, adapt and utilise appropriate technologies for occupational development of rural enterprises in the villages of these countries. In some other developing countries, programmes to build up and operate rural businesses and micro-enterprises with a group of households working together or in clusters have been set off. In this context, the concept of Occupational Community Villages (OCVs) in Vietnam is unique. The OCVs form the country's rural industrial base, in which several households work together to produce a single product. This collective approach allows groups to develop some degree of economy of scale and also greatly improves their sustainability and efficiency. There are similar programmes in many other developing countries aimed at the promotion of cottage and village industries and micro-enterprises and application of S&T for rural industrialisation and income generation in villages. But notwithstanding the success of OCVs in a national setting, they are still not fully equipped to move into higher value markets and do not have enough resources and skills to evolve into small and medium scale enterprises.

In order to deliberate on this issue, the Centre for Science and Technology of the Non-Aligned and Other Developing Countries (NAM S&T Centre) organised an international workshop in Hanoi, Vietnam on 'Application of Science and Technology for Occupational Villages Development' during 2–5 August 2010 jointly with the International Cooperation Department of the Ministry of Science and Technology (MOST), Government of Vietnam and the Centre for Regional Research and Development (CRD) of Vietnam. 15 countries, namely, Cambodia, Ethiopia, India, Indonesia, Iran, Kenya, Myanmar, Nepal, Nigeria, Pakistan, Philippines, Sri Lanka, Tanzania, Uganda and Vietnam participated in this workshop. The 18 overseas participants were from Cambodia [Mr. Hoeung Kimsay, Official, Science and Technology Office, Department of Industrial Techniques, Ministry of Industry, Mines and Energy (MIME), Phnom Penh], Ethiopia [Mr. Yared Awgichew, Agriculture Technology Transfer Expert, Ministry of Science and Technology (MOST), Addis Ababa], India [Dr. Anil Kumar Singh, Scientist-G and Head, Technology and Business Development (TBD), Central Institute of Medicinal and Aromatic Plants (CIMAP), Council of Scientific and Industrial Research (CSIR), Lucknow; Mr. Ammanaghatta Rudrappa Shivakumar, Principal Investigator – RWH, Karnataka State Council for Science and Technology (KSCST), Bangalore; Dr. Brij Mohan Sharma, Secretary, Society of Pollution and Environmental Conservation Scientists (SPECS), Dehradun; and Prof. Vasala Madhava Rao, Head, Centre for Geoinformatic Applications in Rural Development (C-GARD) and Head I/C, CIT, National Institute of Rural Development, Hyderabad], Indonesia [Mr. Darmawan Listya Cahya, Researcher, Agency for the Assessment and Application of Technology (BPPT), Jakarta], Iran [Dr. Hojjatollah Hajihosseini Sefideh, Director, Research Institute for Development Studies, Iranian Research Organization for Science and Technology (IROST), Tehran], Kenya [Dr. Salome Muthoni Guchu, Assistant Director of Research, Ministry of Higher Education Science and Technology, Department of Research Management and Development, Nairobi], Myanmar [Dr. Aye Aye Khai, Associate Professor, Department of Biotechnology, Technological University, Kyauk Se], Nepal [Prof. Dilip Subba, Secretary, Nepal Academy of Science and Technology (NAST), Kathmandu], Nigeria [Mrs. Angela Chineme I. Nwanya, Chief Scientific Officer (Food Division), Raw Materials Research and Development Council, Abuja], Pakistan [Mr. Syed Nawazish Ali Shah, Deputy Scientific Adviser (International Liaison), Ministry of Science and Technology, Islamabad], Philippines [Dr. Ma. Josefina Pormento Abilay, Regional Director, Department of Science and Technology (DOST), MIMAROPA Region, Taguig City, Metro Manila], Sri Lanka [Ms. Jayasamara Gunaratne, Assistant Director, Vidatha Program, Ministry of Technology and Research, Colombo], Tanzania [Mr. Omar Jumanne Bakari, R/O Technology Diffusion and Commercialization, Commission of Science and Technology (COSTECH), Dar Es Salaam], Uganda [Mr. William Balu-Tabaaro, Managing Director, Geo Engineering Technologies (GET), Entebbe], and a representative from the NAM S&T Centre [Mr. M. Bandyopadhyay, Senior Expert]. In addition to these foreign delegates, there were several participants from the host country, Vietnam.

21 scientific papers were presented in three Technical Sessions of the Workshop, which were co-chaired respectively by Dr. Dilip Subba, Nepal and Mr. Vu Quoc Tuan, President, Vietnam Association of Crafts Villages; Dr. Anil Kumar Singh, India and Mr. Luu Duy Dan, Vice President, Occupational Association of Vietnam; and Dr. H. Hajihoseini, Iran and Mr. Vu Quoc Tuan, President, Vietnam Association of Crafts Villages. The present publication is based on the deliberation of the Hanoi workshop and includes the papers contributed by eminent specialists in the field. It comprises seventeen articles and review papers under three sections, namely, I: Science and Technology for Rural Industrialization; II: Case Studies of Varied Approaches to Rural Industrialization; III: Government Policies and Initiatives in Promoting Rural Industrialization.

I acknowledge with gratitude the deep involvement and determined efforts of the editorial team comprising Dr. Seetha I. Wickremasinghe, Principal Scientific Officer and Head, Science and Technology Policy Research Division, National Science Foundation, Sri Lanka; Dr. Ma. Josefina P. Abilay, Regional Director, Department of Science and Technology, MIMAROPA Region, the Philippines; and Ms. Jayasamara Gunaratne, Assistant Director, Vidatha Programme, Ministry of Technology and Research of Sri Lanka in checking the manuscripts and giving their valuable comments to make it a high quality publication. I am also indebted to the entire team of the NAM S&T Centre, particularly to Mr. M. Bandyopadhyay, Dr. V.P Kharbanda, Ms Isha Sehgal, Ms Apeksha Yadav and Mr. Pankaj Buttan in compiling the presented papers, liaising with the authors and editors and giving a shape to this volume.

I sincerely trust that this publication will serve as a priceless reference material not only for the developing countries but also for other countries as well and pave way for the promotion of South-South and North South co-operation in the area of Rural Development.

Prof. Arun P. Kulshreshtha

Director and Executive Head,
NAM S&T Centre

Contents

— Section II —
Case Studies of Varied Approaches to Rural Industrialization

— Section III —
Government Policies and Initiatives in Promoting Rural Industrialization

— *Section I* —

Science and Technology for Rural Industrialization

Chapter 1

Science and Technology Interventions Towards Rural Industrialization in MIMAROPA Region, Philippines

Ma. Josefina P. Abilay

Regional Director, Department of Science and Technology, MIMAROPA Region, 2nd Floor Imelda Building, DOST Complex, Bicutan, Taguig City, Metro Manila, Philippines
E-mail: mjpabilay@yahoo.com

ABSTRACT

DOST-MIMAROPA, an ISO 9001:2008-certified agency and the newest of 16 administrative regional offices (apart from one autonomous region) of the Department of Science and Technology (DOST) of the Philippines, provides science and technology (S&T) interventions mostly to community organizations and agro-based micro, small and medium enterprises (MSMEs) thus promoting rural industrialization in the MIMAROPA region. Its main thrusts are the technology acquisition and upgrading, which encourage firms to adopt technological innovations that boost their overall productivity and competiveness. The core and non-core funded programmes enable MSMES to address technical problems/restraints through technology acquisitions, product development, and product quality enhancement. DOST-MIMAROPA also provides packaging and labeling assistance that underscores the importance of first-rate presentation in a product's consumer recall and marketability. Also offered are comprehensive consultancy services, which serve as an initial effort to build/ improve a firm/farm's productivity and product quality. In most cases, firm/farm-beneficiaries are also endorsed into other S&T programmes that can induce even a faster techno-based business growth. In addition, the DOST also makes available a venue for rural communities to gain/develop skills from relevant S&T training courses, demonstrations, and/or seminar-forums at minimal costs. This programme shoulders all expenses related to an S&T expert/resource person's travel to a distant area in need of assistance. DOST-MIMAROPA also encourages students and out-of-school youths to directly engage in S&T-based entrepreneurial projects where they can learn/develop competencies applicable to real workplace situations. Lastly, numerous scholarships are also made available to intellectually-competent individuals who are already pursuing and/or about to pursue a course in the basic and applied science, engineering, and technology fields. This ensures the presence and involvement of a critical mass of S&T manpower in the country. Collectively these programmes have provided at least 2,489 employments for the past two years and have generated a gross income of about PhP 14 million to the MIMAROPA region.

Keywords: *Dost-Mimaropa, rural industrialization, MSMEs, community-based organizations, agro-based industries, science and technology, technology acquisition and upgrading, employment and income generation*

Introduction

International Scenario

The economy of the Philippines, in comparison with those of its Southeast Asian neighbours, is relatively underdeveloped as has been shown in a number of indicators. While the government has long recognized the importance of science and technology (S&T) to economic growth, it has yet to align its financial priorities according to long-term national research and development (R&D) thrusts. This is reflected in the ratio of R&D expenditures to the nation's gross domestic product (GDP).

The latest survey from the Department of Science and Technology of the Philippines reveals that the country devoted 0.11 per cent of its GDP to R&D in 2007. In contrast to other more developed countries, South Korea allocated 3.47 per cent to R&D in 2007, making it the highest among Asian countries included in the survey. Japan came in second in Asia with 2.64 per cent, while Singapore, with a 2.61 percent-ratio, was next. Other more developed countries had R&D-to-GDP ratios ranging from 1.0 to almost 3.0 per cent. With the exception of Thailand, the rest of the ASEAN countries had R&D-to-GDP ratios of less than 1.0 per cent, although most had ratios higher than that of the Philippines (UNESCO Institute of Statistics, 2007).

Table 1.1: Comparative Figures on Ratio of National R&D Expenditures to GDP

Country	*Per cent*	*Country*	*Per cent*
South Korea (2007)	3.47	Japan (2007)	3.45
USA (2007)	2.67	Singapore (2007)	2.61
Australia (2006)	2.17	Canada (2007)	2.03
New Zealand (2007)	1.26	Russia (2007)	1.12
Hong Kong (2006)	0.81	Malaysia (2006)	0.64
Philippines (2007)	0.11	Indonesia (2005)	0.05

Source: UNESCO Statistical Yearbook and DOST R&D Expenditure Survey.

The government sector of the Philippines accounts for almost 80 per cent of the R&D activities in the country, which is quite the opposite to more developed nations where the private sector undertakes most of the R&D and provides most of the funds for that purpose - typically 60 per cent of the total (UNESCO Institute of Statistics, 2007).

The DOST: Where S&T Thrives

Science and technology change the world rapidly in remarkable ways. The developing countries, having recognized the contribu-tions of such pursuits to economic growth and societal security, have thought deliberately about its innovation systems and have been act-ing to make these systems more effective and efficient in an increasingly inter-reliant and competitive world. Policies that seek to strengthen a nation's education infrastructure, the institutions that perform research and development (R&D), and the innovation environment have now become commonplace. According to the Harvard Report on Global Governance of Technology, national development goals that do not recognize the importance of S&T in economic transformation are likely to fail (Juma, *et al.*, 2001).

In its acknowledgment of the significance of R&D pursuits, the Philippines government has instituted and mandated the Department of Science and Technology (DOST) *"to provide central direction, leadership and coordination of all scientific and technological activities, and to formulate policies, programmes and projects to support national development."*

With fifty-two years into its service, the DOST has dynamically evolved into a self-sufficient S&T body composed of the following: five sectored planning councils responsible for formulating policies, long-term programmes and strategies for S&T development; seven R&D institutes (RDIs) concerned with innovation and engineering of sustainable technologies in differing basic and applied research fields; two collegial bodies that provide assistance, recognition, advisory and establishment of local and international linkages; and sixteen regional offices that identify and address the S&T needs of its seventy-eight component provinces.

DOST in MIMAROPA: A Glimpse

Considering that the Philippines' archipelagic profile causes a highly uneven spatial distribution of economic activity in the regions, the focus-area development approach has become a prevalent national strategy. With application of this strategy, the DOST has established seventy-eight provincial S&T centers (PSTCs) under all sixteen regional offices to make services available and accessible even to the most rural parts of the country. This report tackles how the DOST Regional Office in MIMAROPA (DOST-MIMAROPA), with a unique regional profile to pervade, addresses the delivery of S&T interventions to its five provinces.

DOST-MIMAROPA was established in October 2007 by virtue of the Executive Order (EO) 103 dividing then Region IV comprised of 11 provinces into Region IV-A (CALABARZON) and Region IV-B (MIMAROPA). The EO states that the purpose was to *"accelerate social and economic development and improve public services".*

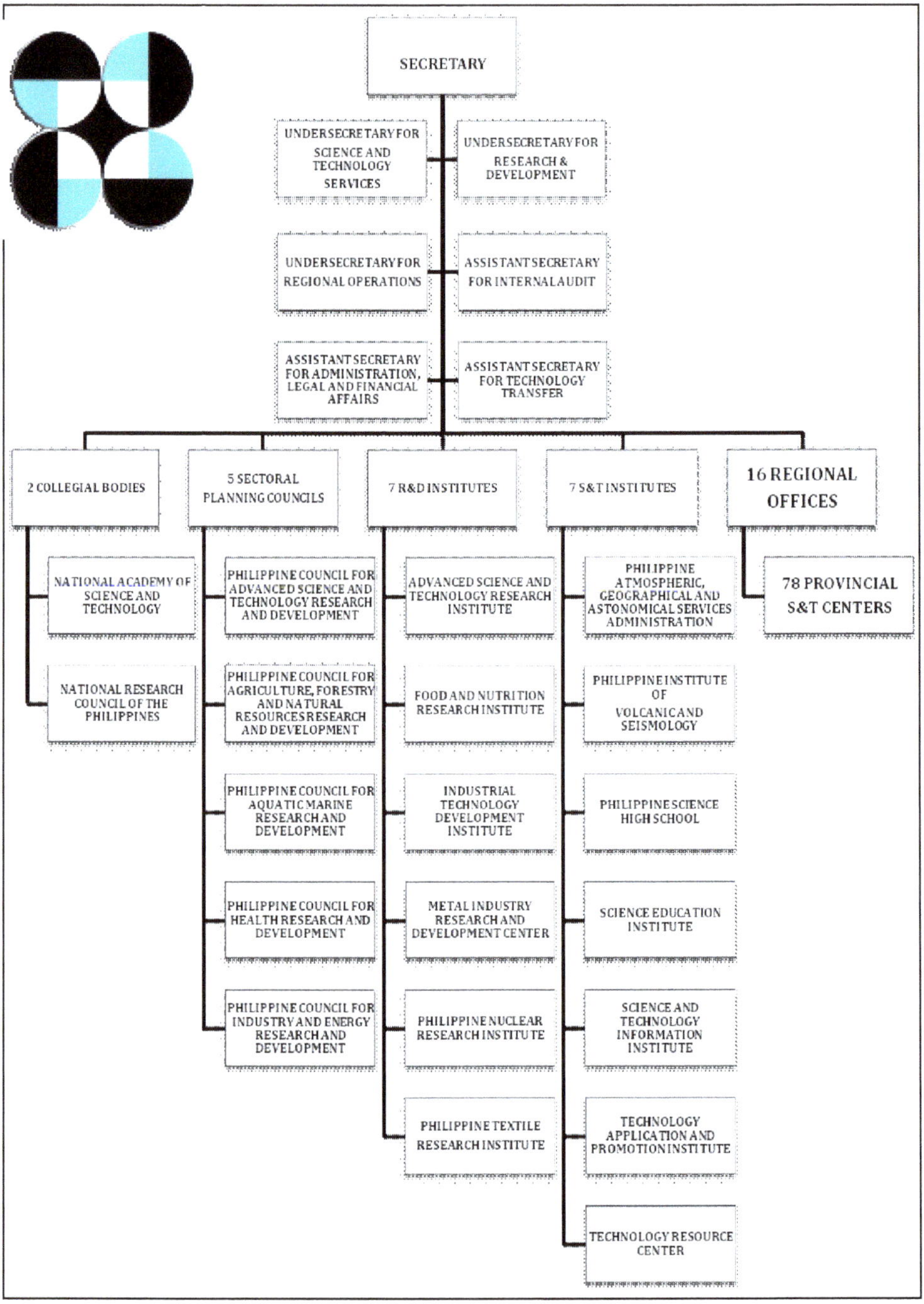

Figure 1.1: The DOST System

Guided by both its vision of "becoming *a virtual bridge of the islands where knowledge, creativity, and innovation drive regional economic growth to improve quality of life*";

And the mission: "to *effectively respond to the social, economic and ecological development challenges of the MIMAROPA Region through S&T interventions*",

DOST-MIMAROPA has since then performed even beyond its clients' expectations.

In the same year, the issuance of EO 161, which requires all government agencies to establish a Quality Management System (QMS) was passed to address the improvement of the country's competitiveness ranking specifically in efficiency-related areas of business and governance. In effect, DOST-MIMAROPA, with full commitment from the top management, has successfully incorporated the ISO 9001:2008 into its system to be effective from October 2009 and has acquired the ISO certification in March 2010. To further strengthen the practice, the regional offices also implemented productivity tools such as 5S, which is a 'walk the talk' strategy since the practice of the same being promoted to assist micro, small and medium enterprises (MSMEs) in the countryside.

The MIMAROPA Region acquired its name from the acronym of its five constituent island-provinces: Mindoro Occidental and Mindoro Oriental (MI), Marinduque (MA), Romblon (RO), and Palawan (PA). It is bounded by the South China Sea to the West, Tayabas Bay to the North, Sibuyan Sea to the East and Sulu Sea to the South putting the region strategically within the radius of urban and industrial influence of the Luzon and Central Philippines super regions. Of the aggregate 2.74 million hectares of mostly old-growth and residual forest land area, MIMAROPA utilizes 542,218 hectares of lands for agricultural use, based on the National Statistics Office (NSO) Census of Agriculture and Fishery. This is equivalent to 19.8 per cent of the region's total land area making agro-industry accountable for a large 38.08 per cent share of the gross regional domestic product (GRDP). The agriculture sector represents a priority area to MIMAROPA as approximately three-quarters of the rural poor depend on agriculture for employment and income (Stads, Faylon, and Buendia, 2007). The primary sectors of agriculture and fishery in the rural areas are critical as they are the economic base of the nation. If these are undermined and weakened, the economy's growth might decline or stop altogether.

DOST-MIMAROPA acknowledges that agro-based industrialization is most fitting in the rural and vastly agricultural region. MIMAROPA's output in the agricultural sector alone takes a 1.1 percent-share of the national GDP. MIMAROPA is resource-intensive, its agricultural industry has backward and forward linkages with others sectors, and it has a very ample labor force. Palawan, the largest of five provinces, is the top cashew nut producer in the country with a total production of 106,353 metric tons (MT) or 90 per cent of the country's total annual yield. Similarly, Mindoro Oriental contributes 60 per cent or 109,183 MT of *Citrus microcarpa* (Calamondin Orange) or locally known as calamansi to the national output. Mindoro Occidental, on the other hand, makes 70 per cent of the total salt produced in the country. The region is the second biggest fishery sector in terms of volume from 2002

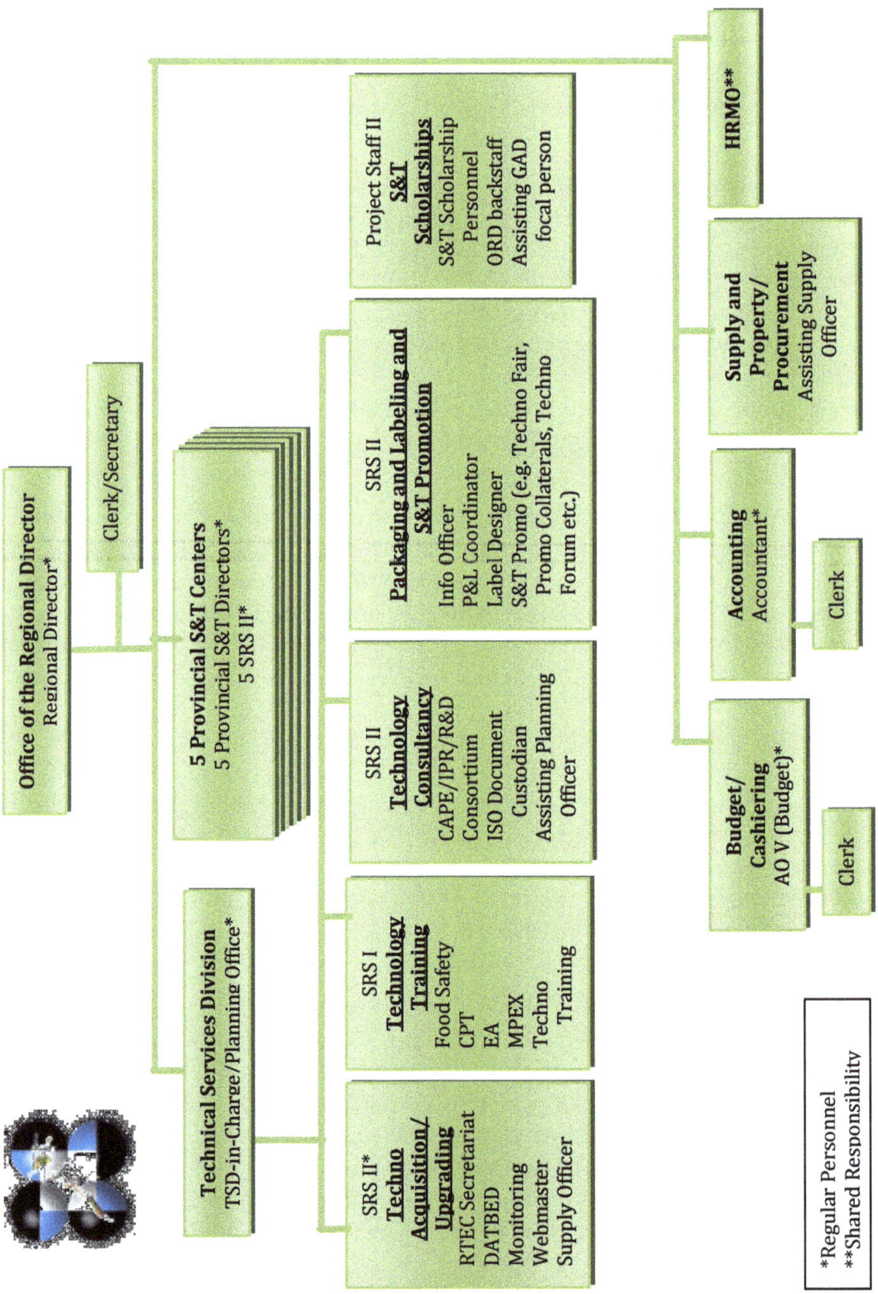

Figure 1.2: DOST-MIMAROPA Organizational Chart

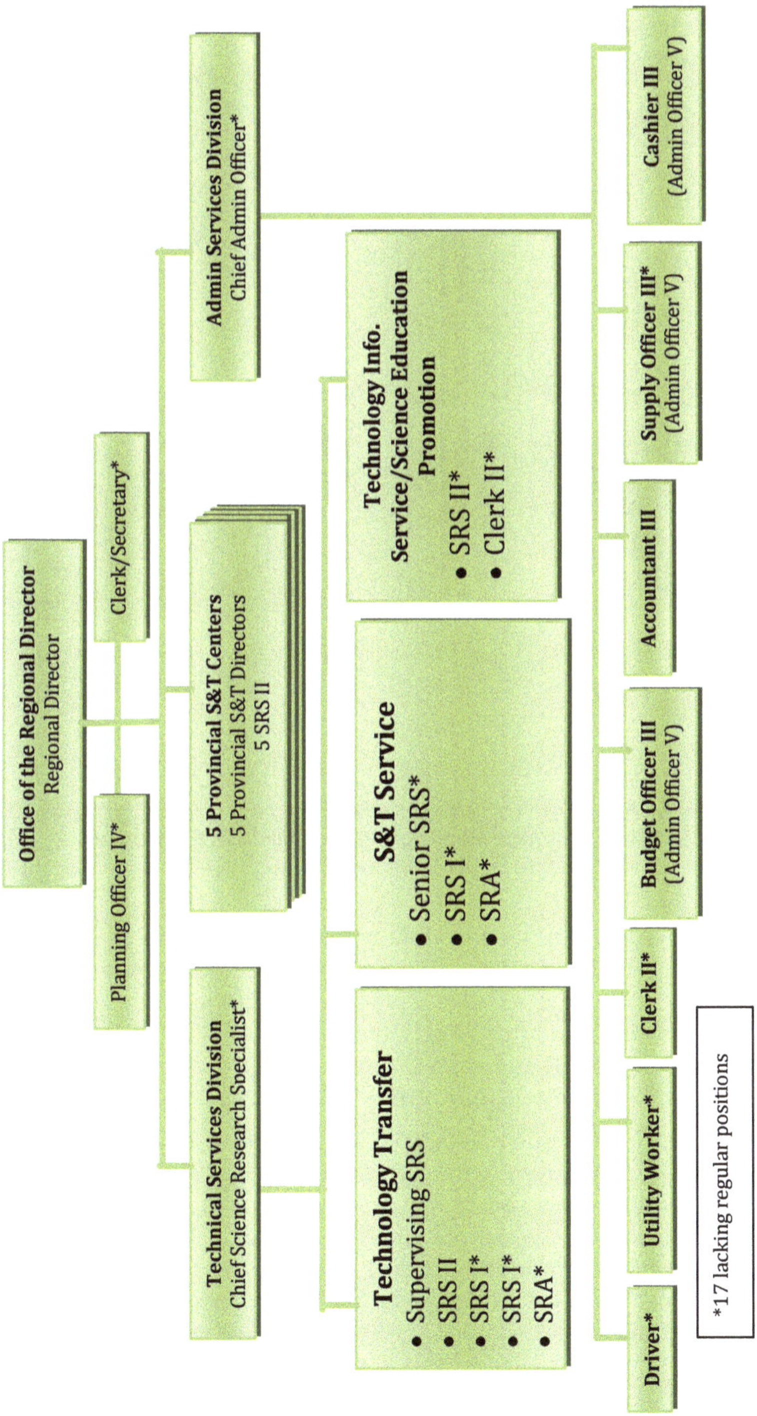

Figure 1.3: DOST Regional Office No. IV-B (MIMAROPA) Organizational Structure as per DOST Administrative Order No. 001, Series of 2004

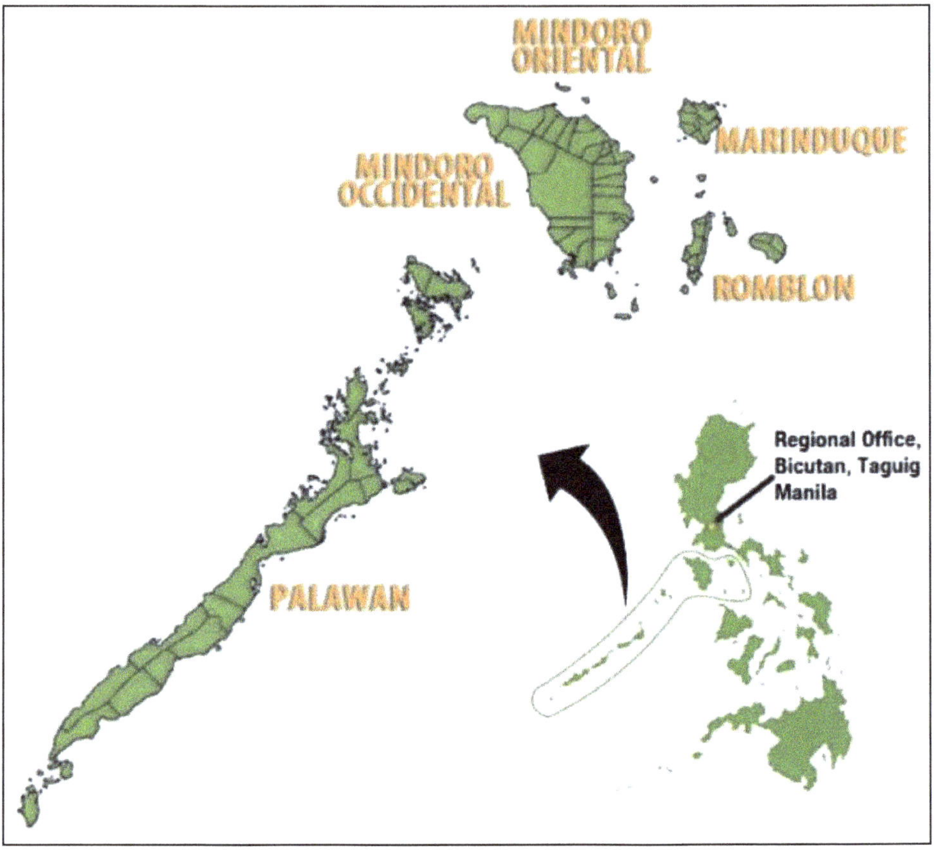

Figure 1.4: The Philippine Map Highlighting MIMAROPA Region and its DOST Regional Office

to 2006 with a 714,162 MT output equivalent to a 17 per cent national share. Other major products include bananas, coconuts, mangos, seaweeds, rice, and corn each of which has a national share of 3 – 17 per cent. Through rural industrialization, the sustainable utilization of these natural and human resources will stimulate economic progress in the provinces. These approaches, through S&T interventions, will further be improved thus, positioning the region and its MSMEs with a competitive advantage.

Through several institutionalized S&T programmes, DOST-MIMAROPA provides interventions that improve the general and technical competitiveness of micro, small and medium enterprises (MSMEs). On the whole, DOST-MIMAROPA identifies the S&T needs and opportunities in the region; delivers S&T services such as technology demonstration and transfer, productivity-oriented training and consultancy services, testing and analysis, S&T promotion and information dissemination; and coordinates the programmes and projects of DOST Councils and RDIs in partnership with LGUs, RLAs, NGOs and community organizations in the region.

Application of S&T for Rural Industrialization in Mimaropa Region

Industrialization, especially in the countryside, requires assistance from the government in order to be mobilized. Hence it is a major thrust of the agency to support the government's initiative to alleviate poverty by providing livelihood opportunities for the people. This objective is attained through provision of S&T interventions to micro, small and medium enterprises (MSMEs) and community organizations. Capacity building mechanisms such as the application of technologies to upgrade processes, improve productivity and quality of products that help firms become competitive thus world-class, are in place through the following interventions:

Technology Acquisition and Upgrading

DOST aims to encourage and assist MSMEs and community organizations to adopt technological innovations to improve their operations and thus boost productivity and competitiveness. It enables firms to address their technical problems through technology acquisition and upgrading, new product development, product quality enhancement including food safety, packaging and label designs, and facility improvement including waste management and energy audit, and skills enhancement through training. Priority sectors covered by the programme are food processing; furniture; gifts; houseware and decors (GHD); marine and aquatic resources; horticulture; agriculture; metals and engineering.

Projects differ in fund sources and scheme of implementation. Core funded projects are funded with repayment while non-core projects are funded under the DOST grants-in-aid (GIA) with complementation of resources from other stakeholders. DOST-MIMAROPA together with its partners provides MSMEs and community organizations with the needed technical assistance and services.

Core Projects

For core projects, DOST-MIMAROPA provides technical and the concomitant financial assistance to MSMEs. Through this, the regional office provides a package of technology interventions such as technology needs assessment (TNA), consultancy, equipment acquisition, and training on current good manufacturing practices (cGMP), food safety, product analysis, and/or improvement on their packaging and label designs.

Non-Core Projects

Non-core projects are another regional priority that was implemented in collaboration with other community organizations, such as local government units (LGUs), regional line agencies (RLAs), and national government agencies (NGAs).

Technology Based Entrepreneurship Development

DOST-MIMAROPA promotes the development of entrepreneurial competencies among students and out-of-school youth by providing funding assistance to implement entrepreneurial projects. Most of the student-beneficiaries gain entrepreneurial skills while already earning from their projects. The state universities and colleges (SUCs), which go through a process of accreditation based on their implementing plan, among other requirements by the DOST, also provides the technical/technology assistance needed by the students.

Packaging and Labelling

Recognizing the utmost importance of appropriate product packaging and labelling to improve product presentation and quality, consumer product recall, and market share, the DOST-MIMAROPA intensified its packaging and labelling assistance to MSMEs.

Consultancy for Manufacturing, Horticulture Crops, and Aquamarine

Consultancy services are provided to MSMEs to ensure the adoption of technologies through technical advices and assistance. These functions come in the form of consultancy for agricultural productivity enhancement and manufacturing productivity improvement.

Productivity consultancy services for agricultural and aquaculture farms aim to increase farm productivity and convert traditional farming activities into agribusiness ventures. The manufacturing firms are likewise given assistance in order to enhance the profitability of MSMEs as well as improve facilities, enhance product quality and safety, and waste minimization and utilization, among others, to make them globally competitive.

Technology Training

Skills training are conducted to complement and enhance various technical and financial assistances provided to the core and non-core clients, manufacturing firms, horticulture and aquamarine farms, and other potential and/or interested firms. Resource persons come from a pool of S&T experts in various agencies of DOST and members of the R&D network, and private sector.

Creating Critical Mass of S&T Manpower

To create a critical mass of manpower, DOST-MIMAROPA administers S&T Scholarship Programmes. The Republic Act No. 7687, also known as the Science and Technology Act of 1994, provides scholarships to deserving students who pursue priority fields of study in the basic and applied sciences, engineering, science and mathematics teaching, technology and technical courses. Part of the administration of scholarship programmes are the accreditation of SUCs who offer these priority courses. In this process, SUCs are required to have graduated at least one batch under the course in order to be accredited. On the other hand, priority courses offered by private higher education institutions (HEIs) are required to be accredited by the Commission on Higher Education (CHED).

DOST scholarship programmes are categorized into two: a) RA7687 scholarship and b) Merit scholarship. There are other categories such as the ladderized programme, BS with bridging programme, or Bridging Education for Science and Technology for indigenous people (BEST for IP). Merit A is formerly known as the NSDB or NSTA scholarship that began in 1958, and is given to qualifiers who exceed the required indicators such as the family income and expenditures during application. Under the ladderized programme, the qualifiers are allowed to take a 3-year technician course first. In the event that the scholars meet all the requirements to continue in a BS course, then they may avail a BS course in the school where they are enrolled. Scholars under the BS with bridging undergo tutorials. The BEST for IP scholarship programme is specific for indigenous people who want to enrol in a technician course.

S&T Promotion

DOST-MIMAROPA promotes S&T through technology forums, technology fairs and utilizes tri-media (print, radio and TV) and the internet, to reach a greater number of prospective clients in the countryside effectively at lesser costs.

Technology Forums

The DOST-MIMAROPA also promotes its S&T services and recent technologies through technology investors' forums. These forums are conducted among different industry sectors and groups to disseminate the available and new technologies and for the audience to gather ideas that may be appropriate for their respective businesses. During these events, DOST funding mechanisms for technology acquisition and upgrading and/or scholarship programmes, in addition to the usual way of disseminating information about it through schools, are also discussed.

Science and Technology Fairs

As the success of S&T programmes and services largely depend on how well the beneficiaries are able to market and obtain profit from their products, S&T fairs are conducted to promote technologies developed by the different DOST networks. At the same time, these fairs provide opportunities for the DOST assisted MSMEs and the products of community organizations to find a market place.

The cooperators' participation in DOST-sponsored fairs such as the National Science and Technology Week (NSTW), Regional Cluster S&T Fairs, and provincial fairs allow marketing opportunities for their products at a competitive advantage.

Likewise, the Regional Invention Contest and Exhibits (RICE) is also conducted, which features the Regional Invention Contest, invention fair and exhibit and other invention development activities to promote the commercialization of inventions. RICE serves as a venue to select regional winners as finalists who will vie for the major awards during the National Invention Contest and Exhibits. The regional contest will select the regional winners/qualifiers in the invention, utility model, industrial design, and creative research and student creative research categories.

Tri-Media

DOST-MIMAROPA uses tri-media coverage to promote various DOST services, information on technologies relevant for MIMAROPA, announce scholarship and activities such as training courses, seminars, symposia, workshop, forums, and call for proposals.

Internet

The DOST, through its eDOST programme provides infrastructure support to all its agencies in terms of information technology capability to keep up with the latest developments in the world. Part of the IT capability building is the provision of IT equipment, software, capability building of IT manpower and hosting the websites of various agencies of the department including the regional offices.

It is geared to improve the DOST's effectiveness in performing its internal operations and delivery of front-line services to the public through competitive and appropriate Information and Communication Technology (ICT) solutions.

Website

True to the vision of being a virtual bridge, DOST-MIMAROPA enriches its website as a tool for providing information services. This measure is a continuing strategy of DOST-MIMAROPA to reach more customers with just a click of a button. The website currently features the various services, directory of regional and provincial S&T centers, success stories, news articles and current activities of the regional office. National information sources from DOST are likewise linked with the website such as the one stop information shop for technologies (osist.dost.gov.ph) where experts as well as technologies in various fields of science and technology can be accessed by the public.

Promo Collaterals

Another way of S&T promotion is the production of promo collaterals such as brochures, leaflets and flyers. To do this, DOST-MIMAROPA came up with project briefs, an information materials which highlights the various S&T projects, the technology intervention given and its impact to the MSME/community organization in terms of productivity improvement, income and employment. The material is used as reference during project visits and as information dissemination during S&T fairs.

Library Services

Student-researchers have free access to mini-libraries stationed in the Provincial Science and Technology Centers (PSTCs) wherein publications and journals from DOST Councils and research and development institutes, as well as other research and information materials about appropriate technologies that may be applicable to the areas of interest, are being maintained.

Results and Outcome of Assistance

Technology Acquisition and Upgrading

DOST-MIMAROPA has been actively involved in supporting the development of different industries in the region by providing technology interventions. The industry players include MSMEs, LGUs, NGAs, cooperatives and other community organizations. The impact of assistance to these organization was measured essentially in terms of increase in the income generated, and the level of employment.

Types of Projects

As shown in Table 1.2, a majority of the industries assisted came from the food sector, which is about 71.6 per cent of all the assisted projects, followed by

Figure 1.5: A Coconut Processing Facility in Gasan, Marinduque

environmental engineering projects receiving 11.32 per cent of the total. The smallest segment was the GHD sector receiving 1.89 per cent share of the total.

Table 1.2: Distribution of Selected DOST-MIMAROPA Clients, by Sector, 2004-2009

Type	*No. of Clients*	*Percentage (per cent)*
Food	38	71.6
Environmental Engineering	6	11.32
Metals and Engineering	5	9.40
Furniture	3	5.66
Gifts, Housewares and Decors (GHD)	1	1.89
TOTAL	53	100 per cent

Assisted projects under the food sector were further classified into specific commodities that the enterprise/organization is involved. There were projects on the processing of coconut, rice, fish, and meat into other products that have a big demand in the market. Other major raw materials being processed into value-added products include nuts (cashew and peanut), *Maranta arundinacea* (arrowroot tubers) and *Citrofortunella microcarpa* (calamansi).

There were also projects that were implemented to provide inputs for the production of the abovementioned commodities. Such projects serve as input support to the livestock industries, and those that ensure productivity in the utilization of aquatic/marine resources. These projects aim at ensuring the sustainability of supply of raw materials for processing.

Arrowroot Cookies

Arrowroot, an abundant tuber crop in Marinduque, is the major raw material used by one of DOST-MIMAROPA's assisted enterprise, Rejano's Bakery, in making its famous arrowroot cookies. DOST-MIMAROPA provided technology interventions through the provision of an entire assembly line that processes raw arrowroot up to its flour form, along with the seminars conducted on Cleaner Production Technologies (CPT) and Good Manufacturing Practices (GMP). The owner was also sent to the Climate Technology Initiative Workshop on Energy Efficiency for Asian Countries for food processing sector in Yokkaichi, Japan. These interventions enabled the firm to increase its production capacity and improve the quality of its products, which, in turn, made it qualified to be displayed in the shelves of the major shopping malls in Manila. A second phase of project assistance was given to this MSME in 2009 for the procurement of packaging materials for its arrowroot cookies. This project alone was able to provide an annual average employment to 260 individuals, which consisted of processors/labourers and local arrowroot farmers. The firm earns an annual average income of PhP3.6 million. For the first quarter of 2010, the firm was able to generate a gross income of about PhP600, 000.00 and directly employed 22 individuals.

Rejano's Bakery received the *Gawad Saka* Outstanding Food Processors Award (regional level) and Outstanding One Town, One Product (OTOP) MSME for the improvement of its arrowroot cookies and increase in gross sales, amounting to PhP5 million in 2009

Noodles

Another food industry in which a number of the MSMEs in the Region are involved is noodle-making. The product, being very affordable and can easily be fortified, provides immediate solution to problems on food security and malnutrition. There are about five of this sort of enterprises in MIMAROPA, which were given DOST assistance to improve their existing noodle processing equipment and procedures. The projects of this food commodity were further classified into firms that use regular noodle ingredients and firms that incorporate vegetable extracts to fortify its products.

Two projects implemented the veggie-noodle technology of DOST involving nutrient fortification. These projects utilize *Cucurbita maxima* (squash) and *Moringa oleifera* (ben oil tree or *malunggay*). A particular project located in Bulalacao, Mindoro Oriental utilizes the locally abundant seaweeds as the major ingredient in noodle-making. This project also supports the local seaweed industry of the province generating livelihood for the fisherfolk members of the community. These five projects were funded with a total investment amounting to P1.9M by DOST-MIMAROPA, which in turn provided direct and indirect employment to 63 individuals and generated income of PhP2.34 million for the first quarter of 2010.

Coconut Products

MIMAROPA has a total coconut area of 161,858 hectares or 29.85 per cent of the total cultivated/agricultural land in the region with an average of 580,000 MT of annual production (BAS, 2008). DOST-MIMAROPA has identified this resource and

Figure 1.6: Noodle-Making Facility in San Jose, Mindoro Occidental

implemented projects that would maximize the utilization of the crop. These DOST projects involve the use of coconut for food and non-food applications.

Marinduqueland Corp., a DOST-MIMAROPA assisted firm in Marinduque, was able to develop a special condiment/food dressing combining fish paste and grated coconut meat called *cocobagoong*, which takes its market appeal from its distinct taste. Aside from this product, the firm also ventured into making cosmetics and health products such as, body lotions, soaps, oils, and massage gels and virgin coconut oil (VCO). The enterprise's VCO product is individually marketed as herbal alternative to generic drugs and is also being used as a major ingredient in manufacturing its cosmetic products. These project alone was able to directly employ 10 and 100 indirect workers in 2009 and generate income of PhP 1.4M from the PhP500,000 assistance provided by DOST-MIMAROPA.

Two projects utilizing coconut for its fiber and alternative energy potential are located in Romblon. The former initially involves the processing of fiber from coconut husks to make household mats using the DOST-developed coco-coir twining technology. This project generated an investment of PhP1.8M from other sources such as the local government unit (LGU) and other national government agencies (NGAs), to complement the PhP400,000 investment provided by DOST-MIMAROPA. The project was also able to generate direct and indirect employment for 45 individuals as it commenced operation.

Another major project was the establishment of a biodiesel processing plant in Romblon making use of the abundant coconut resource in the province. During its operation, the project was able to provide 6 direct and 650 indirect employments for individuals in its value chains and gross sales of PhP 3.4 million from the PhP10.49M generated investment. Thirty-three per cent (33 per cent) of this invested amount came from DOST funds, and the remaining 77 per cent amounting to about PhP8.0M as counterpart from other sources such as the LGU, including the stake of the client.

Other Food Products (Kitchen-Scale Enterprises)

The agency was keen on identifying its clients to qualify for the assistance being provided by the agency. Backyard or cottage-type industries are prominent in the region. Most of them being start-ups, are still in the preliminary stages of their production, with manageable quantities of raw materials/products that would only require a limited space. Furthermore, the capitalization for such industries is of minute amount in comparison with the large industrial and commercial establishments in the Metro.

These backyard industries provide a significant input to the development of the regional economy in terms of its contribution to the regional employment and the GRDP. They are the identified backbone of the regional economy with its growing number and stake in the overall progress of the country.

Six core funded projects by DOST-MIMAROPA fall under this category. Their main products include pastries, refreshments, and sweet/delicacies. With the assistance provided these firms were able to improve their production capacities and enhance the quality of their products. These projects were able to generate 21 direct and 14 indirect employments to workers who mostly perform the processing operations required in these abovementioned kitchen-scale food enterprises. These projects were able to generate an average income of PhP700,000 during their first quarter of operation with the agency's PhP 570,429 investment, which came from its core funds.

Postharvest

DOST-MIMAROPA has aligned its focus in implementing programmes and services by identifying projects that would support the development of the agriculture sector. MIMAROPA produces 921,157 MT of rice, which is about 6 per cent of the total Philippine production and supplies a large volume of the total rice requirements of the mainland Luzon.

Mindoro Progressive Multi-Purpose Co-operative in Mamburao, Mindoro Occidental addressed the need for efficient postharvest equipment necessary to efficiently process the bounteous grain produced. These technologies come in the form of mechanical grain dryers. The dryers are of best assistance to the farmers, especially during the wet season, when drying of the harvest becomes a problem and also at this time, the losses could be countless due to extreme weather conditions if such technologies were not available. These technology interventions address the major issue of postharvest losses, which is estimated to occur during various postproduction operations, (*i.e.*, from harvesting, threshing, drying, milling, storage

Figure 1.7: A Meeting of the Mindoro Progressive Multi-Purpose Co-operative during a Visit to their Grain Drying Facility

handling and marketing), and were estimated to be as high as 37 per cent of the total production (Saunders, *et al.*, 2009). According to the Bureau of Postharvest Research and Extension (BPRE), an estimated P50 billion loss was incurred in 2006 for the agriculture sector alone, excluding the fishery and livestock sector. BPRE added that these postharvest losses affect the food security of the country and annual losses could have been saved if there were proper post-harvest practices. The project provides direct employment to 8 individuals and 550 local farmers and cooperative members were benefited. It is expected to generate an annual average of PhP700, 000 during the first three years of its operation from the PhP850,000 project investment.

Another project in Mindoro Occidental involving postharvest facility upgrading for an agricultural cooperative namely *Kaagapay Kilusang Bayang Tagapagpaunlad* (KAAKIBAT, Inc), was undertaken in 2009 by providing the organization with an improved-design and efficiency of mobile rice milling equipment. The increasing demand for postharvest services being provided by KAAKIBAT to its farmer-members triggered the need to expand/upgrade its operation that will in turn provide additional employment and increase income of cooperative members. The project was able to employ 5 direct and 7 indirect workers and generated an income of PhP231,000 from its first quarter of operation with an initial investment of PhP 850,000 by DOST-MIMAROPA.

Environmental Management

Preservation of the environment has been one of the many benefits to our society derived from the use of technology. DOST-MIMAROPA has been in the frontline in implementing quantifiable measures of addressing the growing concern to conserve the natural resources. This is being achieved by implementing conservation initiatives through the use of mature green technologies in the projects.

The Biogas Utilization Programme of DOST-MIMAROPA is aimed to complement the government's programmes on environment conservation and alternative energy utilization. The continuous dissemination of the technology from as early as the year 2000 to date resulted in the construction and continuous operation of 42 biogas digesters (Table 1.3).

Table 1.3: Number of Biogas Digester Constructed in MIMAROPA, 2000-2009

Year	*No. of Constructed Biogas Digesters*	*Operational Biogas*	*Non-Operational*
2000	3	3	
2001	3	2	1
2002	4	3	1
2003	2	2	
2004	9	3	6
2005	8	6	2
2006	9	8	1
2007	5	4	1
2008	4	3	1
2009	7	7	
Total	55	42	13

There was an increasing trend on biogas digester construction for the period from 2000 to 2009. The rate of construction was from two to four units per year during its introduction in 2000 to 2003. On the fifth year, when potential users were convinced that the technology was working, the construction doubled, with nine digesters per year. There was an observed slight decrease in construction to five and four in 2007 and 2008 respectively, but started regaining momentum to seven constructions in 2009.

Forty-two or about 76 per cent of these constructed biogas facilities are still operational, a majority of which are used for household cooking purposes, with 2 units being utilized for heating purposes in poultry production (Table 1.4). Thirteen or 24 per cent of the projects are non-operational due to change in business of the owners, termination of swine production operation, non-compliance to the DOST design of digester, and scheduled pipe clearing.

Table 1.4: DOST-MIMAROPA Assisted Biogas Projects and Nature of Use, 2000-2009

Use	*Number of Facilities*	*Percent (per cent)*
Cooking	40	95.2
Others	2	4.76
Total	42	100 per cent

Total investment for the construction of biogas facilities from 2000-2009 was PhP3.82 Million from the private sector/owners of the facility. These biogas digester units established in the region minimized household operating costs by eliminating the use of wood or liquefied petroleum gas (LPG). The major impact of this project is climate change mitigation through the use of animal waste as a natural resource to generate alternative energy.

Moreover, a maximum of 2,911 pig heads, the facilities in the region generate 5,822m^3 of biogas per day or 52.398 tons of carbon dioxide. This may be traded in the world carbon market at a rate of eight US Dollar per ton (SURE Inc., 2009) to generate PhP7.5 million development funds for the region, which is enough to construct common service facilities (CSF), technology business incubators (TBI) and other infrastructure facilities needed for regional development annually.

Organic fertilizer processing plant was another project on green technology located in Magsaysay, Mindoro Occidental. This project encourages the use of organically produced fertilizer in place of synthetic fertilizers which, as we know, cause harmful effects to humans and the environment. The technology intervention provided for the project is the acquisition of shredding equipment as replacement to the labor intensive process of grinding and mixing the fertilizer materials. The local government likewise provided counterpart in the form of land and building, composter drums, moisture presser and initial working capital. This project was able to provide direct employment for five individuals and benefit 118 farmer cooperative members, and generate a PhP50,000 income during its first semester of operation from an amount of PhP240,000 investment by DOST and PhP928,875 LGU counterpart.

The incorporation of technology on rice hull ash cement (RHAC) is another project on environmental management in Odiongan, Romblon. Marco Hardware and General Merchandise manufactures concrete hollow blocks from rice hull ash using 50 per cent rice hull and 50 per cent cement. The intrinsic property of rice hull as an insulator allow less moisture penetration thus prevents propagation of moulds and fungi when used as a component in concrete hollow blocks (CHB).

The package of technology provided includes ball mill grinder, hollow block machine, incinerator, motorized concrete mixer, and 2800-liter capacity water tank. The DOST assistance and client counterpart amounting to PhP350,000 and PhP650,000, respectively, resulted in a semi-annual production volume of 3,632 pieces of CHB valued at PhP1,950,000 and 120 beneficiaries including rice farmers, labourers, and construction supply retailers.

For the first quarter of 2010 alone, it has already produced 20,000 pieces of CHB with a gross income of P 119,000 and direct employment to 42 individuals composed of CHB makers and other workers.

Another project in MIMAROPA which aims at developing an alternative energy source to meet the power/fuel requirements is the establishment of a biodiesel (cocomethyl esther or CME) processing facility. Biofuels are renewable energy sources eyed worldwide due to the derived benefit of reducing the emission of greenhouse gases. (Impact data mentioned above under coconut products).

Metals and Engineering

Projects belonging to this sector comprise 9.40 per cent of all DOST assisted enterprises in MIMAROPA. These projects promote locally-developed technologies in the field of engineering, which is vital in making available fabricated equipment to the local enterprises. The projects assisted in regard to these industries are mostly the machine shops or metal work enterprises. Technology interventions come in the form of varied metal working tools and equipment such as hydraulic operated bending machines and welding equipment, which require substantial capitalization.

Other projects involve prototype development and improvement of engineering design of small-scale machines that could be made commercially available upon standardization and pilot-testing of the design. DOST interventions to such projects come in the form of provision of materials needed to develop prototypes and funds for trial testing of the units in field applications.

The total investment generated for these projects amounts to P4.16 million. Seventy-five per cent (75 per cent) of this amount or PhP3.15M came as counterpart funds from the clients. About 104 individuals were provided with employment and PhP3.18 million were the income generated from these metal and engineering projects.

A complete list of projects discussed in this section is available in Appendix 1.

Technology Based Entrepreneurship Development

As of 2009, four out of the six SUCs in the region were already accredited under the programme, namely: Occidental Mindoro State College (OMSC), Mindoro State

Figure 1.8: Hog Fattening, One of the Students' Projects in Mindoro Occidental during a Monitoring Activity

College of Agriculture and Technology (MinSCAT), Romblon State University (RSU), and Marinduque State College (MSC).

Various projects under the programme come from agriculture and other technology based business ventures. These projects were on hog fattening, goat-raising, cattle fattening, piglet production, tomato production, squash production, broiler production, and beekeeping.

Sixty-six students gained entrepreneurial skills while already starting to earn from their newly established projects.

Packaging and Labelling

Improved packaging allowed some firms to penetrate 16 new markets and increase sales by PhP361, 500 during the period of years 2008 and 2009.

For 2009, 35 firms and community organizations were assisted. Improved packaging allows some firms to penetrate other markets and increase their sales.

Figure 1.9a: Some of the Package and Label Designs Developed and Provided by DOST-MIMAROPA

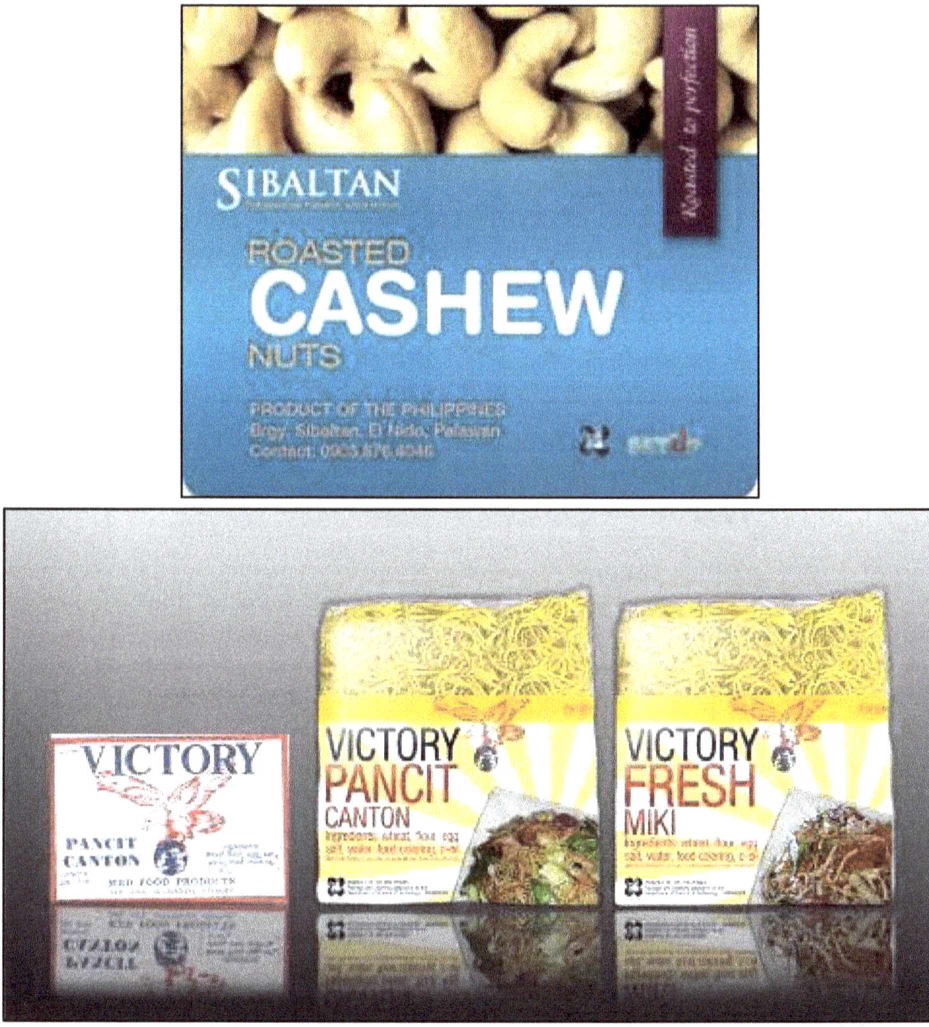

Figure 1.9b: Some of the Package and Label Designs Developed and Provided by DOST-MIMAROPA

Of the 35 firms, seven (7) received the packaging and labelling assistance worth of PhP1, 440,500 to procure packaging materials and execute product label designs. Also, the local GIA project entitled "Development of Innovative Packaging and Labelling for Selected Food Products in MIMAROPA" provided PhP281,500 to 14 firms and community organizations to enable these MSMEs to purchase the recommended packaging materials and implement the recommended package and label designs in order to penetrate the market.

Aside from the financial grant, the DOST-MIMAROPA also provided 38 package and label designs to 22 firms for products such as dried fish, chocolate and cassava cakes, processed and/or frozen meat products, *kalamayhati* (native delicacy made from a concoction of ground sticky rice, sugar and coconut milk) products and other sweets, *pastillas de leche* (carabao milk candy), virgin coconut oil, *sampalok*-based (tamarind) products, *pancit* (noodles), canton/miki, *bukayo* (sweetened young coconut), soap, ginger tea, *ube* (purple yam) cookies, iodized salt, fruit candies, banana chips, rice cakes and peanut butter.

The office was able to assist a large number of package and label designs because it enhanced its internal capability and lessened its dependence on external services in graphic design, product package and label requirements to meet the increasing clients' demand. The staff involved attended relevant training to improve skills in graphic design and product package and labelling.

Packaging and labelling assistance has an increasing demand from micro and small enterprises and will remain as one of the top priorities of DOST-MIMAROPA.

Consultancy for Manufacturing, Horticulture Crops and Aquamarine

Manufacturing

The manufacturing productivity extension was implemented for the first time in the MIMAROPA region in May 2009 with generated funds of PhP 948,636.00. Nineteen (19) firms from Palawan, 13 of which were from Puerto Princesa City, two (2) firms each from Brooke's Point and Roxas, and one (1) each from Narra and Taytay availed of the manufacturing productivity consultancy services in 2009. Of the 19 firms, there were 13 food processing firms and 6 GHD enterprises. These firms passed the prequalification requirements in terms of total number of workers, value of business, annual volume of production, size of the factory, and more importantly on the need for intervention.

The consultancy services include production planning and control, firm lay-out, operational and materials management schemes, 5S training, quality control assurance system, food safety and packaging options, among others.

Horticulture Crops and Aquamarine

For horticulture crops and aquamarine products, the experts were deployed for a period equivalent to one complete farming cycle to help farmers in the implementation of the recommended S&T interventions.

In 2008 and 2009, 46 farms composed of 17 aquaculture farms, 19 mango farms, and 10 banana farms benefitted from consultancy for horticulture crops and aquamarine. Average yield increased from 420 kg/ha to 1163 kg/ha while other mango growers were successful in early flower induction. In general, farm productivity improved except for those farms that were greatly affected by the typhoons during the project period.

The consultancy services resulted in the increase of the net income of three grow-out operators. On the other hand, five grow-out *Oreochromis niloticus niloticus* (tilapia) farmers declared break-even or suffered losses. These five operators could have also experienced increase in income if not because of the typhoons *Jolina*, *Kiko*, and *Ondoy*

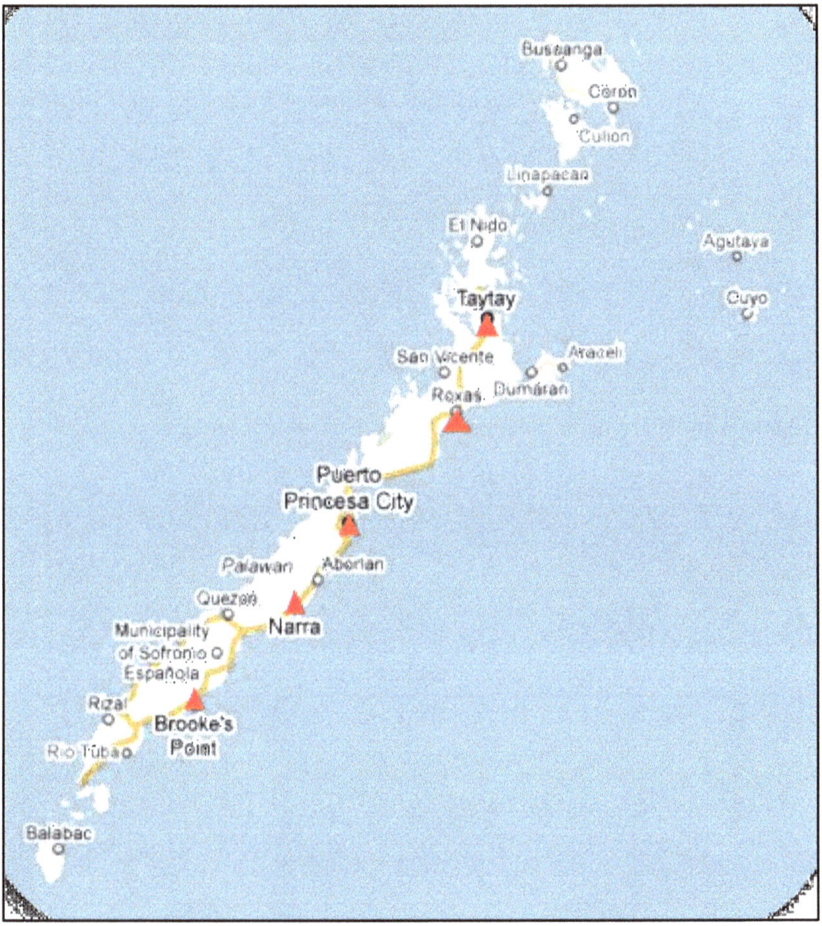

Figure 1.10: Palawan Map Showing Towns with Manufacturing Productivity Extension Clients

during the project period. However, farmers are expected to increase the farm outputs in the forthcoming harvests with the application of improved farming. Aside from augmenting farm income, these agricultural practices taught the farmers on how to reduce pests and diseases incidence.

Ten (10) tilapia farmers, of which eight (8) were engaged in grow-out production and two (2) in hatchery-nursery production from the municipalities of San Jose, Rizal, and Magsaysay in Mindoro Occidental improved their production through various interventions provided under the programme. By clustering these 10 farms into one group, the perennial problem of inadequate and irregular supply of high quality tilapia fingerlings was addressed through the two hatchery-nurseries in the cluster. Factors that contributed to low productivity in grow-out ponds such as improper pond preparation, poor pond water management, inappropriate fertilizer application, and poor quality feeds and feeding management were discussed and studied and recommendations were given.

Before

After

Figure 1.11: Before and after the Adoption of Productivity Recommendation in a Cashew Processing Enterprise

Despite the programme's desire to warrant immediate success, fish farming in brackish water ponds in Mindoro Oriental was faced with problems that may be classified as non-technology related such as poaching, non-compliance on the discussed and agreed technological interventions, and incomplete recording of inputs, outputs, and activities. Thus, while the programme provides fish farmers with technological know-how to improve productivity, it is suggested that fish farmers must be willing and able to implement the suggestions given by the consultants. Likewise, active and hands-on participation of pond operators/owners together with pond caretakers regarding pond management is also needed to ensure compliance with agreed technical interventions.

Figure 1.12: A Visit to Mango and Aquaculture Farms in the Region

The biggest drawback in mango production in the provinces of Mindoro Occidental and Romblon is the presence of unfavourable weather conditions particularly strong winds and rain showers coinciding with the crucial production stages of flowering and fruiting. To negate the ill-effects of weather disturbances and pest incidence, early flower induction and forced flushing were emphasized along with other appropriate crop protection and production management technologies such as proper sanitation, pruning, thinning, and fruit bagging. Farm visits were done regularly to motivate farmers to take necessary activities more seriously and ensure proper dosage applications of flower inducers, fertilizers and pesticides. To maximize farm productivity, agricultural practices such as diversified or integrated farming and organic farming were introduced.

More, participants and other interested mango growers from the municipalities of San Jose, Magsaysay, and Rizal in Mindoro Occidental were also trained on fruit processing and marketing by the consultants.

Technology Training

DOST-MIMAROPA conducted 96 training courses for 2,448 participants in 2008 and 2009.

Technical Courses

The training courses were mostly on food production such as fish processing, mushroom production, and mango processing. Other training courses that were conducted were on organic vegetable production, coffee production, rice processing, lemongrass processing, baking, salt-iodization and washing, candle-making, ceramic production, shielded metal arc wielding (SMAW), oxyacetylene welding (OAW), operation of radio phone/back-up communication, legal metrology, verification of balances, and verification of fuel dispensers.

Others were on product development and management practices, *i.e.*, current good manufacturing practices (cGMP), seminar/lecture on product packaging and

Figure 1.13: Some of the Training Activities Conducted

labeling and project proposal preparation for new and existing entrepreneurs. Moreover, a training course on intellectual property rights (IPR) for technology generators in SUCs and some regional line agencies (RLAs) were also conducted.

Pool of S&T Volunteers

The pool of S&T volunteers has deployed 79 experts and assisted 1,170 clients in providing technical expertise/advisory services, training and seminars to complement the various technical and financial assistances in the past two years.

Food Safety

The DOST-MIMAROPA organized the Food Safety Team (FST) to ensure a quick-response core group for the region's food manufacturing sector. The FST is composed of the five Provincial S&T Directors and one food technologist from the DOST-MIMAROPA, a representative each from BFAR and DOH, and two professors from Western Philippines University (WPU). The FST is tasked to provide the food-based MSMEs with the technical know-how on current food safety practices, regulations and requirements. To enhance their capability, the FST members attended training courses on current Good Manufacturing Practices (cGMP) in February 2009, and the Hazard Analysis and Critical Control Points (HACCP) in August 2009. They conducted their ISO 22000:2005 (Food Safety Management Systems) Training in March 2010. The FST members after having attended these trainings provide technical assistance to the MSMEs/community organizations involved in food processing on various areas where they were trained.

For 2009, the FST conducted food safety seminars, on-site compliance audits and/or trainings to at least thirty-one firms from Mindoro Occidental (5), Mindoro Oriental (6), Marinduque (3), Romblon (7), and Palawan (10).

Creating a Critical Mass of S&T Manpower

RA 7687

Of the 709 fourth year high school students from the MIMAROPA provinces who qualified to take the exam, 109 students or 15.37 per cent passed and were qualified to receive the scholarship. As seen in Table 1.5, there were more male qualifiers (54 per cent) than females (46 per cent) although there were more female examinees (66 per cent) than males (34 per cent). Of the 109 qualified students, 20 were enrolled in HEIs in MIMAROPA. The rest of the qualified students were enrolled in other accredited colleges and universities in the country.

Table 1.5. The 2009 examinees under RA 7687

RA 7687	*Total*	*Male*	*Female*
Examinees	709	241	468
Per cent	100	34	66

Of the scholarship programmes being implemented, BS regular and the ladderized programme are the most popular as reflected in Table 1.6. The two categories also topped in terms of male and female qualifiers, closely followed by Merit A and B.

BS with bridging and Best for IP had the least number of qualifiers as these two programmes were implemented for the first time only in 2009.

Table 1.6: The 2009 Qualifiers Under RA 7687

RA 7687	*Total*	*Male*	*Female*
BS Regular	45	25	20
Merit B	15	8	7
Ladderized programme	23	11	12
BS with Bridging Programme	8	5	3
BEST for Indigenous People (IP)	2	0	2
Merit A	16	10	6
Total	109	59 (54 per cent)	50 (46 per cent)

Figure 1.14: Orientation and Oath-Taking of Scholars

Junior Level Science Scholarship (JLSS) Programme

DOST-MIMAROPA also administers the JLSS. JLSS is a scholarship grant to qualified third year college students who are enrolled in priority fields of study. Unfortunately, while there were 33 examinees, none qualified in 2009.

Philippine Science High School (PSHS)

There were 143 Grade VI students from MIMAROPA provinces who took the entrance exams for the PSHS, of which 11 passed and were qualified to enrol at PSHS. Unfortunately, none of those who qualified actually enrolled in any PSHS campus.

Coordination and Monitoring of On-Going Scholars

For SY 2009-2010, DOST-MIMAROPA coordinated and monitored 48 scholars enrolled in various SUCs in MIMAROPA region.

Table 1.7: DOST Scholars in MIMAROPA

	Total	*Male*	*Female*
No. of Scholars	48	25	23
	100 per cent	52 per cent	48 per cent

Although there were 109 scholarship qualifiers from MIMAROPA for the school year, only 20 freshmen enrolled in the SUCs of the region as most of the students would prefer to enroll in universities in Metro Manila and CALABARZON Region.

S&T Promotion

Technology Forums

There were 28 technology clinics/forums that were conducted in 2009 with 1,306 individuals and 195 firms. The participants were DOST's target serviceable sectors and groups. They were interested in DOST's food processing technologies, and technologies on fiber-based industries, technology acquisition and upgrading assistance, technology and livelihood opportunities for women and packaging and labelling services. Funding mechanisms, HRD opportunities and scholarships are also discussed during forums as side topics.

Science and Technology Fairs

In 2009, DOST-MIMAROPA sponsored two provincial fairs and participated in 10 fairs, including trade fairs where clients are sent to showcase their products and the DOST interventions accorded to them. In the South Luzon Cluster (SLC) S&T Fair, the provinces' ecotourism programmes, destinations and the delectable food products assisted by DOST-MIMAROPA were showcased. This is to attract domestic and foreign tourists who serve as the most promising target market of DOST-MIMAROPAs assisted MSME products.

During the South Luzon Cluster S&T Fair held in SM City Sta. Rosa, Laguna, Dr. Ma. Josefina P. Abilay, DOST-MIMAROPA Regional Director discussed the region's assisted projects in the morning TV show "One Morning Café" of NBN 4, a government channel.

DOST-MIMAROPA will be hosting the SLC S&T Fair in Puerto Princesa City on September 14-18, 2010 with the theme "Filipinnovation: the way forward". The S&T Fair in 2010 became a community event as the city government with five

Figure 1.15: Technology Forums Conducted

Figure 1.16: Regional Director, Ma. Josefina P. Abilay being Interviewed Regarding the Technology Acquisition and Upgrading Programme

Figure 1.17: S&T Fair Opening Programme, Exhibit Booths, Products Display and Centrepiece

non-government organizations (NGOs) and two State Universities and Colleges (SUCs) participated as part of the technical working group (TWG). The NGOs are the City Tourism Council, Local Chamber of Commerce, Specialty Food Association, Hotel and Restaurants, and Travel and Tours association, while the SUCs are the Western Philippines University (WPU) and the Palawan State University (PSU).

Also, the Regional Invention Contests and Exhibits (RICE) will be hosted by the regional offices for the first time in 2010. The South Luzon Cluster (SLC) version will be hosted by DOST Region V in 6-8 September, 2010 in Legazpi City, Albay. At the moment, DOST-MIMAROPA has started to organize inventors in the region to find out who can participate at such events.

Tri-Media

To maximize dissemination of information with no additional cost to DOST, Romblon PSTC hosted a 30-minute radio programme every Thursday at Odiongan and Looc *Radyo Natin* (Our Radio) discussing various science and technology related topics relevant to the province. More, Romblon DOST Provincial S&T Director is a regular contributor to Romblon Sun, a local newspaper that features updates and discussions on the DOST programmes and services. Nutrition Communication Network (NUTRICOMNET) in Mindoro Occidental and Philippine Science Journalist

Figure 1.18: A Press Conference on DOST Services

Association Inc. (SCIJOURN) in Marinduque continue to be regular S&T promoters while other PSTCs rely on the local Philippine Information Agency (PIA) and local cable TV channels.

The DOST Secretary, Estrella F. Alabastro visited Marinduque dubbed as "S&T Projects: A Showcase" on May 20-22, 2009. A press conference was conducted after the project visits, where the technology acquisition and upgrading programme, as well as the on-going/existing assisted projects in Marinduque and the whole region were discussed. Likewise, other successful community-based projects were discussed during the Secretary's project visits around the country.

Library Services

This year, about 1, 572 library and information services users were recorded, most of which were students. Most of the topics being researched were the thesis writing and production technologies.

Intensifying efforts on S&T promotion resulted in increase in the number of clients who availed of the DOST-MIMAROPA services particularly on scholarships and innovation to support technology acquisition and upgrading.

R&D to Support Technology Needs

One of the strategies done to make R&D to be responsive to technology needs is the establishment of the Southern Tagalog Islands Research and Development Consortium (STIRDC) with thirteen (13) founding member organizations composed of the six SUCs and seven regional line agencies in the MIMAROPA Region. STIRDC's main purpose is to develop strategies and implement a dynamic system of research and development attuned to the development needs of the MIMAROPA Region, with

Figure 1.19: A Capability Building Activity of the Researchers of the Southern Tagalog Islands Research and Development Consortium (STIRDC)

the end view of promoting and utilizing research results and information to support regional, as well as national development goals.

Recently, the consortium was accredited by the Philippine Council for Industry and Energy Research Development (DOST-PCIERD). It also started linking with the Engineering Research and Development for Technology (ERDT), to collaborate efforts on applied R&D, initially, on the following areas: coastal engineering, groundwater transport and decontamination, development of food processing technologies for secondary staple foods and postharvest handling.

The health and research consortium is also at the initial stages of exploring its establishment in the region. Recently, a survey on institutions and foundations, SUCs with courses on medicine, health and allied services, and hospitals were conducted to identify the extent of health related research activities in MIMAROPA provinces.

All these efforts are geared towards promoting technology acquisition and upgrading as it provides interventions towards rural industrialization.

Networks and Linkages as Support Mechanisms

With the technology acquisition and upgrading, various efforts and collaborations are entered into by DOST-MIMAROPA to serve as either key player or supporter to the application of S&T to stimulate rural industrialization.

Likewise, DOST-MIMAROPA plays a key role in inter-agency associations in the region promoting the utilization of S&T for the growth of the economy, health, environment and other sectors of society for purposes of complementation of programmes and services including resources in providing government services to the public.

Contribution of DOST Assisted Clients to the Economy

Gross Regional Domestic Product

The gross regional domestic product (GRDP) is an economic tool used to measure the total market value of all the goods and services produced by the region's economy for a given period. It serves as an evaluation mechanism to assess the value of the region's contribution to the growth of the national economy.

Table 1.8: MIMAROPA Contribution to GDP and Growth Rates, 2004-2008

Year	*Philippines*	*Per cent Growth Rate*	*MIMAROPA Region*	*Per cent Growth Rate*	*Per cent Contribution to GDP*
2004	1,152,173,648	-	31,688,281	-	2.75
2005	1,210,497,421	5.06	33,740,765	6.48	2.79
2006	1,276,435,452	5.45	34,526,488	2.33	2.70
2007	1,366,492,669	7.06	37,648,674	9.04	2.76
2008	1,418,952,296	3.84	38,819,577	3.11	2.74
2009*					
Average	1,284,910,297	5.35	35,284,757	5.24	2.75

Source: National Statistical Coordination Board (NSCB)

* Data not yet available.

Table 1.8 shows MIMAROPA Region's GRDP as against the GDP from 2004 to 2008. The region registered an average annual growth rate of 5.24 per cent with a value of Php 35billion for the period covered. The biggest growth in GRDP was noted in 2007 with 9.04 per cent and according to NSCB, this was attributed to the hefty gains of the industry sector–the top contributing sector as illustrated in Figure 1.20, with a 41.25 per cent share of the region's output in the same year. It could be noticed that the region's economy slowed down from a growth of 9.04 per cent in 2007 to 3.11 per cent the next year inflicted by the extensive damage caused by the natural disasters particularly in the agricultural sector which is the next major contributor to the regional economy of about 39.6 per cent in 2007 as shown in Figure 1.20.

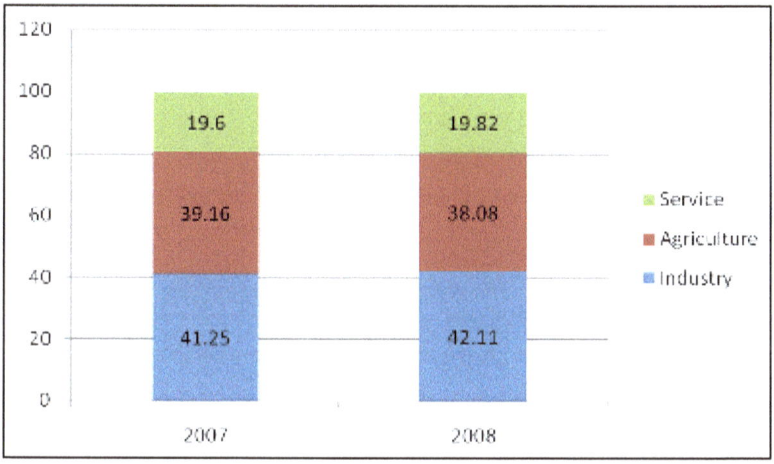

Figure 1.20: GRDP Per cent Distribution by Industrial Origin, MIMAROPA, 2007-2008

The contribution of MIMAROPA region's GRDP to the national account ranges from 2.70 per cent (in 2006) to 2.79 per cent (in 2005) is also presented in Table 1.8. The region's computed average contribution to GDP for the five-year coverage is 2.75 per cent.

Table 1.9: GRDP Distribution by Industrial Origin, 2007-2008

Sector	*2007*	*Per cent Contribution*	*2008*	*Per cent Contribution*
Industry	15,528,237	41.25	16,346,594	42.11
Agriculture	14,742,743	39.16	14,780,661	38.08
Services	7,377,694	19.60	7,692,322	19.82
GRDP	37,648,674	100.00	38,819,577	100.00

Source: NSCB.

In 2008, the industry sector continued to account for the largest share of the regional economy contributing to 42.11 per cent followed by the agriculture sector with 38.08 per cent, and the service sector with 19.82 per cent as illustrated in Table 9 and Figure 1.20.

Table 1.10: Contribution of Industry Subsectors, 2007-2008

Sector/Sector	*2007*	*Per cent Contribution*	*2008*	*Per cent Contribution*
Industry Sector	15,528,237	100	16,346,594	100
Mining and Quarrying	7,603,469	48.97	8,109,350	49.61
Manufacturing	4,391,822	28.28	4,509,880	27.59
Construction	328,659	2.12	362,299	2.22
Electricity and Water	3,204,287	20.64	3,365,065	20.59

Source: NSCB.

Analyzing the industry sub-sectors' performance for 2007 and 2008, mining and quarrying accounts for the biggest chunk of the industry sector's GRDP (50 per cent), followed by the manufacturing sub-sector (28 per cent), then electricity and water (21 per cent) and lastly, the construction sub-sector (2 per cent) as shown in Table 1.10.

Table 1.11: Contribution of DOST-Assisted Projects to GRDP, 2007-2008

	2007 (Php '000)	*Per cent Contribution*	*2008 (Php '000)*	*Per cent Contribution*
GRDP	37,648,674		38,819,577	
Manufacturing	4,391,822	41.25	4,509,880	42.11
Gross Income of DOST-Assisted Projects	25,379.40		15,136.60	
Contribution to GRDP		0.07		0.04
Contribution to Manufacturing		0.58		0.34

Table 1.11 shows the DOST-assisted projects in 2007 posted a gross income of Php 25.4Million, which was 0.58 per cent of the manufacturing sector's GRDP or 0.07 per cent of MIMAROPA GRDP. Whereas, the DOST-assisted projects in 2008 contributed a gross income of Php 15Million which was 0.34 per cent to the manufacturing sector's GRDP or 0.04 per cent of the MIMAROPA GRDP. It could be recalled that it was in 2007 when a notable increase in the performance of the industry sector was observed thus explains the significant difference in the income generated from 2007 to 2008 of DOST-assisted projects.

Investments Generated through Projects in Technology Acquisition and Upgrading

The DOST-MIMAROPA projects generated a total investment of PhP46.3 million in 2008 and 2009 (Table 1.12). Sixty two per cent (PhP28.68 million) came from other sources such as LGUs, RLAs and NGOs. The DOST-MIMAROPA contributed 25 per cent (PhP11.4 million) of the total investment while cooperators shared 13 per cent (PhP6.3 million).

Sixty-six per cent (Php 14.9 million) of the investment in 2008 came from other sources. Only 22 per cent (Php 5 million) came from cooperators. The DOST contributed 12 per cent (Php 2.63 million) of the total investment generated.

In 2009, other sources remained as the major contributor to the investment with a 58 per cent (Php 13.7 million) share. DOST-MIMAROPA's share increased to 37 per cent (Php 8.79 million) and cooperator's share decreased to 5 per cent (Php 1.15 million).

Comparing the investment generated as against the budget based on the General Appropriations Act (GAA) of DOST-MIMAROPA in 2008 and 2009, which are Php 8.3 million and Php 8.7 million respectively, this regional office was able to generate Php 2.73 in 2008 and Php 2.72 in 2009 for every peso invested by the agency from the total investment generated from projects (Table 1.13). The investment generated for every peso provided by DOST-MIMAROPA was almost the same with only a slight difference in the GAA of the two consecutive years.

Table 1.12: Investment Generated Through Technology Acquisition and Upgrading Projects, Core and Non-core, 2008-2009

Year		*Investments*			
		DOST	*Co-operator*	*Other Sources*	*Total*
2008	Core	2,354,383	618,430	6,998,380	9,971,193
	Non Core	274,400	4,471,540	7,941,620	12,687,560
	Total	2,628,783	5,089,970	14,940,000	22,658,753
2009	Core	6,912,500	800,000	10,640,000	18,352,500
	Non Core	1,879,600	350,000	3,100,000	5,329,600
	Total	8,792,100	1,150,000	13,740,000	23,682,100
Both	Core	9,266,883	1,418,430	17,638,380	28,323,693
	Non Core	2,154,000	4,821,540	11,041,620	18,017,160
	Total	11,420,883	6,239,970	28,680,000	46,340,853

Table 1.13: Investments Generated by DOST, 2008-2009

Investment Generated	*2008*	*2009*
Against DOST-MIMAROPA GAA	2.73	2.72
Against core-funded (DOST)	0.26	0.11
Against non-core funded (DOST)	45.2	1.84

The total investment generated from technology acquisition and upgrading core projects in 2008 amounted to Php 9.9 million. DOST invested Php 2.3 million, while cooperators invested Php 618, 430 as shown in Table 1.12. This implies that for every peso invested by DOST, Php 0.26 served as counterpart investment of the cooperators in 2008.

An amount of Php 18.3 million total investments was generated from core funded projects in 2009. Out of the total investment, cooperators invested Php 800,000 and DOST invested Php 6.9 million. This means that cooperators invested Php 0.11 as counterpart investment for every peso invested by DOST. It is observed that the amount of DOST assistance increased as a result of more clients availing the service, although the counterpart fund provided per client for every peso from DOST decreased since most of them were kitchen-scale.

For 2008, the non-core funded projects generated a total investment of Php 12.6 million. From this, DOST invested Php 274,400 while the remaining came from cooperators and other sources, implying that for every peso invested by DOST, a Php 45.2 counterpart was invested by cooperators and other sources such as LGUs, RLAs and NGOs.

In 2009, the DOST invested Php 1.8 million for its non-core funded projects while the cooperators and other sources contributed a total investment of Php 3.4 million. The cooperators and other sources invested counterpart investment of Php 1.84 for every peso invested by DOST. The decrease in the amount of counterpart funds from 2008 to 2009 could be accounted for the increase in the DOST-MIMAROPA investment. The office provided more non-core funds in 2009 to address the need of non-core clients, mostly community organizations, for government assistance.

4.3 Regional Employment

Employment generation is an established indicator measuring the level of economic growth for an area at a specified period or duration. It is also being used as a device in determining the extent of impact of any intervention to a particular project.

MSMEs and community-based enterprises are the identified core clients of DOST-MIMAROPA. The technology interventions provided enable them to enhance their productivity and increase production, hence, creating additional manpower requirement to sustain their level of operations. These enterprises operate within a value chain system linking independent units that require different sets of labor force. The forward and backward links in this supply and demand network operates with an end result of income and employment generation.

Food industry related projects have the largest percentage in terms of its contribution to employment generation with an average of 75.35 per cent. This result was expected given that about 79.7 per cent of the overall recipients of technology assistance come from the food sector. This is followed by the metals and engineering, and environmental management projects, both contributing a 10 per cent share to the total pie.

Table 1.14: Employment Generated by DOST-MIMAROPA Assisted Projects by Sector, 2007-2009

Sector	*Employment Generated*	*Distribution of Employment Generated*
Food	1,875	75.35
Metals and Engineering	259	10.42
Environmental Engineering	257	10.32
Furniture	37	1.5
Gifts, Housewares and Decors (GHD)	36	1.45
Total	2,489	100 per cent

Employment generation is a function of the number and type of assisted enterprise, investment generation, and the scale of operation of the individual business units. Based on its CY 2007-2009 projects on S&T application, DOST-MIMAROPA was able to provide employment to a total of 2,489 individuals coming from the different sectors (Table 1.13). These generated jobs include enterprise-level employment and those created from the entire value chain during the course of project implementation.

Table 1.15: Contribution of DOST-MIMAROPA Assisted Projects to Regional Employment, 2007-2009

Year	*DOST-MIMAROPA Assisted Projects Employment Generated*	*Labor Force* in MIMAROPA*	*Per cent Contribution*
2007	535	1.67 M	0.032
2008	840	1.17 M	0.071
2009	1,114	1.22 M	0.091
Total	2,489	4.06 M	0.064 per cent (Average)

**Source*: Bureau of Labor and Employment Statistics, Department of Labor and Employment (DOLE).

In 2007, a total of 535 individuals were provided with employment, which is equivalent to about 0.032 per cent of the 1.67M total labor force in the region per the Philippines' DOLE data. A majority of these employments came from the metals and engineering sector which was the emerging industry during the year.

In 2008, DOST-MIMAROPA assisted projects contributed 0.071 per cent or 840 employment generated out of the regional labor force of 1.17M individuals. This is a 55 per cent increase in regional labor force contribution since 2007 that could be attributed to the increasing number of assisted enterprises. The employment generated during this period mostly came from the food sector, which began to grow with the continued rise of backyard food industries in the region.

In 2009, the assisted projects' share on the total regional labor force was 0.091 per cent, which was a 22 per cent increase compared to that of 2008 employment data. This is equivalent to 1,114 additional jobs in the region, which covered projects mostly from the food sector.

Conclusion

Lessons Learned

According to the new DOST Secretary Mario G. Montejo:

> *"The use of S&T for the betterment of society in terms of better products, services and solutions is a sound business model".*

The DOST-MIMAROPA has been doing this by actively promoting and orchestrating the use of S&T to increase income and employment in order to improve the quality of life of the people.

Likewise, positive lessons are learned from the past as stated by the former DOST Secretary, Estrella F. Alabastro in a message for the 2008 annual report of the regional office:

> *"Let me congratulate the Regional Director, Dr. Ma. Josefina P. Abilay, and the staff of DOST-MIMAROPA for their dedication to meeting the day-to-day challenges of delivering quality service to the people of the region. In this time of global economic slowdown, we hope that they will continue to focus their efforts on the development of a forward-looking, knowledge-intensive economy through innovation, hard work, and an unwavering commitment to excellence."*

Assistant Secretary and Officer-in-Charge of the Office of the Undersecretary for Regional Operations Carol M. Yorobe supported this by saying:

> *"Despite the limited human resource, infrastructure, funding and other resources, DOST-MIMAROPA has lived up to its targets for the year. As a newly established Regional Office, my hope is that the seeds of hard work in transferring DOST technologies and in promoting S&T services will progress to bring more benefits to the lives of the people we serve".*

The learnings encountered in bringing S&T services to the rural communities is a way to making improvements so that every accomplishment becomes a sweeter success and successes are brought to greater heights.

Towards a More Effective and Responsive S&T Services Delivery

DOST-MIMAROPA recognizes that lessons learned should give way towards a more effective and responsive S&T services delivery. A study made by the author in 1997 examined the factors associated with which technologies were adopted in the Philippines (Abilay, M.J.P., 1997). The study was conducted to develop marketing strategies for new technologies generated by the DOST System based on actual survey of adopters and non-adopters of the technologies of DOST. The influence on their decision to adopt the DOST technologies were carefully noted. Based on the study's empirical data and confirmed by the actual experiences of DOST-MIMAROPA, the following were identified:

- ☆ Conduct of TNA to determine appropriate technology; and the technical advantage of the technology itself.
  - TNA is an effective tool to ensure appropriate technological intervention for proposed projects
  - Benchmarking could also be done to find out the best technology available in the class and then evaluate where the DOST's available technologies would fall. This would allow for a more appropriate positioning of the available new technologies. Moreover, benchmarking would give information on what kind of new technology could be generated next which is simultaneously based on the needs of the market.
- ☆ Good prospects for economic benefits such as the marketability of the product or service and overall viability of the business is an effective mechanism for convincing MSMEs and community based organizations to avail of the S&T interventions that is crucial for rural industrialization.
- ☆ Entrepreneurial client with absorptive capacity to adopt the technology
  - Identification of entrepreneurial people as technology transfer beneficiaries is important.
  - Successful adopters making testimonies about the DOST- generated technology is another way of improving its marketability. Success stories can attract the potential users to avail and benefit from the assistance of DOST.

- Technology investment forums if done with focus on intended users are also vital tools in technology utilization as they are made aware of the technologies offered in the forums. With direct interaction between the technology generator and the technology user, specific concerns from the potential use of the technology are sorted out.
- "Focused publicity" also helped as those seriously interested in acquiring new technologies and related services monitor such publications.

☆ Support Services

- SUCs/local researchers to conduct quick production shop floor applied R&D for MSMEs to respond to the immediate needs of the industry
- Continued/sustained linkages between the DOST-MIMAROPA and the LGUs, NGOs and other pertinent organizations that provide services to the adoptors of technology.

☆ Technology transfer unit

- The Regional Offices/Provincial Science and technology Centers (PSTCs) play a crucial role as technology transfer units for DOST-generated and other commercially available technologies being at the forefront of the undertaking. Personnel with background in certain technical areas as well as in business and entrepreneurship helped since technology transfer is a business activity.
- Furthermore, it is important and necessary to keep S&T personnel updated of new information and new technologies generated by the DOST research network and to provide the regional Office/PSTCs personnel a periodic seminar/training to keep them effective in their technology promotion and commercialization role.
- Cooperation within and outside the technology transfer unit and support of top management in all activities are very crucial and provides indicators for successful transfer of technology from the agency to its users. Regional Offices/PSTCs also need to be in regular touch with the existing users of the new technologies of DOST and with those who have stopped using them to get feedback in order to improve the technology and the delivery of related services.
- Continuous competency improvement of S&T personnel, provision of adequate logistical support such as manpower and transportation shall be put in place to support/sustain the delivery of S&T services in the countryside.

Acknowledgement

I am heartily thankful to the Department of Science and Technology Secretary, Mario G. Montejo; Undersecretary for S&T Services, Fortunato T. Dela Pena; Undersecretary for R&D, Graciano P. Yumul; and OIC for the Office of Undersecretary for Regional Operations, Assistant Secretary Carol M. Yorobe, whose encouragement, guidance and support enabled me to succeed in all my undertakings as a Regional Director, including this endeavour.

I also offer my appreciation to all my staff that helped me in the directions set for the regional office under my leadership and also during the completion of the paper. To my family, thank you for all the love! And to God be the glory!

References

1. Abilay, M.J.P, 1997. Factors Influencing the Decision to Adopt New Technologies, Thesis (DBA) University of the Philippines, Quezon City.
2. DOST Administrative Order No. 001, Series 2004. Implementing Guidelines for the Operationalization of DOST-CALABARZON and DOST-MIMAROPA in Accordance with Executive Order No. 103 and DBM National Budget Circular 484. Taguig City, Metro Manila
3. DOST-MIMAROPA, 2007. Performance Review Report. January 2006 – June 2007. Taguig City, Metro Manila.
4. DOST-MIMAROPA, 2009. Annual Report 2008. Taguig City, Metro Manila.
5. DOST-MIMAROPA, 2009. Annual Report 2009 (Unpublished). Taguig City, Metro Manila.
6. Executive Order No. 103, 2003. Dividing Region IV into Region IV-A (CALABARZON) and Region IV-B (MIMAROPA), Transferring the Province of Aurora to Region 3 and for Other Purposes. Manila
7. Executive Order No. 161, 2007. Institutionalizing Quality Management System in Government. Manila.
8. Juma, C., Fang, K., Honca, D., Huete-Perez, J., Konde, V., Lee, S. H., *et al.*, 2001. Global Governance of Technology: Meeting the Needs of Developing Countries. Int. J. Technology Management, Vol. 22, Nos. 7/8, pp. 629 - 655.
9. National Statistical Coordination Board (NSCB), 2009. Regional Accounts of the Philippines. Makati City
10. Stads, G.-J., Faylon, P. S., and Buendia, L. J. 2007. Key Trends in Agricultural R&D Investments in the Philippines. Agricultural Science and Technology Indicators, pp. 1-2.
11. UNESCO Institute of Statistics, 2007. *Statistics Data Centre* http://stats.uis.unesco.org/unesco/TableViewer/tableView.aspx?ReportId=1782 (retrieved July 08, 2010).
12. Viava, J.C.A. 2009. Application of Biogas Technology: Carbon Market Potential, Presentation. Solutions Using Renewable Energy (SURE Inc.).

Appendix 1
List of Selected Ongoing Projects of DOST-MIMAROPA by Commodity/Sector

Sl.No.	Commodity	Project Title	First Name	Amout (Php)
		FOOD		
1.	Arrowroot	Facility Upgrading of Arrowroot Starch Production and Product Quality Improvement (*Phase 1*)	Rejano's Bakery	466,430.00
2.	Noodles	Upgrading of Noodle Production	JKarlo Trading	70,000.00
		Upgrading Noodle Processing Facility for the Production of Fortified Products	Fresh Miki Making Enterprise/ Ms. Margie F. Faderanga/ Dapawan, Odiongan, Romblon	275,000.00
		Upgrading of Noodles Processing Center	MBD Food Products/Ms. Macrina B. Delfin/ Labangan-Poblacion, San Jose, Occidental Mindoro	530,000.00
		Upgrading of Food Processing Facility	Provincial Nutrition Office Boac Marinduque	230,000.00
		Productivity Improvement of Seaweed Canton Noodles	Samahan ng Mga Mangagawa sa Balatasan	264,000.00
3.	Cashew and Peanuts	Upgrading of Peanut Processing Enterprise	Glowens Food Products/Ms. Preciosa Ilao Fojas/Bansud, Oriental Mindoro	240,000.00
		Improving Peanut Processing Facility of MamasYamiDeli	MamasYamiDeli/Ms. Ellen E. Muros/ Brgy.Ligaya, Odiongan, Romblon	300,000.00
		Upgrading the Production of Peanut Butter and Coco Vinegar	BUBOG SELF-EMPLOYMENT ASSISTANCE KAUNLARAN, MPC/Bubog, San Jose	113,200.00
		Sibaltan Community Cashew Enterprise Project	Sibaltan, El Nido, Palawan	150,000.00
4.	Coconut	Upgrading of Coconut Production facility for the Production of refined Bleached Deodorized Oil	DBN Manufacturing Enterprise/ Mr. Nolan G. Ngo/Porvenir St., Poblacion, Looc, Romblon	950,000.00
		Facility Upgrading of MarinduqueLand Processing Plant	MarinduqueLand Corp.	508,000.00
		Development of Packaging for Coco Water Vinegar	Catubosan Multi-Purpose Cooperative	160,000.00

Contd...

Appendix 1–Contd...

Sl.No.	Commodity	Project Title	First Name	Amout (Php)
		Livelihood Program for Community Based Enterprise	Samahang Palawano, Brookes Point, Palawan	150,000.00
		Development of Coco Coir Industry	Arya Coconut Farmers Multi Purpose Cooperative, Calatrava, Romblon	400,000.00
5.	Processed Fish	Upgrading of Equipment for Bagoong Production and Product Quality Improvement	DCM Bagoong	142,000.00
		Upgrading the Fish Paste and Sauce Production Process of Brader Food Products	Brader's Food Products/Mr. Dionisio Jiminez Julian Brgy. Concepcion, Calintaan, Occidental Mindoro	216,874.00
		Establishment of Milkfish Processing Center	Marias Marine Products	231,000.00
		Establishment of a Lamayo (Semi-dried)' Fish Processing Enterprise	APLAYA Women's Association, Bulalacao, Mindoro Oriental	150,000.00
6.	Ice	Improving the Production Process of d'Icehouse	D'Icehouse Food Products/Ms. Leticia F. Santiago/San Roque II, San Jose, Occidental Mindoro	995,000.00
7.	Salt	Upgrading of Salt Production Enterprise	Mr. George dela Cruz, Mindoro Oriental	10,000.00
		Upgrading of Salt Production Enterprise	Mr. Bobby dela Cruz, Mindoro Oriental	
		Upgrading of Salt Production and Iodization	Madrigal Salt Farm; Duque Salt Farm; Magsino Salt Farm; JALD Inc. Corp.	
8.	Rice and Corn	Upgrading the Mobile Rice Milling Facility of KAAKIBAT	KAAKIBAT Cooperative/Mr.Juanito M. Masib/ Labangan-Poblacion, San Jose, Occidental Mindoro	850,000.00
		Upgrading of the Grain Drying Facility of JALIRI Farms	JALIRI Farms	990,000.00
		Production of Organic Fertilizer	Mindoro Christian MPC	163,000.00

Contd...

Appendix 1–Contd...

Sl.No.	Commodity	Project Title	First Name	Amout (Php)
9.	Input Support	Input Support to Meat Processing Enterprise	CNVS Feed Mill/Mr. Cesar C. Bagtas/ Adrialuna, Naujan, Oriental Mindoro	990,000.00
		Productivity Enhancement of Meat Processing Enterprise	Bansud Livestock Marketing Cooperative (BLMC)	150,000.00
		Animal Feed Mill Establishment	KAD Bayan MPC	511,000.00
10.	Aquatic Resources	Enhancing Productivity of Fisheries Resources through Appropriate Technologies in Marinduque	Seaweed Farmers Association	340,000.00
11.	Kitchen-scale Food Processing Enterprise	Modernizing the Baking Equipment and Tools of Chariz Cakes	Chariz Cakes and Party Balloons/ Ms. Jennifer S. Cometa/Lakandula St., San Jose, Occidental Mindoro	188,500.00
		Improving the Production Process of Dhel's Food Products	Dhel's Food Products/Mr. Felipe M. Cajayon/ Mabini Ext., San Jose, Occidental Mindoro	97,359.00
		Upgrading the Soya Hon Beans Milk Production of Soya, Hon!	Soya, Hon! Soy Products/Ms. Estela P. Lilagan/Labangan, San Jose, Occidental Mindoro	105,550.00
		Improving the Cooking Process and Product Packaging/Labeling of Richblitz Sweets	Richblitz Sweets/Ms. Grace D. Dinglasan/ San Roque II, San Jose, Occidental Mindoro	86,200.00
		Improving the Production Process of NICAH Food Products	NICAH Food Products/Ms. Gloria M. Sabay/ Pag-asa, San Jose, Occidental, Mindoro	121,500.00
		Improving and Expanding the Production of Jelotin Crispy Potato	Jelotin Crispy Potato/	159,820.00
12.	Calamansi	Calamansi Processing	Tugdaan Center for Human Resources Development, Inc., Naujan, Mindoro Oriental	34,000.00
		GIFTS, HOUSEWARES and DECORS (GHD)		
13.	Marble Crafts	Upgrading of Equipment for Production of Marble	Encarnacion Marble Craft/Ms. Estella P. Encarnacion/Caplacan, Romblon, Romblon	200,000

Contd...

Appendix 1–Contd...

Sl.No.	Commodity	Project Title	First Name	Amout (Php)
		METALS and ENGINEERING		
14.	Metal/ Machine Works	Enhancement of Product Quality and Service of Metal	Ulip Welding Shop/Mr.Jaime A. Ulip/ Pangalaan, Calapan City, Oriental Mindoro	400,000
		Expansion of a Metal Working Shop	Darnel's Iron Works	250,000
	Construction	Production of Hollow Blocks from Rice Hull Ash Cement Materials	Marco Hardware and General Merchandize/ Mr. Eddie M. Fabila/Brgy. Anahao, Odiongan Romblon	350,000
	Prototype Development	Development of Improved Banana Chipper	WPU, Aborlan, Palawan	55,280.00
		Development of Improved Cashew Sheller	WPU, Aborlan, Palawan	44,165.00
		ENVIRONMENTAL ENGINEERING		
15.	Green Farming	Production of Organic Fertilizer	Mindoro Christian MPC	163,000.00
		Organic Fertilizer Processing Plant (Ecological Recycling and Composting Center)	Local Government of Magsaysay/ Allium Growers' MPC	250,000
		Greenhouses in Support of the RSC Provincial Tissue Culture Laboratory	Romblon State College, Odiongan, Romblon	300,000.00
	Biofuels	Utilization of CME as Diesel Fuel Substitute in Romblon - 3rd Phase	SVFPMPC, Romblon	
		Process Improvement for CME Production of the Romtron Biodiesel Plant	St. Vincent Ferrer Parish MPC	390,000
		Biogas as Alternative Energy Source and Waste Abatement System	IMPLEMENTED IN ALL PROVINCES	

Appendix 2a: Briefs of Selected Projects

Rejano's Bakery is known for its cookies made from first grade high-fiber starch extracted from arrowroot tubers, organically grown in Marinduque. These crispy and crunchy cookies are ideal for picnics, snacks or parties.

PROJECT DESCRIPTION: Improvement of production capacity through upgrading of facility, and consultancy on the sustainability of appropriate raw material supply to achieve desired product quality

DOST INTERVENTIONS: R & D of the process of making arrowroot flour then assistance in the acquisition of appropriate equipment to make the process right;

The following set of equipment was acquired: tubers washer, dryer with stainless tray, filter centrifuge for pulp expulsion, starch extractor-extruder, fabricated drying shelves with pan, platform and table scale;

cGMP seminar; Cleaner Production Technologies; Product package and label design; product promotion

DATE OF IMPLEMENTATION: March 2006
DOST ASSISTANCE: P 466,400
CLIENT COUNTERPART: P 1,500,000

STATUS: Actual production capacity is 1,170.83kg per month, valued at PhP58, 641.58; completed payment on May 2009

IMPACT: Improved product quality; Increased flour output by 5.02%; 263 direct and indirect employment generated

MARKET: SM, Landmark, and Robinson's Supermarkets

CONTACT:
Ms. Carmelita Reyes
Rejano's Bakery
Banahaw, Sta. Cruz
Marinduque
Phone: 042-311-1069

Appendix 2a: Briefs of Selected Projects

PROJECT DESCRIPTION: Upgrading of noodle production enterprise

DOST INTERVENTION: Technical assistance on plant layout, cGMP and Food Safety , Noodle Formulation; Productivity improvement through provision of furnace, stainless vat and production accessory to increase production efficiency of noodles; Renovation of noodle factory for GMP compliance based on the plant layout recommended by FNRI in 2006

DATE OF IMPLEMENTATION: September 2007 - August 2008 (being monitored)

DOST ASSISTANCE (SETUP):	PhP 530,000
CLIENT COUNTERPART:	PhP 780,000

STATUS: Production capacity of 260 kgs. of miki noodles per day

IMPACT: 30% increase in productivity of the enterprise from 200kgs/day to 260kgs/day production capacity; gross income of P 1.9M in 2009; 60 direct and indirect employment per year

MARKET: Provincewide

PROBLEMS ENCOUNTERED:

RECOMMENDATION:

CONTACT:
Mr. Duane Delicana
JKARLO Trading
SMV, Calapan City
Mindoro Oriental

Chapter 2

Science and Technology for Rural Industrialization and Employment Generation in India

M. Bandyopadhyay

Senior Expert, Centre for Science and Technology of the Non-aligned and other Developing Countries (NAM S&T Centre), New Delhi, India

E-mail: bandodst@gmail.com

ABSTRACT

Policies and programmes of the Government of India have been designed with the aim of alleviating rural poverty based on increasing the productive employment opportunities. One of the consistent objectives of the development policies in India has been the rapid and widespread development of micro and small enterprises including rural and village industries. In pursuance of these policies, several support services and infrastructural facilities have been created, and a number of promotional and protective measures taken to facilitate the growth of rural, micro and small industries in the country. Several programmes for the development and introduction of technology for application in rural areas and promotion of rural enterprises have been initiated by the relevant Ministries and Departments. A network of technology extension and demonstration centres, and training institutions provide technology information and training to rural artisans and industrial workers in skill up-gradation, which has helped enriching the rural industrial sectors in the scope and content.

Though considerable work has been done by the field level institutions, efforts are confined largely to relatively urbanized centres and the rate of diffusion of new technologies to rural enterprises has been slow. Organizations under various S&T Ministries and Departments are engaged in the development of technologies relevant to rural applications, but there is no mechanism for creating an appropriate linkage between the technology providers and the entrepreneurs for dissemination of such technologies in rural areas. Due to technological backwardness, the rural industry is facing the problems regarding viability of production and stiff competition in the market due to inconsistent quality.

Keywords: *Agriculture, Employment, Income generation, Industry, Micro and small enterprises, R&D, Rural areas, Skill development, S&T, Unorganized sector, Village.*

Introduction

Agriculture is the primary occupation of the people living in villages and would continue to remain so in the future also. However, the rural economy cannot exclusively depend on agriculture alone. The increasing aspirations of the people for a better quality of life and the need for reduction in rural-urban disparities necessitate the development of the secondary and tertiary sectors in rural areas. In order to reduce pressure on land and diversify rural economy, fresh avenues for employment have to be found in non-agricultural vocations such as rural industries.

The experience of many countries proves the fact that technology is an important factor of production besides land, labour and capital. For an all round progress of the village economy, it is necessary to adopt an integrated approach towards rural development by maximization of production and generation of employment through the introduction of Science and Technology (S&T). People in rural areas need technical skills to start and maintain income generating ventures, and also to sustain the production systems, such as, repair of agricultural appliances, maintenance of irrigation systems, use of low cost building materials etc. All these activities require concerted efforts to generate, transfer, adapt and utilize appropriate technologies. Various demand driven needs for the introduction of science and technology in rural areas include: providing minimum basic needs and improving quality of life of the people (through the provision of food, clothing, housing, road and transportation, energy, communication, healthcare, drinking water, consumer durables, entertainment and other urban amenities), employment generation, enhancement of income, poverty alleviation, empowerment of rural women, population control, minimizing rural to urban migration, development and utilization of local resources, providing market for industrial products, and relocation of industries from cities to rural areas (Bandyopadhyay, 1994).

In the planned process of growth initiated after the independence of India, high priority has been given to the welfare of rural population with a clear objective of eradication of poverty, reduction of inequality and creation of productive jobs. One of the consistent objectives of the planned development in India has been the rapid and widespread development of micro and small enterprises including rural and village industries. In pursuance of the policy of the Government, several support services and infrastructural facilities have been created and a number of promotional and protective measures taken for facilitating the growth of rural micro and small industries in the country.

There is a vast network of Research and Development (R&D) and academic institutions spread throughout the country. Several programmes for the development and introduction of technology for application in the rural areas and promotion of rural enterprises have been initiated by the relevant Ministries and departments of the Government of India. A number of extension and demonstration centres and experimental field stations provide technology information and training to rural artisans and industrial workers in skill up-gradation.

This paper presents an overview of the policy initiatives, institutional framework and promotional measures for introducing S&T for industrialisation and employment generation in rural areas and discusses the current status of various sectors of rural industries in India. In the concluding part, the major problems and constraints experienced by the rural industries have been examined, and a few recommendations to address such problems have been made.

Concepts and Definitions

Enterprise and establishment

In India, the terms 'enterprise' and 'establishment' are used interchangeably. An 'enterprise' is an undertaking, which is engaged in the production and/or distribution of some goods and/or services meant mainly for the purpose of sale, whether fully or partly. It may be owned or operated by a single household or by several households jointly, or by an institutional body. Enterprises are divided into following three classes based on the total number of workers and employment of hired workers:

- *Own Account Enterprise (OAE)* - an undertaking run by household labour, usually without any hired worker employed on a 'fairly regular basis'[1].
- *Non Directory Establishment (NDE)* – an establishment, which employs less than six workers (hired and household) on a 'fairly regular basis'.
- *Directory Establishment (DE)* – an establishment, which employs six or more workers (hired and household) on a 'fairly regular basis'.

An 'establishment' is an enterprise, which has got at least one hired worker on a 'fairly regular basis'.

Rural Areas and Villages

It is difficult to demarcate rural areas from semi-urban and urban centres. In rural areas, the smallest area of habitation is the village, which generally follows the limits of a revenue village that is recognised by the district administration. According to the current definition, an area comprised in any village, and including the area comprised in any town, the population of which does not exceed 20,000 or such other figure that the Government of India may specify from time to time, is considered as a rural area.

---

1 Fairly regular basis major part of the period of operation(s) of the enterprise during the last 365 days.

India is the second largest populous country in the world having a population of 1.18 billion (estimate for April 2010). As per the Census of India 2001, about 72.2 per cent of the population lives in rural areas in 138 million households spread over 593,731 inhabited villages (total no. of villages: 638,387, including 44,656 uninhabited villages). The size of these villages varies considerably. 219,063 villages have a population less than 500, while only 3,961 villages have a population of more than 10,000. The Census of India regards most settlements of less than 5,000 as a village. Most Indian villages are small; nearly 80 per cent having less than 1,000 inhabitants (GOI, 2001).

Rural Industry

The term 'rural industry' has no clear definition and is basically referred to artisan-based traditional household economic activity based in rural areas. Rural industries generally include agro-based enterprises, *khadi*, handloom, handicrafts, sericulture, coir, other village industries, and also tiny and service units situated in the rural areas. However, due to progressive introduction of technological inputs in such activities and initiation of programmes for manufacturing other products, the scope of rural industries has become larger to encompass all productive activities in rural areas relating to effective utilization of locally available resources.

Khadi

Khadi is Indian handspun and hand-woven cloth using handspun yarn only. The raw materials may be cotton, silk or wool, which are spun into threads on a spinning wheel called a *charkha*. The production activity is carried out in rural areas.

Village Industry

Village industry in India is currently defined as "any industry located in a rural area, which produces any goods or renders any service with or without the use of power and in which the capital investment per head of artisan or worker does not exceed Rs.[2] 100,000 (Rs. 150,000 for hilly areas) or such sum that may be specified by the Government".

Micro and Small Enterprises (MSE)

There is no generally accepted definition of micro, small and medium-sized enterprises. The distinction among the micro, cottage, tiny, small and informal sectors is often blurred. The current definition of micro and small enterprises in India is as under (MSME):

Classification	*Manufacturing Enterprise*	*Service Enterprise*
	Limit for investment in plant and machinery	Limit for investment in equipment
Micro enterprise	Rs. 2.5 mn./US$ 50,000	Rs. 1.0 mn./US$ 20,000
Small enterprise	Rs. 50 mn./US$ 1.0 mn.	Rs. 20 mn./US$ 0.4 mn.)

---

2 1 US$ = ~ Rs. 50

Self-employed

Persons who operate their own farm or non-farm enterprise or are engaged independently in a profession or trade on own-account or with one or a few partners are deemed to be self-employed in household enterprises. The essential feature of the self-employed is that they have autonomy (*i.e.* how, where and when to produce) and economic independence (*i.e.* market, scale of operation and money) for carrying out their operation. The remuneration of the self-employed consists of a non-separable combination of two parts – a reward for their labour, and a profit from their enterprise.

Unorganized Sector

The National Commission for Enterprises in the Unorganized Sector (NCEUS) defines 'unorganized sector' (or informal sector) as the sector *consisting of all unincorporated private enterprises owned by individuals or households engaged in the sale and production of goods and services operated on a proprietary or partnership basis with less than 10 workers* (NCEUS, 2008). A large section of the self-employed consists of those who are engaged in non-farm activities. This definition applies to all sectors of economy, both in urban as well as rural areas, and including agriculture but excluding plantation and other types of organized agriculture. The unorganized sector predominantly consists of own-account enterprises, in which there are no hired workers, but run by self with or without the help of unpaid family members.

Artisans in the villages such as blacksmiths, carpenters, weavers, potters, sculptors and makers of handicraft, metal ware, farm implements, hand tools etc., form a major segment of the unorganized sector that provides a great potential for rural employment and income generation. These artisans play an important role in the village life besides being the custodians of heritage of rural India. A unique feature of these artisans is that they largely live and operate in clusters.

Policy Initiatives for Rural Industrialization and Employment Generation

Various policies adopted in India to promote rural industrialization and employment generation in the rural areas are discussed below:

Policies on Rural Development and Rural Employment Generation

The Ministry of Rural Development is the nodal Ministry of the Government of India, which has the objectives to alleviate rural poverty and ensure improved quality of life for the rural population especially those below the poverty line. These objectives are achieved through formulation, development and implementation of programmes relating to various spheres of the rural life and activities, from income generation to environmental replenishment. In order to ensure that the fruits of economic reforms are shared by all sections of societies, some schemes and programmes were initiated for the five elements of social and economic infrastructure, critical to the quality of life in rural areas *viz.*, health, education, drinking water, housing and roads. The present strategy of rural development in India mainly focuses on poverty alleviation, better livelihood opportunities, provision of basic amenities and infrastructure facilities through innovative programmes of wage and self-employment. In pursuance of these policy directions, a time bound plan for 2005-09, *"Bharat Nirman"* was initiated for creation of rural infrastructure by the Government of India in partnership with the state governments and *'Panchayatiraj institutions'*[3], in the areas of irrigation, roads, housing, water supply, electricity and telecommunication connectivity with specific targets for each of these goals (MRD).

Policies on Rural Industrialization

The planners and policy makers in India consistently laid emphasis on the development of small and rural industry. Rapid and widespread development of small industries including cottage, village, household and tiny industries has been almost a regular feature in the objectives of the planned development in the country. The Industrial Policy Statement, 1980, inter-alia stated – *"The Government is determined to promote such a form of industrialization in the country as can generate economic viability in the villages. Promotion of suitable industries in rural areas will be accelerated to generate higher employment and higher per capita income for the villagers in the country without disturbing the ecological balance. Handlooms, handicrafts, khadi and other village industries will receive greater attention to achieve a faster rate of growth in the villages"* (GOI, 1980). After the economic liberalization in 1991, a new industrial policy was announced wherein also the importance of small and rural industries was not undermined, and a liberal and congenial environment was sought to be provided for the development of industry in the country including micro, small and rural industries.

India today operates the largest programme for the development of micro and small enterprises (MSEs) in any developing country. The government of India has taken a number of steps for the promotion, protection and development of this sector. A separate policy statement for the promotion and strengthening of small, tiny and rural industries was made in 1991 (GOI, 1991). Subsequently, the Micro, Small and Medium Enterprises Development (MSMED) Act 2006 was adopted in October, 2006 (GOI, 2006).

The government of India's encouragement to the MSE and rural industry sectors is reflected in the fact that a separate Ministry of Small Scale Industries (SSI) and Agro and Rural Industries (ARI) was established in October 1999 as the nodal Ministry for policy formulation, and implementation and coordination of schemes and programmes on promotion and development of these sectors of industries. This Ministry was bifurcated into two separate identities *viz.* Ministry of SSI and Ministry of ARI in September 2002. Later, all the activities were again brought under a single Ministry of Micro, Small and Medium Enterprises (MSME) in May 2007. The Office of the Development Commissioner, MSME (earlier DCSSI) functioning within the Ministry of MSME is entrusted with the responsibilities of providing techno-economic and managerial consultancy, common facilities and extension services, and facilities for technology upgradation, modernization, quality improvement and infrastructural support to MSMEs.

---

3 Local governments

Science and Technology Policies

The country's commitment to science, technology and innovation was evidenced with the adoption of the Scientific Policy Resolution (SPR) in 1958 and the Technology Policy Statement (TPS) in 1983. Later, keeping in view the need of the liberalized economy in the country, a comprehensive Science and Technology Policy was adopted in 2003. The objectives of the S&T policy include, encouraging research and innovation in areas of relevance for the economy and society, particularly by promoting close and productive interaction between the private and public institutions in science and technology and strengthening enabling mechanisms that relate to technology development, evaluation, absorption and upgradation from concept to utilization. The National S&T Entrepreneurship Development Board (NSTEDB) was established in 1982 and several science and society related programmes were initiated by various science related Ministries, departments and agencies, which complement the efforts of the Government for rural development, micro and small enterprises creation and employment generation.

Skills Development

Another major policy initiative, which has a direct bearing on employment generation in rural areas, is the national policy on skills development or the national skill development mission, which was approved by the Union Cabinet in February 2009. This policy aims at empowering all individuals to enable them to get access to decent employment and to promote inclusive national growth. Further, the policy promotes the public-private partnership to ensure that the needs of the industry are met.

Credit and Financial Assistance

The credit in the form of a) term loan and working capital and b) financial incentives and subsidies are crucial factors for the development of rural industries. In India, a multi-agency approach is followed for providing rural credit. The credit provided by banks to the micro-enterprises is treated as assistance to priority sector. About 40 per cent of 79,933 bank branches in the country are located in the rural India, and another 19,057 branches are in the semi-urban centres. Further, commercial banks have set up Regional Rural Banks (RRBs) or *Grameen* Bank, which have penetrated every corner of the country and complements other initiatives for the development of rural India. At present, there are 82 RRBs with 15,200 branches across 593 districts, and about 90 per cent of which are located in the remote rural areas. Schemes have been drawn up in the nationalised banks to extend credit facilities to small borrowers in rural areas for activities on agriculture, rural industry, retail trade, small businesses, roads and water transport. The institutional structure for the flow of credit to the micro, small and village industries comprises two apex level institutions *viz.*, Small Industries Development Bank of India (SIDBI) and National Bank for Agriculture and Rural Development (NABARD). Also, Khadi and Village Industries Commission (KVIC) extends financial assistance to its implementing agencies in the form of grants and loans.

Institutional Support and Promotional Measures for Introduction of S&T for Rural Industrialization and Employment Generation

Due to the vastness and diversity of the country, there is no single coordinated organization or programme for technological capacity building and introduction of technology in the rural areas. Different ministries and departments of the Government have instituted schemes and programmes, and set up a large number of institutions and promotional agencies for the development and transfer of technology, providing technical consultancy, and conducting skill upgradation training and demonstration programmes in their respective fields for rural development, employment generation and industrialization in rural areas. Important schemes and programmes being implemented by various Ministries and departments of the Government of India (Federal Government) are highlighted in the following sections.

Ministry of Agriculture

Department of Agriculture and Co-operation

The Department has launched the Agri -Clinics and Agri- Business Centre Scheme to take better methods of farming to each and every farmer across the country. This programme aims to tap the expertise available in the large pool of agriculture graduates. The Department is also providing the start-up training to graduates in agriculture and allied subjects such as horticulture, sericulture, veterinary sciences, forestry, dairy, poultry farming, and fisheries, etc.

Department of Animal Husbandry, Dairying and Fisheries (DADF)

The DADF of the Ministry of Agriculture is responsible for matters relating to livestock production, preservation, protection and improvement of stocks, dairy development, and matters pertaining to fishing and fisheries development boards. There are 36 subordinate offices and institutions on animal husbandry, dairy development and fisheries under the above department located in various parts of the country.

Indian Council of Agricultural Research (ICAR)

ICAR is one of the largest national agriculture research systems in the world. There are 97 institutes and 45 agricultural universities under ICAR. Some of the important programmes of ICAR are: Institution-Village Linkage Programme; National Agriculture Technology Project (NATP) (1998); National Agriculture Innovation Project (NAIP) (2005); and Krishi Vigyan Kendras (KVKs)[4]. Agricultural universities and research institutes under ICAR have developed a number of technologies on post-harvest operations, agro-processing and animal-based products. Several value-added products from horticultural, livestock and fishery produce have been developed, which include jelly, leather, powder and toffee from Aonla[5], sweet dish from jackfruit, coated trout fillets, curried products from freshwater fish, canned trout, prawn soup

---

4 Agriculture Science Centres.

5 Aonla or amla (*Emblica officinalis*), Indian gooseberry.

powder, smoked fish, tuna biryani, and ready-to-serve mahaseer (Fish-*Barbus mosal*) curry. Linkages between agricultural research, education and extension have been established for the direct benefit of the farming community, and through such initiatives many agro-based industries have come up (ICAR).

Ministry of Micro, Small and Medium Enterprises (MSME)

The Ministry of Micro, Small and Medium Enterprises (MSME), Development Commissioner, MSME and other organizations of the Ministry operate several schemes and programmes for providing techno-economic and managerial consultancy, common facilities and extension services, and facilities for technology upgradation, modernization, quality improvement and infrastructural support for MSMEs including rural enterprises as given below (MSME, 2010):

- *Micro and Small Enterprises Cluster Development Programme (MSE-CDP)* - for the development of clusters for capacity building, technology upgradation, quality standardization and testing, and also to promote skill development, marketing and export support, improved credit delivery, setting up of common facility centres, etc.
- *Credit Linked Capital Subsidy Scheme (CLCSS)* - for facilitating technology upgradation.
- *Technological Services* - offered through 30 technology resource centres and 28 small industries service institutes; 6 product-cum-process development centres; 4 regional testing centres; 7 field testing stations and mini tool rooms.
- *Product Development, Design Intervention and Packaging (PRODIP) scheme* - for quality improvement of khadi products and to diversify into new products.
- *Scheme of Fund for Regeneration of Traditional Industries (SFURTI)* - for comprehensive development of traditional industry clusters of *khadi*, village industry and coir products.
- *Prime Minister's Employment Generation Programme (PMEGP)* - to provide training and financial support to educated unemployed youth, both in rural as well as in urban areas, for setting up of self-employment ventures through industry, service and business routes.
- *Rajiv Gandhi Udyami Mitra Yojana (RGUMY)* - to provide handholding support and assistance to potential first generation entrepreneurs, who have already successfully completed entrepreneurship or skill development programmes, or vocational training through selected lead agencies called 'Udyami Mitras'[6], in the establishment and management of a new enterprise, in dealing with various procedural and legal hurdles and in completion of various formalities required for setting up and running of the enterprise.

---

6 Means: 'Friends of entrepreneurs'.

Khadi and Village Industries Commission (KVIC)

KVIC is charged with the planning, promotion, organisation and implementation of programmes for the development of *Khadi* (ref. section-2.3, Page-4 for definition) and other village industries in the rural areas in coordination with several agencies engaged in rural development. Its functions also comprise building up of a reserve of raw materials and equipments for supply to entrepreneurs. KVIC undertakes R&D projects with the help of national level institutes such as Indian Institutes of Technology (IITs) and National Institutes of Technology. It has also set up Mahatma Gandhi Institute for Rural Industrialization (MGIRI) and in-house testing laboratory for the promotion of technology support for rural industry. KVIC implements its programmes through state level Khadi and Village Industries Boards (KVIBs), 5000 registered institutions, 30,129 co-operative societies and a large number of banks and financial institutions (KVIC).

KVIC has set up a Rural Industry Service Centre (RISC), which is a common facility unit for Khadi and Village industry (KVI) to provide backward and forward linkages and services such as raw material support, skill upgradation, training, quality control, testing facilities, marketing promotion, design and product development etc., in order to strengthen the rural industry clusters. Training facilities are also provided for upgrading skills of the artisans in order to increase their earnings. In addition, other facilities are provided for developing new designs, new products, and diversified products in consultation with experts/agencies for value addition of rural manufacturing units; raw material support; and preparation of product catalogue.

Coir Board

The functions of the Coir Board include promotion and development of coir industry, fixing grade standards and inspection of fibre, yarn and coir products. The S&T activities are: modernization, standardization and distribution of looms and equipments, extension services and imparting training to people from industry. A new scheme on rejuvenation, modernization and technology upgradation of coir industry has been started in 2008 with a total outlay of Rs. 2.43 billion (MSME, 2010).

S&T Related Ministries, Departments and Agencies

Department of Science and Technology (DST)

The DST is implementing a programme on Science for Equity, Empowerment and Development (SEED) with broad objectives of providing opportunities to motivated scientists and field level workers to take up action oriented and location specific projects aimed at socio-economic upliftment of the poor and disadvantaged sections of the society through appropriate technological interventions, especially in the rural areas. Under this programme, efforts have been made to associate national laboratories and other specialist S&T institutions with each major project for expert inputs, utilizing national S&T infrastructure and link it up with the grassroots S&T interventions (DST). Relevant projects under the SEED programme include, among others, holistic and integrated development of 50 villages based on S&T needs; design and development of farm and off-farm machinery/implements; bio-integration of farming activities and resource management; organic cultivation and processing of medicinal plants; rural energy programme for non-grid areas; programme on animal husbandry and IT for disadvantaged sections; rural or women technology parks in remote areas; coordinated programme on non-edible oils; pilot demonstration and dissemination of rural technologies etc.

S&T Entrepreneurship Development (STED) scheme of the National Science and Technology Entrepreneurship Development Board (NSTEDB) of DST has been implemented in 48 districts to create micro-enterprises in rural areas by utilizing local resources and application of rural technologies. Under the scheme, 2600 micro-enterprises have been created and 7000 persons were provided with jobs. Further, NSTEDB organizes skill development programmes for training rural people in S&T skills for setting up self-employment ventures, under which 35,000 persons have been trained (NSTEDB).

Department of Biotechnology (DBT)

The DBT implements Biotechnology based programmes for the society aiming at women, under-privileged sections and the rural community. The activities in rural areas have been designed for creating awareness on biotechnology packages for income generation and better health care and for dissemination of the use of biotechnological process and tools for creating new avenues of employment in the rural areas through training and demonstration projects. More than 25,000 rural people have been trained in various areas such as wasteland utilization, vermi-culture and vermi-composting, food technology, sericulture, mushroom cultivation, floriculture, biofertilizer, aquafarming, poultry farming, rabbit rearing for quality wool production, medicinal and aromatic plants cultivation and formulation of herbal medicines (DBT).

Council of Scientific and Industrial Research (CSIR)

Various institutions under CSIR have provided S&T based solutions to mitigate the vulnerability and to improve the quality of life especially for the rural society. CSIR institutions having a direct bearing in the sector of societal goods and services reported several achievements in developing technologies for rural applications in food and agro-based technologies; building and construction technologies; drinking water, environment and sanitation; cultivation and processing of economic plants; cottage and village industry etc,. for example: solar power operated community; reverse osmosis desalination unit for sea water; light weight instant house for disaster victims; bio-village approach in widespread cultivation of aromatic and herbal plants etc. (CSIR).

National Research Development Corporation (NRDC)

NRDC promotes technologies that are particularly suited to rural needs to reduce the drudgery of women in their daily lives, to provide literacy aids, safe drinking water, low cost housing, sanitation, better agricultural implements and post-harvest technologies, solar pumps, heaters, solar cookers and solar dryers, cement from paddy husk ash. It also provides employment by the utilisation of farm wastes. NRDC has also set up 52 rural technology demonstration cum training centres in various parts of the country to give hands-on training and demonstrations to artisans and technicians living in rural areas (NRDC).

National Innovation Foundation (NIF)

The NIF was set up by the Department of Science and Technology as an autonomous organization with the objective of providing institutional support in scouting, spawning and sustaining grassroots innovations and helping their transition towards self-supporting activities. The NIF has network partners, the Honeybee Network and the Society for Research and Initiatives for Sustainable Technologies and Institutions (SRISTI), which have been scouting innovations by farmers, artisans, women, etc. at the grassroots level and the Gujarat Grassroots Innovations Augmentation Network (GIAN) that scales up innovations. So far, more than 100,000 innovations and traditional knowledge practices have been scouted from 500 districts of the country. The areas covered in the innovation database include plant variety, utilities and general machinery, farm implements, energy devices, agricultural and traditional knowledge practices, livestock management, herbal remedies, biodiversity examples, innovation concepts and ideas (NIF).

Ministry of New and Renewable Energy (MNRE)

The MNRE is supporting application-oriented non-conventional energy technologies throughout the country. The programmes that are useful to the rural people are given below.

1. Remote village electrification programme under which 4,997 households in remote villages and 1,257 remote hamlets have so far been provided solar home-lighting systems.
2. National biogas and manure management programme under which 4.18 million family type biogas plants were set up providing the benefit of clean cooking and lighting fuel for over 4.2 million rural households.
3. Solar Photovoltaic (SPV) demonstration, under which over 1.4 mill SPV home and street lighting systems were installed, mostly in rural areas, and a large number of solar lanterns were distributed while solar pumps and solar power plants were set up.
4. Development or up-gradation of water mills.

Further, the Government has recently launched a National Solar Mission with the aim to set up an enabling environment for solar energy penetration in the country with a target to develop and deploy 20,000 MW solar powers, including 20 million solar lighting systems in the rural areas by 2022. In order to utilize various sources of non-conventional energy like biomass, solar energy, wind energy etc., renewable energy development agencies at the state level and the Indian Renewable Energy Development Agency (IREDA) at the national level have evolved several programmes while providing subsidies for producing and using non-conventional energy devices to meet the energy needs of the rural people (MNRE, 2010).

Ministry of Rural Development (MRD)

Council for Advancement of People's Action and Rural Technology (CAPART)

The objectives of CAPART include, encouraging, promoting and assisting voluntary action for the implementation of projects for enhancement of rural prosperity including introduction of new technological inputs and the development of technology that are appropriate for rural areas. CAPART supports the projects under the Advancement of Rural Technologies Scheme (ARTS) for the development and dissemination of innovative technologies relevant for rural areas.

Swarnjayanti Gram Swarozgar Yojana (SGSY)[7]

This programme was initiated by MRD after merging the erstwhile programmes *e.g.* Integrated Rural Development Programme (IRDP), Training of Rural Youth for Self-employment (TRYSEM) etc. The objective of SGSY is to create self-employment opportunities for the rural poor by organising them into Self Help Groups (SHGs), their training and capacity building and provision of income generating assets through a mix of government subsidy and bank credits. The programme aims at establishing a number of micro-enterprises in rural areas based on the ability of the beneficiaries and the potential of each area. Identification and management of appropriate technology including issues of technology transfer, upgradation and absorption have been given due importance (MRD).

Mahatma Gandhi National Rural Employment Guarantee Act (MGNREGA)

The Programme under the Act was launched in 2005 under the aegis of MRD, which aims at enhancing the livelihood security of people in rural areas by guaranteeing hundred days of wage-employment in a financial year to a rural household whose adult members volunteer to do unskilled manual work. The programme does not aim at any technological intervention, and emphasizes only on ensuring some minimum income to the rural poor.

Ministry of Textiles

The Government adopted the National Textile Policy in 2000, which laid the major thrust on technological upgradation, enhancement of productivity, quality consciousness, strengthening of raw material base, product diversification, etc. The textiles industry gets technology support under several programmes *viz.*, Technology Upgradation Funds Scheme (TUFS), Technology Mission for Cotton (TMC), Jute Technology Mission (JTM), Powerloom Schemes, Jute Manufacturers Development Council (JMDC), Integrated Textiles Parks etc.(MOT, 2010).

Further, the Ministry administers the Development Commissioner (Handicrafts), the Development Commissioner (Handlooms) and the Central Silk Board for development and promotion of the respective sectors. For the handicraft sector, assistance is provided under design and technology upgradation scheme, which aims at upgradation of artisans' skills, improvement and diversification of products,

---

7 Golden jubilee village self-employment programme.

development of new design of prototypes, supply of improved/modern equipments to the craft persons, preservation of traditional art and crafts of high aesthetic value, etc. For handlooms, the research projects directed towards technological upgradation are supported. The Central Silk Board has established several R&D and training institutes, a seed technology laboratory, germplasm resources centre etc. in order to develop new varieties of mulberry silkworm to suit various agro climatic conditions, and to develop value added products and services, methodologies and practices to increase productivity, quality and profitability of sericulture.

Table 2.1: R&D Organizations and Technology Extension and Demonstration Centers in Various Sectors of Rural Industry*

Sector	*Type/Name of Institutions*	*No.*
Animal Husbandry	Institutes under ICAR, DADF	30
Coir industry	R&D and training institutes	3
Dairying	National Dairy Research Institute, Karnal	1
Fisheries	Central institutes on various sectors of fisheries, Postharvest Technology training, Coastal Engineering and Fisheries, Nautical and Engineering training, etc.	12
Handloom, Textiles and Handicrafts	Weavers' Service Centres	25
	Indian Institute of Handloom Technology	5
	Indian Institute of Carpet Technology	1
	National Textile/Fashion Institutes and Textile Research Associations	11
	Textile Testing Houses	19
Jute and Fibers	National institutes *e.g.* Central Research Institute for Jute and Allied Fabrics, Indian Jute Industries Research Association, Institute of Jute Technology, National Institute of Research on Jute and Allied Fibre Technology etc.	8
Leather	R&D, design and training institutes	18
Postharvest Technology (crops)	Postharvest Technology Centre (IIT, Kharagpur); Indian Institute of Crop Processing Technology (Thanjavur); Central Institute of Post Harvest Engineering and Technology, PAU, Ludhiana	3
	Regional extension service centres (for Rice)	12
	Testing centres (for Rice)	4
Poultry	Central Poultry Development Organizations	5
Sericulture/Silk	Central institutes	10
	Regional research stations	20
	Regional extension centres	64
	Sub-regional extension centres	12
	Demonstration-cum-service centres	16

* The list is not complete. There are several other institutions and agencies at the national, state and regional levels which are engaged in technology demonstration, extension and training in various sectors.

Besides the above, several research organizations, regional technology centres, and technology extension and demonstration centres have been set up in various parts of the country including rural areas and semi-urban centres for carrying out product development, transfer of technologies, skill development and self-employment training, and providing field level services to assist the farmers and rural entrepreneurs in different sectors. The institutional infrastructure for R&D, and technology extension and demonstration in various sectors of rural industry has been summarized in Table 2.1. Also, the institutional support for training and extension for rural development, entrepreneurship development and skill formation that has been created in the country is shown in Table 2.2.

Table 2.2: Training and Extension Centres for Rural Development and Human Resource Development for Entrepreneurship and Skill Formation*

Function	*Name of Institutions/Organizations*	*No.*
Agricultural extension	Krishi Vigyan Kendras (KVKs)	571
	Kisan Call Centres	25
	Agriculture Technology Extension Centres	44
	National Institute of Agricultural Extn. Management	1
Rural development	District Rural Development Agencies (DRDAs)	626
	National Institute of Rural Development (NIRD)	1
	State Institutes of Rural Development	28
	Institute of Rural Management, Anand	1
Industrial Development at District level	District Industries Centres (DICs)	626
Entrepreneurship	National level Institutes for Entrepreneurship Development	4
development, Skill	Udyami Mitras	355
development,	Community Polytechnics	617
Vocational training	Industrial Training Institutes (ITIs) – Govt.- run	1931
for self-employment/	Industrial Training Centres (ITCs)- privately-run	6003
wage employment	Rural Self-employment Training Institutes. (RSETIs)	99
	Technology Service Centres and Technology Transfer Centres of National Small Industries Corporation	8

* The list is not complete. There are several other institutions and agencies at the national, state and regional levels which are engaged in similar activities.

The Government has also been encouraging participation of non-governmental organizations (NGOs) for integrated rural development.

These organizations are playing a significant role as intermediaries between technology providers and the beneficiaries in the rural areas, which undertake activities on skill training and dissemination of technological information. For example, M S Swaminathan Research Foundation (MSSRF) is a non-profit research organization established in 1988, which has been developing and following a pro-nature, pro-poor, pro-women and pro-sustainable on-farm and non-farm livelihoods through appropriate eco-technology and knowledge empowerment. By one estimate, there are more than 10,000 NGOs in the country.

Technology Areas, Product Varieties and Status of Rural Industry

The rural industry in India is primarily based on local and regional agricultural produce. There are activities like plantation agriculture, modern dairying, animal rearing, etc. which require an industrial organization but are agricultural in the sense that they are related to plant and animal life. One of the most important areas of rural industrialization is agro-based industries which use agricultural output as an input and convert the same into products by value addition. In addition to the farm-based enterprises, the non-farm sector constitutes an important segment of the country's economy and is considered as a useful means for decentralization of economic activities to rural India. The current status of various sectors of rural enterprises is given under the following groups:

Farm-based Enterprises

Important occupations in this category include establishment and management of modern nursery, apiculture (bee keeping), sericulture, mushroom cultivation, kitchen gardening, vermi-composting, floriculture, agri-businesses etc. The current status of some selected occupations under this category is given in Table 2.3.

Table 2.3: Status of Important Farm-Based Enterprise Sectors

Sl.No.	*Occupation and Status#*		*Remarks*
1.	**Apiculture (Bee keeping)**		
	Production of Honey (tonnes)	65,000	Unorganized activity. Provides income to rural and tribal people. Has potential to provide self-employment to ~6 million.
	No. of bee-keepers	150,000	
2.	**Floriculture**		
	Area cultivated (ha)	166,000	A labour intensive activity. Provides direct employment to 20 workers per ha and indirect employment to another 40 per ha.
	Production–Loose Flowers (tonnes) Cut Flowers (mill. nos.)	868,000 4365.4	
3.	**Mushroom Cultivation**		
	Production (tonnes)	48,000	Commonly grown species: white button, paddy straw and oyster mushroom.
4.	**Sericulture (Silk)**		
	No. of villages engaged	53,814	An important labour intensive cottage industry in rural and semi-urban areas. India 2nd largest producer of silk in the world
	Production (tonnes)	18,475	
	Employment (mill.)	6	

Contd...

Table 2.3–Contd...

Sl.No.	*Occupation and Status#*		*Remarks*
5.	**Spices**		
	Area cultivated (mill. ha)	2.617	A source of livelihood and employment for large number of rural people
	Production (mill. tonnes)	4.357	
6.	**Plantation Crops**		
	Area cultivated (mill. ha)	3.190	Provides direct and indirect employment to large number of rural people.
	Production (mill. tonnes)	11.300	

Sources: NHB; MOT, 2010.

Animal Husbandry, Dairying and Fisheries

The animal husbandry and dairy development play an important role in the socio economic development of India. Besides providing cheap nutritional food to millions of people, they are helpful in generating gainful employment in the rural areas, particularly among the landless labourers, the small and marginal farmers and the women by supplementing their family incomes. Livestock are the best insurance against the vagaries of nature like drought, famine and other natural calamities. According to the National Sample Survey Organization (NSSO), the estimate of employment in animal husbandry sector (2004-05) was 11.44 million in the principal status and 11.01 million in the subsidiary status.

Dairying has become an important secondary source of income as well as an employment source for millions of rural families. India is the largest producer of milk in the world, almost entirely from the output of millions of smallholder farms. The cooperative movement (Operation Flood) has been important in dairy development in different parts of the country. Most of the milk is produced by small, marginal farmers and landless labourers who are grouped into cooperatives at the village level. The current status of the animal husbandry and dairy development sectors is given in Table 2.4.

Non-Farm Enterprises

The rural non-farm sector encompasses all non-agricultural activities that include mining and quarrying; household and non-household manufacturing; handicrafts, food processing; repair and construction; trade and commerce; transport and other services in villages and rural towns. As such, this sector contributes significantly to provide supplementary employment to small and marginal farm households, reduce income inequalities and rural-urban migration. The distribution of selected non-agricultural establishments in rural areas is shown in Table 2.5.

Table 2.4: Status of Animal Husbandry, Dairy and Fisheries Sectors

Sl.No.	*Sector and Status#*		*Remarks*
1.	**Dairy Development**		
	Milk production (mill. Tonnes)	104.8	India Largest producer of milk (15 per cent) in the world, mainly from millions of smallholder farms and cooperatives.
	No. of farmers (mill.)	13	
	No. of village cooperatives	128,800	
2.	**Livestock Rearing**		
	Total livestock (mill. No)	485.012	Provides empl. to landless labourers, small/ marginal farmers,rural women.
	Employment (mill.)	18	
3.	**Wool**		
	No. of Sheep (mill.)	64.269	India 7th largest clean wool producer, 1.8 per cent of world production.
	Production (mill. kg)	45.0	
	Employment	2.7	
4.	**Meat and Meat products**		
	No. of Slaughter houses Authorized/Unauthorized	2700/ 12000	India contributes 2.4 per cent of global meat production.
	Production (mill. Tonnes)	6.5	
5.	**Fisheries and Aquaculture**		
	No. of fishing villages	3332	India is the third largest producer of fish and the second largest producer of fresh water fish in the world.
	Production (mill. Tonnes)	.7.12	
	Seed production (mill. Fry)	31688	
	Employment (mill.)	14.49	
6.	**Poultry**		
	No. of birds (mill)	489.0	An important contributor to rural economy. India is the 5th largest producer of eggs (3.6 per cent) and 9th largest producer of poultry meat in world.
	Egg production (bill. No.)	53.532	
	Broiler meat production (mill. Tonnes)	2.2	
	Employment (mill.)	3.0	

Sources: DADF, 2010; DADF, 2003; GOI, 2010a.

As per the Fifth Economic Census – 2005, there were a total 41.83 million establishments in the country engaged in different economic activities other than the crop production and plantation, out of which around 25.54 million (61.05 per cent) were in rural areas. 18.11 million establishments in rural areas were own account establishments and the remaining were with hired workers. Further, out of the total 25.54 million establishments in rural areas, 5.57 million (22.36 per cent) were engaged in the agricultural sector, rest 19.83 million (77.64 per cent) were in the non-agriculture sector. Also, 52.07 million persons were reported to be working in the rural establishments, out of which 41.89 million worked in the non-agricultural rural establishments (CSO, 2008). The status of a few selected sectors of non-farm enterprises is given in Table 2.6.

Table 2.5: Distribution of Selected Non-Agricultural Establishments in Rural Areas

Sl.No.	*Types of Non-agricultural Establishments in Rural Areas*	*Per Cent of Total*
1.	Manufacturing	26
2.	Community, social and personal service	8
3.	Hotel and restaurant	4
4.	Transport and storage	4
5.	Health and social work	2
6.	Construction	1
7.	Sale, maintenance and repair of motor vehicles	1
8.	Wholesale trade	2
9.	Retail trade	39

Source: CSO 2008.

Table 2.6: Status of selected Non-Farm Enterprise Sectors

Sl.No.	*Sector/Product Group and Status#*		*Remarks*
1.	**Food Grain Milling**		
	No. of rice hullers	91,000	Mostly in rural and semi-urban areas.
	No. of small flour mills	300,000	Largely in unorganized sector.
	No. of oil expellers	250,000	Unorganized sector, mostly animal operated.
	No. of mech. pulse mills	11,000	Remaining use small flour mills.
2.	**Jute and Jute Textiles**		
	Production(mill Tonnes)	1.36	Supports 4 million farm families. Also provides livelihood to another 140,000 in tertiary sector.
	Direct employment	260,000	
3.	**Coir Industry**		
	Fibre production (tonne)	437,800	An important cottage industry contributing significantly to the economy of the major coconut growing States.
	Coir products (tonne)	172,000	
	Employment	650,000	

Contd...

Table 2.6–Contd...

Sl.No.	Sector/Product Group and Status#		Remarks
4	**Handloom**		
	No. of handloom clusters	470	2nd largest employment generating occupation after Agriculture
	Production(m.Sq. mtr.)	6677	
	Employment (mill)	6.55	
5.	**Handicrafts**		
	Production (Rs. bill)	215	Unorganized sector, contributes substantially to employment generation.
	Employment (mill.)	6.77	
6	**Leather Industry**		
	Total output (Rs. bill.)	250	75 per cent of output contributed by tiny, cottage and small artisans. India 3rd largest producer, contributes 10 per cent of global requirement.
	Employment (mill.)	2.5	

Sources: MOT, 2010; MSME, 2010.

Khadi and Village Industries

Many rural industry sectors are also categorized as Khadi and Village Industries (KVI), which comprise both the farm-based and the non-farm enterprises. Industries covered under Khadi and Village Industries Commission (KVIC) programmes include agro-based and food processing (pulse and cereal processing), fruits and vegetable processing, village oil industry; mineral based (pottery and lime; polymer and chemical - leather, non-edible oils, soaps and detergents, cottage match industry, plastics etc.; forest products – medicinal plants, apiculture, hand-made paper and fibre; and rural engineering and biotechnology. Total production in the KVI sector in 2007-08 was Rs. 166.78 billion (*Khadi*– Rs. 5.43 billion and Village industry – Rs. 161.34 billion) and the employment generated was 9.92 million (0.91 mill in *Khadi* and 9.01 million in village industries) (MSME, 2010; KVIC).

Micro and Small Enterprises

The micro and small enterprises (MSEs) which are located both in rural and urban areas have played a vital role in the economic progress of India. The sector has grown rapidly over the years and produces a range of over 6000 products. The contribution of some major sub-sectors of MSEs in terms of manufacturing output is shown in Table 2.7.

As per the quick estimates of the Fourth All India Census of micro, small and medium enterprises (2006-07), the number of MSEs is estimated to be 26 million with an estimated 60 million jobs created by them. 28 per cent of the units are manufacturing enterprises while remaining 72 per cent are service enterprises. Further, 13.5 million enterprises are in the rural area, out of which 12.8 million are unregistered (MSME, 2010; GOI, 2010).

Table 2.7: Contribution of Various Product Groups to Manufacturing Output of MSEs

Sl.No.	*Product Group*	*Manufacturing Output (per cent)*
1.	Food products	18.97
2.	Textiles and readymade garments	14.05
3.	Basic metals	8.80
4.	Chemical and chemical products	7.55
5.	Metal products	7.52
6.	Machineries and equipments	6.35
7.	Transport equipments	4.50
8.	Rubber and plastic products	3.90
9.	Furniture	2.62
10.	Paper and paper products	2.03
11.	Leather and leather products	1.98

Source: MSME 2010.

Unorganized Sector

Various surveys carried out by the National Sample Survey Organization (NSSO), the Economic Census and the Census of MSMEs indicate that an overwhelming majority (more than 94 per cent) of enterprises in India come under the category of unorganized sector as per the definition given by the NCEUS (ref. section-2.6, Page-4). According to a report of the NCEUS, there are 58 million enterprises in the unorganized sector (both in rural and urban areas) that provide employment to 104 million persons. About half of these enterprises are classified as the self-employed and own-account enterprises (NCEUS, 2007)· A sector-wise break-up of non-farm unorganized enterprises (2006-07) are given in Table 2.8.

Table 2.8: Sector-wise Break-up of Non-farm Unorganized Enterprises (Rural and Urban)

Sector	*No. of Units (million)*
Micro-enterprises (under Development Commissioner MSME)	12.6
Prime Minister Rozgar Yojana (PMRY)	2.8
Swarnjayanti Gram Swarozgar Yojana (SGSY)	7.5
Khadi and Village Industries (KVI)/Rural Employment Guarantee Programme (REGP)	0.5
Swarnjayanti Sahari Swarozgar Yojana (SJSSY)	0.6
Agro-related	8.2
Handloom	6.5
Handicrafts, sericulture, coir, wool, manufacturing, retails trade, small businesses etc.	19.3
TOTAL (Non-farm unorganized enterprises) (2006-07)	58.0[¥]

Source: NCEUS, 2007.

Also, according to the National Sample Survey (NSS) 62nd Round report, a total of 17.071 million unorganized manufacturing enterprises were estimated to be operational during 2005-06 providing jobs to 36.44 million persons. Out of these, 12.128 million (71 per cent) enterprises were in rural areas. 98 per cent of the enterprises were proprietary, and 90 per cent were not registered (NSSO, 2007). Further, according to the NSS 63rd Round report, 16.5 million service sector enterprises were estimated to be working during 2006-07, while providing employment to 33.5 million persons. 60 per cent of the service sector enterprises *i.e.*, 9.9 million were in the rural areas. 90 per cent of the rural service enterprises were own-account enterprises (NSSO, 2009).

Observations and Conclusions

The growth and performance of the rural industrial sector in India, particularly the small scale industries has by and large, come up to the expectations with respect to number of units, value of output and generation of employment. The industry has proved to be an important instrument for generating productive employment and minimizing disparity in income of rural people, and has become a crucial issue in the process of planning for poverty alleviation and economic progress of the country.

The rural non-farm sector occupies a significant partner of the Indian economy, and has all the potential of becoming more important in the future. Even with the emergence and rapid growth of the modern large scale industries, the micro, small, traditional and decentralized sectors of industry continue to play a very important role in the national economy.

Success Factors

The Government has always made special provisions for the growth and development of the micro and small industries including the rural industries in the industrial policies. The Federal government and a few state governments have set up dedicated Ministries/departments on micro, small, cottage and rural industries in order to give special and coordinated attention to these sectors. Efforts made through programmes like Integrated Rural Development Programme (IRDP) and Training of Rural Youth for Self Employment (TRYSEM) (now combined as SGSY) have resulted into a significant shift in the occupational pattern of the rural people from agriculture to the secondary and tertiary sectors, and mobilization of funds through these programmes has resulted in a certain degree of commercialization in the rural areas. Earlier, the rural entrepreneurs mainly depended upon the informal credit institutions and money lenders to raise funds. But with the creation of a large number of rural bank branches and through several liberal measures, the rural enterprises are now being provided credit facilities, subsidies and incentives. This has resulted in a substantial improvement in the position regarding availability of capital resources for rural industrialization and employment generation.

On the infrastructure front, some facilities that were available in the rural areas earlier had helped in the growth of rural industry considerably. In the recent past, a time bound plan, "Bharat Nirman" has been initiated by the Government for creation of rural infrastructure in partnership with the state governments and *Panchayatiraj institutions*#, in the areas of irrigation, roads, housing, water supply, electricity and telecom connectivity. The programmes under this plan, such as Pradhan Mantri Gram Sadak Yojana[¥] to provide road connectivity up to remote villages, Rajiv Gandhi Grameen Vidyutikaran Yojana (RGGVY)[*] to electrify all un-electrified villages, increasing rural tele-density and broadband coverage in villages, and Indira Awaas Yojana (IAY)[©] to provide houses to the rural poor, would further boost the process of rural industrialization.

A large number of support services and institutional mechanisms have been put in place for helping the rural and small industry. A huge network of technology extension and demonstration centres and training institutions have been set up by the Government which has helped enriching the industries in the scope and content. Also, some efforts have been initiated by the S&T institutions for providing technological services to the industry, results of which are gradually diffusing into rural areas.

Another important factor influencing the success of rural industry is the involvement of a large number of voluntary agencies and non-governmental organizations (NGOs) in the activities on awareness creation, human resource development, dissemination of information, skill upgradation and transfer of technology, which has greatly complemented the efforts of the Government.

Problems and Constraints

The major problems faced by rural entrepreneurs and village artisans include the non-availability and exorbitant prices of raw materials, inadequate and exploitative market, outmoded and traditional technology, widely scattered industries, and illiteracy and inability of artisans to organize themselves in solving their problems. The rural non-farm sector has been suffering due to inadequate infrastructure, such as roads, electricity and communication facilities, lack of sufficient skilled labour and adequate access to credit, information, training facilities etc. There has been stagnation in the case of many traditional and small scale industries in the rural areas. Despite the increased plan outlays, priority and concessional financing, the small and village units have been working below their potential capacity and a large number of them have become sick. Some areas of concern are discussed hereunder.

Lack of Coordination

In spite of a reasonable success in the rural, micro and small industries in terms of production, growth and employment generation, they suffer from various problems, which arise largely due to the lack of coordination, both at the grass root and the nation level, since the subject is looked after by different departments and organizations. Though several coordinating agencies have been set up at the centre and the states, the rural industrialization programmes have suffered due to differences in the inter-agency perceptions. The top-down approach in the Government hierarchy has resulted in delay in decision making and lack of coordination among various implementing bodies.

Managerial Weaknesses

Though it is believed that a small enterprise is easily manageable because it is small, there is a variety of problems that a small unit has to face. Generally, one or two individuals manage such a unit with little or no expertise in various areas of management such as finance, production, manpower, marketing, liaisoning with regulatory agencies, etc. This causes managerial and operational weaknesses for the units in the long run. The situation in case of rural enterprises, which are run only by the owner-manager, is still worse.

Technological Inadequacies

In spite of favourable policy measures and availability of vast institutional set up and resources, the rural industry is facing the problems of technological obsolescence, low productivity and competition in the changing market on the basis of product quality. As per the Third All India Census of Small Scale Industries (SSI) (2001-02), 85.38 per cent of the total registered SSI units do not have any access to technical know-how.The traditional handloom, jute spinning and weaving, and handicrafts industry suffer due to lack of technological upgradation and competition from the organized units such as power looms and large mills. Labour productivity in rural enterprises is usually very low indicating low skills and lack of training of workers in production techniques, maintenancc of quality and standards, etc.

The rural enterprises suffer from a number of disadvantages compared with the larger firms in responding and adapting to new technological developments. The micro and small units are often not aware of the recent scientific advancements in the absence of a proper information dissemination mechanism. There is no motivational input and incentive for them for innovation and technological upgradation. They lack suitable qualified technical specialists and financial resources and are unable to support a formal R&D activity required for pursuing technological innovation.

Assistance and training on technology have so far been provided through the Government run extension centres, and the academic and research community in the country has not been able to contribute much in this direction in the absence of an appropriate mechanism of linkage among S&T institutions, implementing agencies and entrepreneurs. There is a lack of proper coordination between the S&T related Ministries and agencies on the one hand, and the user Ministries and the industry on the other. Although the organizations under various S&T related Ministries and departments are engaged in the development of technologies that are often relevant to rural applications, dissemination and introduction of such technologies in rural areas is not under the mandates of these organizations. There is no mechanism for providing forward and backward linkages through which the experts in the respective fields may address specific technological problems of rural and micro enterprises and suggest solutions to the same. Various S&T institutions and agencies seem to be working in isolation, thereby leading to multiplicity and overlapping of efforts, and dissipation, and poor and inefficient utilization of funds (Bandyopadhyay, 2011).

The Working Group for the Eleventh Plan on effective rural technology delivery in its report observed that often technologies that were developed in the laboratories but not tested and proven under field conditions were disseminated on a large scale, leading to poor results, non-acceptance by intended beneficiaries, under-performance and unsuitability to rural conditions. Further, experiences on successes, and more importantly of failures, have been poorly documented and inadequately shared among stakeholders leading to unnecessary repetitions, infructuous expenditure and considerable demotivation among both technology providers and users (PSA, 2006).

An overwhelming proportion of the enterprises and workers in India are in the unorganised sector. These enterprises are present both in the rural and urban areas but their presence is significantly more in the rural areas. The productivity and income of the unorganised sector enterprises are hampered due to obsolete technology and lack of technical skills among their workers. The National Commission for Enterprises in the Unorganized Sector in its report on technology upgradation for enterprises in the unorganized sector (NCEUS, 2009) have identified the following constraints:

- Inadequate attention to rural technologies
- Lack of grass-root contacts
- Poor technology delivery
- Problem of affordability
- Inability to achieve the economies of scale
- Lack of coordination
- Poor financial condition and lack of technological orientation of non-governmental organizations
- Poor data bank
- Lack of appreciation of changed environment
- Inadequate patenting facility
- Lack of linkage between large enterprises and unorganized enterprises.

Lack of Funds

In the absence of adequate information on credit facilities provided by banks, the rural entrepreneurs often fall into the trap of private money lenders who exploit them by way of high rate of interest. By and large, the banks and other lending institutions tend to treat the small rural entrepreneurs in the same manner as they treat the large scale commercial borrowers. Availability of institutional finance has not been commensurate with the need of rural and small industry and not in proportion to their contribution to the economy. Further, the rural, micro and small scale industries are clubbed together for the purpose of providing incentives, subsidies and other promotional and productive measures, due to which, the benefit of such facilities goes largely to the modern small scale units located in urban areas and the rural units are put at a disadvantage in this regard.

Recommendations

Technology Choice

The first priority for the progress and sustenance of rural industry should be to concentrate on formation and expansion of less capital-intensive projects involving maximization of the existing indigenous technological capabilities. More and more emphasis should be laid on the upgradation and appropriateness of technology rather than investing on the high-tech and sophisticated areas. The initial approach should be on building up the techno-economic infrastructure and development of ancillarization for large industrial houses.

An important policy issue for rural development should be on evolving a technology based product mix. Technologies selected should be such that would help creating the physical resources and assets for creation of wealth on a continuing basis leading to self-sustained development of the rural areas. Appropriate technologies developed and introduced in any given area should not only lead to increase in productivity, production and income but also enable accumulation of surplus for capital formation in the rural areas. The solution to the problems of low labour productivity and inefficient utilization of resources lies with the introduction of appropriate technology in products and processes, and improvement of skills of artisans and workmen. A programme of developing a comprehensive resource inventory for each of the geographical units should be initiated by utilizing the expertise of S&T institutions of the locality and such information should be used for converting the resources into commercially viable projects by application of S&T. A number of new technology areas have been identified by various scientific agencies, which have a great potential for establishment of production units and generation of jobs in the rural areas. An inventory of appropriate technologies developed at the national and state level R&D and academic institutions should be prepared, which could be transferred to the rural areas

It is imperative that the rural industry must keep pace with the technological developments elsewhere in the world. The industry must strive for introducing technological innovations in its products and processes by careful selection of those technologies, which are appropriate for a particular location. Such technology should be thoroughly tested at the field level through demonstration of pilot projects and field trials and then transplanted commercially at a reasonably low opportunity cost. This necessitates establishment of some infrastructural facilities in the rural areas.

In order to find solutions to the problems being experienced by the sectors of the traditional handloom, jute processing and handicrafts, suitable programmes for introduction of efficient looms and use of modern hand tools or semi merchandized tools should be initiated and supported. Dedicated low cost computerized integrated design and sample making systems for handlooms and handicrafts should also be evolved to improve production and competitiveness of goods manufactured by these sectors.

Linkage between Technology Providers and Rural Entrepreneurs

There should be a proper coordination and linkage between the user departments such as Ministries of Rural Development, MSME, Social Justice and Empowerment, etc. on the one hand, and the S&T Ministries and institutions on the other for dissemination of proven technologies for application in the rural areas, and utilization of the expertise and facilities of the latter by the rural entrepreneurs. Academic institutions and other national research laboratories should be involved in the process of technology diffusion, creation of new technology-based enterprises, technical skill formation and employment generation.

According to a case study carried out on the application of science and technology for rural employment generation through the erstwhile scheme of the Department of Science and Technology *viz.*, Mass Employment Generation through Science and Technology (MEGSAT), it was possible to infuse technological skills into various trades through S&T methods and practices to the beneficiaries with a much higher success rate and to create sustainable jobs through such processes either by wage employment or self-employment. As an example, a Mini Jute Mill developed by the joint efforts of the Government (both at the federal and state level), a research laboratory (Jute Technology Research Laboratory under ICAR) and a non-governmental organization were found to be extremely successful in demonstrating how S&T could help in generating employment for the people in the rural areas. It was estimated that at least six million people could be benefited if the model was replicated, and such mini jute mills are already operating in a few jute growing areas in the country (Bandyopadhyay and Thiagarajan, 1993).

Academic and Research institutions should set up the Rural Technology Development and Extension Centres (RTDECs) (a few IITs and other institutions have already set up similar centres) for providing necessary advice, technological information and assistance to farmers and entrepreneurs. The programme on the establishment of Rural Technology Parks (RTPs) initiated by the Government should be expanded to cover all the major districts to provide central facilities, work stations, and guidance on technology transfer, product development, scale up, quality control, etc. The RTPs should act as intermediaries between the RTDECs on the one hand and the MSE/rural enterprise clusters on the other, so as to provide linkages between the technology providers and rural entrepreneurs.

Training Needs

In order to minimize the gap between the demand and the availability of skilled manpower in the country, vocational education to train the rural youth in various technical skills should be expanded. Training programmes for entrepreneurs and workmen suitable to various fields and appropriate transfer mechanisms need to be designed and implemented. Orientation programmes for coordinators and trainers should be organized from time to time in order to update their knowledge and to enable them to design and implement training and demonstration programmes suitably. Teachers of science and engineering institutions should be motivated to take up the work related to rural technology and rural industrialization.

Ancillarisation

The rural areas have a great potential for ancillarisation or sub-contracting of the large public and private sector industry. Through suitable policy measures, large industry should be motivated to either give ancillary rights to entrepreneurs in rural areas or start their own small units in those areas for manufacture of spares, components and intermediates required for the mother units.

Conclusion

The Government of India has laid a great emphasis on rural industrialization concomitant with rural development through an integrated approach. There has been a tremendous growth in the rural industry sector along with agricultural production. The rural industry in India has a great diversity and strength in terms of the variety of products manufactured and the impact it has made on the rural economy through generation of income and employment, value of output and value addition is significant.

It is an inescapable requirement that science and technology inputs must be introduced into the rural industry and the skills of rural workers and artisans should be upgraded so as to make the products competitive and units financially attractive to the entrepreneurs. There is an urgent need to consolidate the initiatives already taken and undertake a coordinated programme on introducing technological innovation in rural areas by establishing appropriate linkages between academics and research institutions, and the rural industry.

List of Abbreviations

ARI: Agro and Rural Industries
ARTS: Advancement of Rural Technologies Scheme
CAPART: Council for Advancement of People's Action and Rural Technology
CLCSS: Credit Linked Capital Subsidy Scheme
CSIR: Council of Scientific and Industrial Research
CSO: Central Statistical Organization
DADF: Department of Animal Husbandry, Dairying and Fisheries
DBT: Department of Biotechnology
DIC: District Industries Centre
DRDA: District Rural Development Agency
DST: Department of Science and Technology
GIAN: Gujarat Grassroots Innovations Augmentation Network
GOI: Government of India
IAY: Indira Awaas Yojana
ICAR: Indian Council of Agricultural Research
IIT: Indian Institute of Technology
IRDP: Integrated Rural Development Programme
IREDA: Indian Renewable Energy Development Agency
ITC: Industrial Training Centre
ITI: Industrial Training Institute

JMDC: Jute Manufacturers Development Council
JTM: Jute Technology Mission
KVI: Khadi and Village industry
KVIC: Khadi and Village Industries Commission
KVK: Krishi Vigyan Kendra
MEGSAT: Mass Employment Generation through Science and Technology
MGIRI: Mahatma Gandhi Institute for Rural Industrialisation
MGNREGA: Mahatma Gandhi National Rural Employment Guarantee Act
MNRE: Ministry of New and Renewable Energy
MOT: Ministry of Textiles
MRD: Ministry of Rural Development
MSSRF: M S Swaminathan Research Foundation
MSE: Micro and Small Enterprise
MSE-CDP: Micro and Small Enterprises Cluster Development Programme
MSME: Micro, Small and Medium Enterprises
NABARD: National Bank for Agriculture and Rural Development
NAIP: National Agriculture Innovation Project
NATP: National Agriculture Technology Project
NCEUS : National Commission for Enterprises in the Unorganized Sector
NDE: Non Directory Establishments
NGO: Non-governmental Organization
NHB: National Horticultural Board
NIF: National Innovation Foundation
NIRD: National Institute of Rural Development
NSS: National Sample Survey
NSSO: National Sample Survey Organization
NSTEDB: National S&T Entrepreneurship Development Board
OAE: Own Account Enterprise
PAU: Punjab Agricultural University
PMEGP: Prime Minister's Employment Generation Programme
PMRY : Prime Minister Rozgar Yojana
PRODIP: Product Development, Design Intervention and Packaging
RGGVY: Rajiv Gandhi Grameen Vidyutikaran Yojana
RGUMY: Rajiv Gandhi Udyami Mitra Yojana
RISC: Rural Industry Service Centre
RRB: Regional Rural Bank
RSETI: Rural Self Employment Training Institute.

RTDEC:	Rural Technology Development and Extension Centre
RTP:	Rural Technology Park
SEED:	Science for Equity, Empowerment and Development
SFURTI:	Scheme of Fund for Regeneration of Traditional Industries
SGSY:	Swarnjayanti Gram Swarozgar Yojana
SHG:	Self Help Group
SIDBI:	Small Industries Development Bank of India
SPR:	Scientific Policy Resolution
SPV:	Solar Photovoltaic
SRISTI:	Society for Research and Initiatives for Sustainable Technologies and Institutions
SSI:	Small Scale Industries
STED:	S&T Entrepreneurship Development
TMC:	Technology Mission for Cotton
TPS:	Technology Policy Statement
TRYSEM:	Training of Rural Youth for Self-employment
TUFS:	Technology Upgradation Funds Scheme

References

1. Bandyopadhyay M., 1994, Introduction and Utilization of Industrial Technologies in Rural Areas – Case Study in India, Submitted to the Economic and Social Commission for Asia and the Pacific (United Nations) (UN-ESCAP), Bangkok for the Regional Report, December 1994.
2. Bandyopadhyay M., 2011, Policies and Programmes for Technological Capacity Building of Small and Medium Enterprises in India, in "Technology Transfer and Small and Medium Enterprises in Developing Countries" by (Ed) K. Ramanathan, Keith Jacobs, and M. Bandyopadhyay, Daya Publishing House, New Delhi.
3. Bandyopadhyay M. and Thiagarajan R., 1993, Science and Technology in Rural Employment Generation – An Experiment in West Bengal; Journal of Rural Development, Hyderabad, India, Vol. 12, No. 4, July 1993.
4. CSIR, Council of Scientific and Industrial Research (CSIR), Government of India, New Delhi Website: http://www.csir.res.in.
5. CSO, 2008, 5th Economic Census All India Report 2005, Central Statistical Organization (CSO), Ministry of Statistics and Programme Implementation, Government of India, January 2008.
6. DADF, 2003, 7th Indian Livestock Census, 2003, Department of Animal Husbandry, Dairying and Fisheries, Government of India.
7. DADF, 2010, Annual Report 2009-10, Department of Animal Husbandry, Dairying and Fisheries, Ministry of Agriculture, Government of India.

8. DBT, Department of Biotechnology (DBT), Ministry of Science and Technology, Government of India, New Delhi Website: www.dbtindia.nicin
9. DST, Department of Science and Technology, Ministry of Science and Technology, Government of India, Website: www.dst.gov.in.
10. GOI, 1980, Industrial Policy Statement, Department of Industrial Development, Ministry of Industry, Government of India, July 1980.
11. GOI, 1991, Policy Measures for Promoting and Strengthening Small, Tiny and Village Industries, Department of Small Scale Industries, Agro and Rural Industries, Ministry of Industry, Government of India, August 1991.
12. GOI, 2001, Census of India 2001, Office of the Registrar General and Census Commissioner, Ministry of Home Affairs, Government of India, 2001.
13. GOI, 2006, The Micro, Small and Medium Enterprises Development (MSMED) Act, 2006, Ministry of Micro, Small and Medium Enterprises, Government of India, October 2006.
14. GOI, 2010, Report of Prime Minister's Task Force on Micro, Small and Medium Enterprises (2006-07), Government of India, January 2010.
15. GOI, 2010a, Indian Economic Survey 2009-10, Ministry of Finance, Government of India, February 2010.
16. ICAR, Indian Council of Agriculture Research (ICAR), Ministry of Agriculture, Government of India, Website: www.icar.org.in
17. KVIC, Khadi and Village Industries Commission, Website: www.kvic.org.in
18. MNRE 2010, Annual Report 2009-10, Ministry of New and Renewable Energy (MNRE), Government of India (www.mnre.gov.in).
19. MOT, 2010, Annual Report 2009-10, Ministry of Textiles, Government of India.
20. MRD, Ministry of Rural Development, Government of India, Website: http://rural.nic.in/.
21. MSME, Ministry of Micro, Small and Medium Enterprises (MSME), Government of India, Website: www.msme.gov.in.
22. MSME, 2010, Annual Report, 2009-10, Ministry of Micro, Small and Medium Enterprises (MSME), Government of India.
23. NCEUS, 2007, Report on Financing of Enterprises in the Unorganized Sector and Creation of a National Fund for the Unorganized Sector (NAFUS), National Commission for Enterprises in the Unorganized Sector, Government of India, November 2007.
24. NCEUS, 2008, Report on Definitional and Statistical Issues relating to Informal Economy, National Commission for Enterprises in the Unorganized Sector, Government of India, November 2008.
25. NCEUS, 2009, A Report on Technology Upgradation for Enterprises in the Unorganized Sector: Status, Constraints and Recommendations, National

Commission for Enterprises in the Unorganized Sector, Government of India, April 2009.

26. NHB, National Horticulture Board, Website: http://nhb.gov.in
27. NIF, Website: www.nif.org.in
28. NRDC, National Research Development Corporation (NRDC), New Delhi Website: http://www.nrdcindia.com
29. NSSO, 2007, Operational Characteristics of unorganized manufacturing enterprises in India, NSS 62nd Round Report (July 2005 – June 2006), Report No. 524 (62/2.2/1), National Sample Survey Organization (NSSO), Ministry of Statistics and programme Implementation, Government of India, December 2007.
30. NSSO, 2009, Service Sector in India - Operational Characteristics of Enterprises, NSS 63rd Round Report (July 2006 – June 2007), Report No. 528 (63/2.345/1), National Sample Survey Organization (NSSO), Ministry of Statistics and Programme Implementation, Government of India, February 2009.
31. NSTEDB, National S&T Entrepreneurship Development Board, Government of India, Website: www.nstedb.com
32. PSA, 2006, Report of the Working Group on Effective Rural Technology Delivery (including partnership with Voluntary Organizations) for the Eleventh Plan (2007-12), Office of the Principal Scientific Adviser to the Government of India, November 2006.

Chapter 3

The Role of Science and Technology Towards Small and Medium Enterprises and Rural Development in Iran

H. Hajihoseini

(Director of Technology Development Studies Institute, IROST)
Ministry of Science, Research and Technology (MSRT),
Iranian Research Organization for Science and Technology (IROST)
E-mail: hajihoseini@irost.org

ABSTRACT

The Islamic Republic of Iran has pursued a development strategy of self-reliance with some degree of success. Iran is a middle-income developing country, with a broad industrial base, a relatively well developed science and technology infrastructure and skilled manpower. Iran's economy is largely dependent on the primary sector. It needs to have promotional policies for small and medium enterprises (SMEs) and related measures to encourage entrepreneurship in Iran. SMEs have an important role in job creation, economic and industrial development in Iran. The user-producer relationship is weak and the innovation activities in Iran are not very much demand-driven. A unique feature of Iran's innovation system is the marginal role played by foreign companies bringing in new technologies in the form of new products, processes and management techniques. At present, there are many key actors in Iran's innovation policy system. In this direction the author discusses the role of the Supreme Council of Cultural Revolution (SCCR) and the Iranian Research Organization for Science and Technology (IROST). IROST provides technical and financial support to SMEs acts as a granting council receiving proposals from researchers in SME to develop prototypes, peer reviewing the proposals and making the awards. The Ministry of Science, Research and Technology (MSRT) has also a strong position in scientific research and technological development. In Iran SMEs are helping create employment and are dynamic in adapting to economic changes relatively quickly and are more innovative than the large enterprises. Unfortunately, Iranian statistics relating to SMEs are scattered and incomplete. Concludes that In spite of huge industries, medium and small business plants and factories have not been able to launch R&D centres or even laboratory to test their products and raw materials for their technology development. They still need to have access to some accredited laboratories or technology centres.

Keywords: *Rural Industrial development strategy, Innovation systems, Rural development, Small and medium enterprises, Innovation policy, Clusters.*

Introduction

The Islamic Republic of Iran has pursued a development strategy of self-reliance with some degree of success. Endowed with abundance of oil and natural gas resources, Iran did not face any balance-of-payment constraints vis-à-vis its import. Yet, it adopted an import substitution policy that allowed it to use its oil revenues to acquire foreign technologies for industrialization. Iran is a middle-income developing country, with a broad industrial base, a relatively well developed science and technology infrastructure and skilled manpower.

However, Iran still remains largely a natural-resource-based economy. Diversification is an imperative, not only because natural resources become more accessible but also because export success in world markets increasingly demands knowledge-intensive production and innovation-based competition. Above all, there is a need to provide quality jobs for 800,000 skilled work force that enter the labour market every year.

The shift towards a more knowledge-based economy will require creating a national innovation system based on science and technology that would not only merely transfer ready-made technologies, but also engage in re-invention, developing new technologies and diffusing them economy-wide. There is a need to better link the science and technology infrastructure to the requirements of the productive sector generally, and in particular, building up capabilities in the high technology areas.

Iran's Economy

Iran's economy is largely dependent on the primary sector. The agriculture and the oil and gas sectors together accounted for 32 per cent of GDP during 1991-2003 while the industrial sector accounted for 20 per cent (including water supply, electricity and gas). This figure has changed to different mixture in 2009 (Table 3.1). Iran's manufacturing industry has been built up mainly through licensing of technology from abroad and in some cases through the reverse engineering. The State-owned enterprises, which are mainly large

Table 3.1: Average Contribution of Economics Sectors to GDP (per cent)

Sector	*2009*
Agriculture	10.9
Industry	45.2
Services	43.9

Source: Central Bank of the Islamic Republic of Iran and the World Bank data.

enterprises, continue to account for much of the industrial sector. It is necessary to have promotional policies for small and medium enterprises (SMEs) and related measures to encourage entrepreneurship in Iran, such as the provision of seed and venture capital and the establishment of science and technology parks and business incubators with adequate funding and support services to assist start-up enterprises.

Since 1990, Iran's economic plans have emphasized a gradual move towards a market-oriented economy and development of the private sector. Recently, many reforms were approved by the government and they are as follow:

- ✰ Approval of the Foreign Investment Promotion and Protection Act, aimed at simplifying the inflow of foreign capital and easiness of technology transfer from abroad;
- ✰ Liberalization of foreign trade through the elimination of non-tariff barriers and regulations for contract deposit;
- ✰ Equalization and reforming the system of exchange rates, aimed at increasing transparency in the government budget and subsidies;
- ✰ Direct Tax Law reform, consisting in the reduction of corporate taxes from 54 to 25 per cent and personal income tax rates from 54 to 35 per cent;
- ✰ Banking System reforms, through the establishment of non-banking credit institutions and private banks.

Iran's National Innovation System (NIS)

The main actors in the Iranian national innovation system are the government, the research institutes/universities and the enterprises. The irony is that all of the research institutes/universities and an overwhelming majority of the enterprises are also state-owned. It is said that due to this government ownership, there are close links between the research institutes/universities, enterprises and government. Other actors such as business corporations, business supporting organizations and consumer groups are very weak and play almost no role in the system. As a result, the user-producer relationship is weak and the innovation activities in Iran are not very much demand-driven. Absence of the private enterprises that base their innovation strategies on conditions of demand and competition makes it difficult to derive larger economic benefits from innovation. Such larger benefits that Iran has not presently realized include opportunities for commercializing new products, emergence of spin-off enterprises and new entrepreneurs, etc.

Competition

Competition is the key driving force for innovation and technological change and the industrial sector of Iran lacks effective competition. The system of licenses and resource allocation (subsidies) ensures that there is only limited competition (and mainly based on price) in the vast majority of industries. This lack of limited competition does not motivate companies to develop new products or product features. Recognizing this, the Government is gradually opening up the economy to competition, but the process is very slow.

Foreign Direct Investment (FDI)

A unique feature of Iran's innovation system is the marginal role played by foreign companies (Table 3.2). The foreign companies bring in new technologies in the form of new products, processes and management techniques. The local operations of foreign companies lead to spill over effects and diffusion of new technologies into the wide economy. They also spur competition and motivate domestic companies to upgrade their technologies and to innovate in order to compete. The Government has established free zones where foreign companies can locate operations, but the wide diffusion of technologies into the whole economy seems to be a small move.

Table 3.2: Comparison of Foreign Direct Investment (FDI)

Country	*FDI, Millions of US$, 2008*
Iran	1,490
Republic of Korea	2,200
Malaysia	7,380

Source: World Bank (2008)

Science, Technology, and Innovation Policy System

At present, there are key actors in Iran's innovation policy system namely the Supreme Council of Cultural Revolution (SCCR); the Ministry of Science, Research and Technology (MSRT); the Ministry of Industry and Mines (MIM); the Ministry of Agricultural Jahad; the Ministry of Health, Treatment and Medical Education (MOH); and the Technology Cooperation Office (TCO). Prior to Islamic Revolution in 1979, knowledge generation through the higher education and the scientific researches were the core functions of the Ministry of Science. In 1985, the medical education and research was assigned to a new Ministry of Health, Treatment and Medical Education (MOH), which also licenses the technology imports for the pharmaceutical sector although the Ministry of Industry and Mines nominally has responsibility for the latter.

The Supreme Council of Cultural Revolution (SCCR) is the highest policy making and the legislative body for all stages of pre-university and academic education. Its resolutions do not require parliament's approval and become the law automatically. Members of the SCCR include heads of the three powers of the state *viz.*, Ministry of Education (MOE), Ministry of Science, Research and Technology (MSRT) and Ministry of Health and Medical Education (MHME), as well as several other cultural experts. The Ministry of Education (MOE) is responsible for all stages of pre-university education. Within the MSRT, technology development falls under a separate Vice Ministry from that of research.

Besides, some work in the area of technology policy is underway, mainly through the activities of the Iranian research organization for science and technology (IROST). It provides technical and financial support to SMEs. To some extent, therefore, IROST acts as a granting council receiving proposals from researchers in SME to develop prototypes, peer reviewing the proposals and making the awards. This is part of a joint programme MSRT/IROST/Vice President for Strategic Planning/private sector/ institutes to support innovation projects up to the experimental development scale and encourage public-private partnerships in the process.

The MSRT, therefore, plays a relatively small role in the funding of innovation projects, but it has a strong position in scientific research and technological development (Abbasi, Hajihoseini, 2004).

Technological Capacity in the Industrial Sector

The two specific policies introduced during the period of reconstruction have attempted to reorient research more towards the industry. The first was to promote applied research with a view to making university research applicable to industrial needs. During the 1990s, the Ministry of Industry also sought to directly promote research and development activities in large and medium-sized companies, most of which were state-owned. By 1996, 158 companies had received official operating licenses for their R&D centres. The MSRT's Report on Private Sector Research Institutes (2003) notes that, by 2000, 76 technical and engineering research institutes have been created in the enterprise sector (including public and private enterprises). The earliest was established in 1971, followed by three others in the 1970s, 17 in the 1980s, 52 in the 1990s and two in the year 2000.

Relevance to the role of research in industrial innovation is the relatively small proportion of researchers located in private research centres. In 1996, the total number of researchers, research assistants and technicians amounted to 68,385, of which 82 per cent were employed in public sector institutes and 18 per cent in private sector research centres.

Apart from building up capacities in the knowledge system, Iran has also built up substantial technological capacity in the productive sector. It has a fairly well developed manufacturing capacity in the automotive industry, telecommunications and pharmaceuticals but the knowledge system has not percolated into the production system properly. Enterprises only undertake production, and do not perform innovation activities. While such a strategy was sufficient to cater to an import-substitution economy, it does not result in a dynamic capability for sustainable development. Moreover, even large manufacturing enterprises rely on imports for inputs. This is mainly because of the absence of strong support industries (supplier networks in the form of small and medium-sized enterprises) in Iran. Small and medium-sized enterprises make a relatively very small contribution to the national product.

The large enterprises (which are mainly state-owned), can build up a strong supplier industry within the country by developing technologies relating to the components and parts and diffusing these technologies to SMEs. Several other developing countries such as India and China have followed such a strategy. SMEs create employment and they are dynamic in adapting to economic changes relatively quickly. SMEs are also observed to be more innovative than the large enterprises. Also, the experiences of many countries have shown that the SMEs can make a substantial contribution to industrial and economic development.

Unfortunately, Iranian statistics relating to SMEs are scattered and incomplete. The number of small and medium sized *industrial* SMEs are, however, around 345,000 formally registered businesses, of which 96.1 per cent belong to the category of micro-enterprises (with a workforce of 1-9 employees), 3.3 per cent to the category of small enterprises (with 10-49 employees), 0.3 per cent to medium-sized enterprises (with 50-99 employees), and 0.4 per cent to businesses larger than 100 employees. As the present study shows, these industrial SMEs provide approximately 1.3 million jobs, out of the total employed labour force of 15.6 million.

The Tables 3.4 and 3.5 shows that 98.4 per cent of all businesses are micro enterprises with 1-9 employees, where as the total of small businesses with 10-49 employees' amounts to only 1.42 per cent. Obviously, there is an imbalance between the large number of micro enterprises and the marginal number of small and medium sized businesses. It may be noted that the absence of a reasonable number of medium-sized enterprises that amounts to only 0.1 per cent of the total number of enterprises, is negatively affecting Iran's ability to produce for the export market.

The Agricultural Jihad Ministry is responsible for the small agro-based and rural industries. The number of enterprise by type, size and employment is summarized in Tables 3.3 and 3.4.

Table 3.3: Number of Enterprises by Size and Sector

The Type of Enterprise	*1-5 Employees*	*6-9 Employees*	*10-49 Employees*	*50-99 Employees*	*More than 100 Employees*
Services	878774	5631	3478	231	150
Productive	334630	17125	13236	1055	1207
Mine	454	355	413	-	-
Total	1213858	23111	17127	1286	1357
Per cent of Total	96.6	1.8	1.4	0.1	0.1

Source: UNIDO, 2003.

Table 3.4: Number of Enterprises by Size and Employment

The Type of Enterprise	*1-5 Employees*	*6-9 Employees*	*10-49 Employees*	*50-99 Employees*	*More than 100 Employees*
Number	16735	12418	22318	2022	1584
Per cent	30.4	22.5	40.5	3.7	2.9
Number of Workers	62778	89572	42363	133315	588944
Per cent	4.8	6.9	32.6	10.3	45.4

Source: UNIDO, 2003.

The state-sponsored service providers include the Vocational Training Organization, the Industrial Management Institute, the Industrial Estates Corporation, and the Science and Industry Service Companies run by the provincial universities. However, vigorous private sector business services for consultation, training, and assistance in information and technology are yet to emerge.

The rationale for promoting small and medium-size enterprises in Iran is not one of scale but rather a scope. They usually cover heterogeneity of products and services based on a variety of skills, materials and markets, while creating a range of benefits for local and national economies.

In Iran, the small businesses suffer from similar obstacles as has been experienced by other developing countries. These include: poor macro-economic environment of high inflation (about 20 per cent, with a fluctuation of 9–10 per cent) and high interest rates; burdensome regulations; adverse labour and tax laws; lengthy and arbitrary procedures for securing bank loans; foreign currency shortages; lack of competent business development services; and an overall sense of discrimination against small enterprises. The main barriers to SME development in Iran is the lack of access to various kinds of information, including:

- Marketing information (on domestic and foreign markets, price structures, packaging requirements, etc);
- Information on the financial and technological standing of SMEs to enable investors to select healthy businesses for their investment;
- Technical and scientific information;
- Information on raw material suppliers and buyers.

This situation is exacerbated by the dominance of the oil sector in the overall economy and of the large state enterprises in industrial production. Thus, the potential for subcontracting to SMEs by the large automotive, transport, home appliances and oil industries is yet to be realized.

The Technology and Business Services to SMEs

Iran Small Industries Organization (ISIO)

Iran Small Industries Organization (ISIO) was established according to Parliament's legislation with the aim of formulating policies and guidelines for small enterprises development and supporting them to enhance their employment rate in the industrial sectors. ISIO was formed in by merging two previous organizations IIEC and ISIO. The Iran Industrial Estates Corporation (IIEC) was established in 1985, in order to establish the industrial estates (parks) and provide the necessary infrastructures and services for industrial investors.

ISIO's Main Policies

ISIO has some key policies as follows:

1. To enable the business environment for small industries by providing and recommending necessary proposals for policy-makers
2. To promote entrepreneurship and develop human resources
3. Assisting firms in increasing productivity and quality
4. Facilitating the transition to a market economy
5. To Develop information technology and e-commerce in small industries

6. To provide access to financial and capital resources
7. To support start-ups, giving priority to those in undeveloped regions
8. To develop technology, R&D activities and innovation in small industries
9. Construction/improvement of regional infrastructure by establishing industrial parks and prefabricated workshops.

Roles of Industrial SMEs on Employment and Economy in Iran

As below the tables identifies the role of industries, number of their workers and their added values. As it is shown, the small factories which have workers between 1 to 9 have the most important role in employing candidates and economic development. They have more than 50 per cent share of industrial economy and employment in Iran. Hence, SMEs have an important role in job creation, economic and Industrial development in Iran. This will be so important if we consider the roles of companies in the field of services and production, which most of them are SMEs companies.

Industrial Development Strategy in Iran and the SMEs

Iran has had various industrial development strategies in different years in the past. Before the Islamic Revolution in 1979, Iran did not have a very detailed and systematic strategy for industrial development. The consumer goods, machinery and industrial equipments, and other needs were imported mainly from the west due to the oil production and export. This was the main motive and lever for the industrial and economic development of Iran. After the Revolution (1979), due to the political crises, for example 8 years war between Iraq and Iran and some limitations on export and import, there was a need for a systematic industrial development strategy, what was called *self sufficiency* of needs. Iranian officials decided to develop the capability for domestic production of a variety of products and services by Iranian firms about 15 years. This was a strategy already used by some socialistic countries. By the end of war and beginning of the third and forth socio-economic development plan, due to some renovations in political, economy and also WTO policies, Iran's industrial development strategy changed to become more export-oriented.

Nowadays Iranian industrial development strategy again due to some foreign policies is changed and is faced with some limitations on importing and exporting equipments and commodities. However, a clear industrial strategy and technological road mapping is a challenge for Iran's industrial development. This situation makes industries to be inefficient, facing with different problems such as limitation of importing their needs and the difficulty in the flow of knowledge, etc. Under such conditions, due to changes of law and regulation in Iran, the SMEs could be more compatible to the industrial strategy of government of Iran and this makes them uncompetitive compared to other international firms in all around the world.

Agriculture and SMEs in Iran

Iran works on different and various kinds of production in the area of agricultural and industrial fields. Most of Iranian industries are government-funded (directly or indirectly), this is also the same for agricultural entities at the large scales. Even

Table 3.5: Growth in the Number of SMEs (Less than 10 employees) within the Food Industry (UNCTAD, 2005)

Item	*Industry Sector*	*No. in 2000*	*No. in 2006*	*Per cent Annual Increase*
1.	Food and Beverage in general	1954	2705	5.57
2.	Fish	514	722	5.83
3.	Cooking oil	31	48	7.56
4.	Meat	93	130	5.74
5.	Packaging (pistachio)	21	46	13.96
6.	Treatment	208	248	2.97
7.	Dairy products	126	233	10.97
8.	Animal food	62	93	6.99
9.	Sugar	57	73	4.21
10.	Pasta	178	135	4.50
11.	Bread	65	110	9.16
12.	Biscuit and cookie	289	368	4.11
13.	Tea	135	184	5.30
14.	Non-alcoholic Beverages	62	97	7.74

Table 3.6: Growth in SME's Employment (Less than 10 Employees) within the Food Industry (UNCTAD, 2005)

Item	*Industry Sector*	*No. in 2000*	*No. in 2006*	*Per cent Annual Increase*
1	Food and Beverage in general	129457	165075	4.12
2	Fish	38355	47351	3.57
3	Cooking oil	11020	10302	1.12
4	Meat	3610	5297	6.60
5	Packaging (pistachio)	1360	1107	3.37
6	Treatment	10574	13063	3.59
7	Dairy products	9671	20563	13.40
8	Animal food	2012	3350	8.87
9	Sugar	20936	23223	1.74
10	Pasta	3514	3394	0.58
11	Bread	1197	2189	10.58
12	Biscuit and cookie	17472	17853	0.36
13	Tea	3767	5285	5.81
14	Non-alcoholic Beverages	13500	12145	1.75

though most of agricultural firms are private which are in small level. So, Interaction between industrial SMEs and agricultural firms from employment points of view are very important. The writer notes that enhancement of capability buildings of SMEs on employment in agricultural field especially for young and university graduated should be considered and pay more attention to Iranian authorities. The following tables show growth in the number of SMEs as well as SME's employment within the food industry:

Industries, Industrial Towns, Research and Technology development and SMEs in Iran:

Most of small and medium factories are in industrial cities. Some details of Iranian industrial cities are shown in Table 3.7. Table 3.8 also provides some information about the role and importance of Iranian industrial cities. It is shown that the technology Parks are yet to be established, and developed. Technology Parks and centres give different services to some SMEs such as; technical information, research and laboratory tests and other services.

Table 3.7: Some industrial information about Iran in 2007 (www.iraniec.ir; www.foodtechpark.com)

	No. of Factories	*Workers 10^3*
Small factory (1-9 workers)	16,057	10714
Medium factory (10-50 workers)	12.151	2666
Factory (more than 50 workers)	3906	8038
All the factories	32,114	21428

Table 3.8: Some Information About Iranian Industrial Cities in 2008 (www.iraniec.ir)

Number of industrial towns	393
Number of industrial company in the industrial cities	20594
Workers	442163
Number of technology cities	2
Number of technology centres in the industrial cities	8
Number of IT centres	3

Conclusion

In spite of huge industries, medium and small business plants and factories have not been able to launch R&D centres or even laboratory to test their products and raw materials for their technology development. They still need to have access to some accredited laboratories or technology centres. Such kinds of firms are very important to economy and employment, and establishing of these types of centres would be very effective to industrial development at the national level. Therefore, using standard methods of design and production such as ISO/IEC 17025 will be more useful on technology development in SMEs. So the author suggests nine recommendations as follow:

- To provide a careful and comprehensive strategy which contributes to long-term policy-design and policy-making
- It is suggested for many industrial towns have technical services centres and incubators and also accredited laboratory to give more qualified services to SMEs in Iran.
- It is recommended that universities and research centres use standard methods and technology management systems in their research and laboratory activities. This will raise the confidence of SMEs.
- To create a conductive environment in order to stimulate the development of knowledge-based SMEs.
- To provide support in management, engineering and technical areas for tenant SMEs.
- To encourage R&D and innovation in SMEs.
- To promote innovation growth in the development of technology- driven SMEs that are competitive.
- To provide first class infrastructure and services for SMEs.
- Engaging Iranian embassies across the globe in these issues.

References

1. Abbasi, F. and Hajihoseini, H (2004): Evaluating Iranian National Innovation System (NIS), 2004 STEPI International Symposium on S&T Policy, 30 November-1 December, 2004.
2. Central Bank of the Islamic Republic of Iran
3. http://amar.sci.org.ir/Detail.aspx Ln=F&no=261504&S=TP, Iranian Statistics Center, population of Iranian 2006, Tehran, Iran
4. Industrial Research Centers, ministry of industry, deputy or research and training, 2007, Tehran, Iran
5. Iranian Industrial Cities, Iranian Industrial Cities Company, 2003, Tehran, Iran.
6. Iranian Organization for Science and Technology (IROST) yearbook 2008, Forsat St. 71, Tehran, Iran.
7. Iran Fourth Five-Year Development Plan 2005-2009.
8. MSRT, IROST, and RITDS (2003a)."The macro survey of the Iranian economy, part 1:Growth,value added, productivity, inflation rate, exchange rate, foreign debt, and population.
9. Statistical Yearbook of Iran, 2000.
10. UNIDO (2003): Strategy Document to Enhance The Contribution of an Efficient and Competitive Small and Medium-Sized Enterprise Sector to Industrial and Economic Development in the Islamic Republic of Iran, February 2003.
11. UNCTAD (2005): Country Reports: Science and Technology Promotion, Advice and Application for the Achievement of the Millennium Development Goals,

Commission on Science and Technology for Development, Eighth session, Geneva, 23-27, May 2005.

12. UNCTAD (2004).
13. World Bank (2003).
14. www.iraniec.ir
15. http://www.foodtechpark.com/
16. www.iraniec. com, Iranian Industrial Cities Company, Tehran, Iran.
17. www.min.gov.ir, Iranian Ministry of Industry Portal, Tehran, Iran.
18. www.amat.sci.org.ir, Iranian Statistics Center, Tehran, Iran.

Chapter 4

Rural Uplift through Science and Technology

Syed Nawazish Ali Shah

Deputy Scientific Adviser (International Liaison),
Ministry of Science and Technology (MoST)
Government of Pakistan, Pakistan
E-mail: anbarshah@yahoo.com

ABSTRACT

Pakistan is a country, which is thickly populated having ~ 170m population with growing industrial base but mainly an agricultural economy. More than 60 per cent population live in rural areas and like other developing countries, are less privileged/poor. The Government of Pakistan is trying through various programmes for reduction of poverty and provision of basic necessities of life with the aim to improve/increase per capita income. There are a number of diverse programmes initiated by different Ministries/Departments of the government for uplifting the standards of life of the people of the country though, this paper accounts for some of the salient ones.

The Aik Hunar Aik Nagar (AHAN) programme based on the "One Village One Product" concept aims to alleviate poverty in the rural and peri-urban areas of Pakistan by supporting rural micro and small enterprises, which are engaged in non-farm products. Thus, AHAN aims at providing the much needed non-traditional work opportunities in rural areas by adopting and indigenizing the said concept.

Pakistan though bestowed with abundant natural resources of all kinds, is still an energy deficient country and the people as well as the industry faces severe energy shortage problems. The government is striving to start a number of projects in the energy sector to meet the public and private sector demand. In order to facilitate the rural population, energy saving and renewable energy production programmes, which have been initiated in the area of Solar (Solar Thermal and Photovoltaic), Wind Energy, Micro hydro Power and Biogas Production besides introduction of fuel efficient technology thereby designing fuel efficient cooking stoves for the rural population, which save up to 25 per cent fuel and are environment friendly, being produced less or no smoke while burning the wood.

Keywords: *Aik Hunar Aik Nagar (AHAN)/One Village One Product, Khes (Sheets), Ceramics, Carpets, Silver jewelry, Renewable technologies.*

Introduction

Pakistan, officially the Islamic Republic of Pakistan, is a country in South Asia. It has a 1,046-kilometre (650 mi) coastline along the Arabian Sea and Gulf of Oman in the south, is bordered by Afghanistan and Iran in the west, and India in the east and the China in the far northeast. Tajikistan also lies very close to Pakistan but is separated by the narrow Wakhan Corridor. Thus, it occupies a crossroads position between South Asia, Central Asia and the Middle East. The region forming modern Pakistan was at the heart of the ancient Indus Valley Civilization and then later was the recipient of Vedic, Persian, Indo-Greek, Turco-Mongol, Islamic,Hindu and Sikh cultures. The area has witnessed invasions and/or settlements by the Indo-Aryans, Persians, Greeks, Arabs, Turks, Afghans, Mongols and the British.

While the Indian independence movement demanded an independent India, the Pakistan Movement (led by Quaid e Azam Muhammad Ali Jinnah of the Muslim League) sought independent states for the majority Muslim populations of the eastern and western regions of British India as well. The British granted independence and also the creation of one Muslim majority state of Pakistan that comprised the provinces of Sindh, North-West Frontier Province, West Punjab, Balochistan and East Bengal. With the adoption of its constitution in 1956, Pakistan became an Islamic republic. In 1971, a civil war in East Pakistan resulted in the creation of Bangladesh.

Pakistan is a federal parliamentary republic consisting of four provinces and four federal territories including a federal capital territory. It is the most populous sixth country in the world and has the second largest Muslim population after Indonesia. It is a Muslim-majority state that possesses nuclear weapons and is designated as a major non-NATO ally of the United States. Pakistan is a founding member of the Organization of the Islamic Conference and a member of the United Nations, Commonwealth of Nations, Next Eleven economies, the G20 developing nations and the Non aligned Group.

Programmes

AHAN - Introduction

Pakistan, in order to meet its goal of reducing poverty, needs to expand non-farm economic opportunities in addition to agriculture and livestock in the rural areas and small towns. Therefore, Government of Pakistan has launched a project titled "Aik Hunar Aik Nagar" (AHAN)/Rural Enterprise Modernization Project based on the One Village One Product concept. Primary objective of this project is to alleviate poverty in the rural and peri-urban areas of Pakistan by supporting rural micro and small enterprises engaged in non-farm products. Thus, AHAN aims at providing the

much needed non-traditional work opportunities in rural areas by adopting and indigenizing the One Village One Product concept.

AHAN Project was formally initiated in July 2006 after its approval by the Planning Commission. During a short period of one year it has actively started work in all four provinces in collaboration with Provincial/District Governments, RSPs (pl. expand), microfinance institutions and other stakeholders. The project has adopted two pronged approach. On the one hand, it has started pilot projects in different clusters of artisans and poor producer groups while on the other hand it has also started research and analysis.

The exercises discover the potential in this area and to see how a long term strategy can be developed for creating sustainable income generation opportunities to a large segment of our society, living in semi urban and small cities and towns of Pakistan. The idea is to use a holistic approach at the grass roots level by creating synergies through network of rural enterprise initiatives. The project is catering to both supply side and demand side: On the supply-side, it aims at building the capacity of crafts persons, involving well-known designers, to produce contemporary products while assuring quality; and on the demand-side, arranging/participating at exhibitions nationally and internationally, displaying set ups, designers' outlets, developing linkages with exporters, etc.

AHAN – Areas of Intervention

Product Development and Design

AHAN assists the crafts persons by providing capacity building services to the cluster in product design, quality improvement and quality sustenance through technical assistance, production management training and product design training.

Marketing

The intervention includes development and implementation of marketing strategy including market analysis, product selection, and brand development, promotional campaign, testing marketing and distribution system for rural non-farm products of Pakistan.

Enterprise Development

Crafts persons and artisans are introduced to effective ways of enterprise development of their products. It aims in supporting employment generation activities by providing awareness and exposure to entrepreneurs in marketing of their products and managing their businesses.

Linkages with Microfinance Institutions

AHAN is developing networks between the rural entrepreneurs and the microfinance institutions to provide support to clusters in obtaining easy credit facilities for their growth and development.

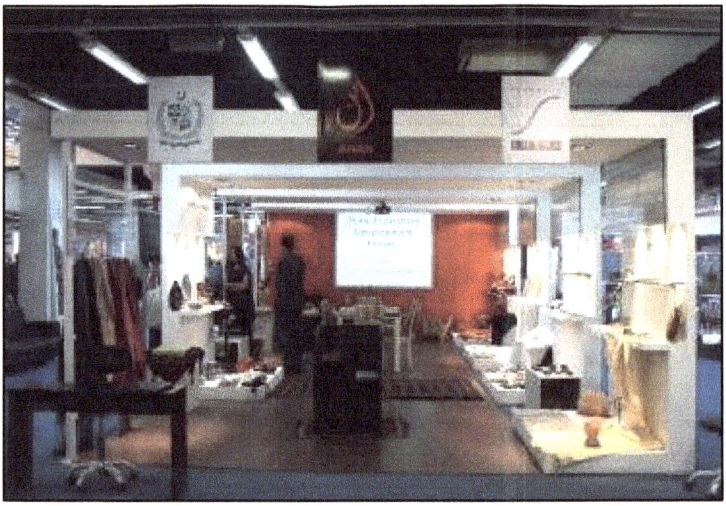

Sustainability of Projects

AHAN has designed its interventions in clusters with the objective of ensuring sustainability of the benefits achieved through pilot projects. Effective exit strategies are designed to allow continued cluster development in the post-intervention phase.

Baseline Studies and Impact Assessment

Great emphasis is laid on conducting initial assessment surveys of clusters under baseline study phase to not only measure effectiveness of pilot projects but also to document current status of identified regions/products for intervention. On culmination of the project, AHAN would conduct the impact evaluation in collaboration with respective partner organizations to evaluate benefits received by the crafts persons. In this regard, consultants have also been hired to help AHAN augment this area.

Monitoring and Evaluation

AHAN realizes the importance of effective monitoring systems for both internal evaluation and pilot project assessment. Consultants have been hired to assist in modification of monitoring mechanisms to ensure maximum efficiency of AHAN; both as a project and a team.

AHAN – Pilot Projects' Statistics

AHAN has recently initiated 25 pilot projects after completion of Phase-I of 4 pilot projects. Another 14 pilot projects are in pipeline. Hence, AHAN intends to provide assistance directly to more than 6,832 beneficiaries through its network of 44 pilot projects in all four regions. 58 per cent of the beneficiaries are craftswomen, thereby ensuring adequate support to females of rural areas.

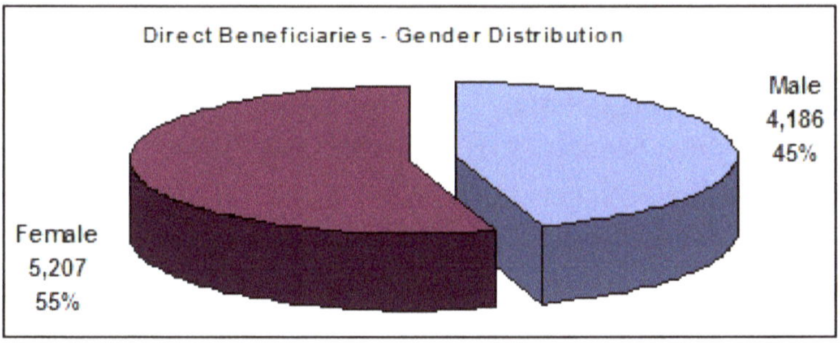

AHAN: Partner Organization Share

AHAN encourages participation from stakeholders of the clusters. Budgets of ongoing pilot projects are being shared with respective partner organizations on an average of 40:60. This not only ensures ownership of the partners for pilot projects but also helps in sustainability of the activities after the project period is over.

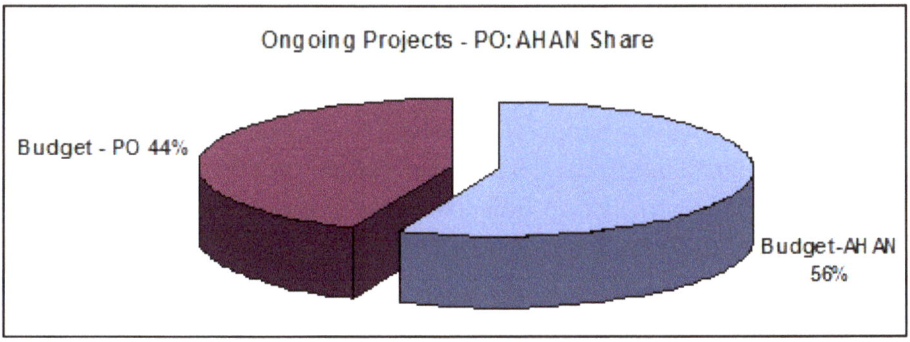

AHAN – Partner Organizations

AHAN has collaborated with private sector partners for incorporating their expertise in products development and design innovation. In addition to this, some partners have also provided reasonable businesses and entrepreneurial opportunities to crafts persons of various clusters.

Private sector partners

- ☆ Deepak Perwani
- ☆ Amir Adnan
- ☆ Rizwan Baig
- ☆ Irum Zia
- ☆ Ruby Jewelers
- ☆ Taneez
- ☆ Noorjehan Bilgirami

AHAN has involved Educational Institutions to provide capacity building services such as consultancy on product design, product range development and respective trainings.

Educational Institutions

- ☆ National College of Arts
- ☆ Indus Valley School of Art and Architecture
- ☆ Beacon House National University
- ☆ Centre of Excellence in Arts and Design (CEAD)
- ☆ Sardar Bahadur Khan Women University (SBKWU)

AHAN has engaged assistance from all the listed Government Organizations, Associations, Chambers and Non Profitable Organizations to maximize–its outreach in rural areas.

Supporting Organizations

- ☆ District Governments
- ☆ Chambers of Commerce and Industries
- ☆ Punjab Small Industries Corporation
- ☆ All Pakistan Handloom Association
- ☆ Bahawalpur Rural Development Project
- ☆ Balochistan Rural Support Programme
- ☆ Sindh Small Industries Corporation – Sindh
- ☆ Sarhad Industries Development Board– Kyber Pukhtoon Khwa
- ☆ Rural Support Program Network (RSPN)
- ☆ Balochistan Small Industries
- ☆ Directorate (BSID)

NGOs/RSPs

- ☆ Karvaan Crafts Development Empowerment Women Association (DEWA) APWA
- ☆ Mashal Skill Development Center

- ✰ National Rural Support Programme
- ✰ Sindh Rural Support Organization
- ✰ Thardeep Rural Development Programme
- ✰ Islamic Relief Al-Falah Foundation
- ✰ Lead Pakistan Catholic Relief Services
- ✰ Cholistan Development Council Behbud Association
- ✰ Strengthening Participatory Organization (SPO)
- ✰ Shaur Development Organization

AHAN – Punjab Region

Punjab is the country's most populous province which may be divided into different zones on the basis of its cultural richness such as Central, Eastern, Northern and Southern Punjab. Each zone is abundant in both highly and semi skilled craftsmen. These craftsmen and women are proficient in making articulate hand made products. The range of these products includes fabric, wood carved decoration pieces, footwear items, carpets, curtains (traditionally known as chiks), sport equipments and surgical instruments etc. The following list transcribes the range of products according to the zones mentioned above:

1. *Central Punjab*: Khes, ceramics, furniture, earthenware, carpets and electronic appliances etc.
2. *Eastern Punjab*: Wood products, kites, music instruments, foot ware, lacquer work pieces and chiks etc.
3. *Northern Punjab*: Khussas, traditional chappals, furniture, Taxila stone work and embroidery etc
4. *Southern Punjab*: Hand embroidery clothes, silver jewellery, chunnari, carpets, prayer mats, zari work.

Project Profiles

The following is the list of initiatives under taken/under consideration by AHAN in Punjab:

Sl.No.	Product Name	District	Partner Organization	Direct Beneficiaries
1.	Silver Jewellery	Bahawalpur	Beacon House National University (BNU), Ruby Jewellers, BCCI	10
2.	Hand Embroidery Project	Bahawalpur	Rizwan Beyg	42
3.	Mathera Pilot Project-Cholistan	Bahawalpur	Cholistan Development Concern (CDC)	15
4.	Roehi Handloom fabric	Bahawalpur	National Rural Support Programme (NRSP)	600
5.	Multan Handloom Cluster	Multan	Multan Association	90

Contd...

Table–Contd...

Sl.No.	*Product Name*	*District*	*Partner Organization*	*Direct Beneficiaries*
6.	Lacquer work Pilot	Sargodha	Strengthening Participatory Organization (SPO)	80
7.	Janu Wala Pilot Project	Bahawalpur	Bahawalpur Rural Development Programme (BRDP)	20
8.	Sandan Pilot Project	Bahawalpur	BRDP	20
9.	Kaarvan Stiching	Lahore	Kaarvan Crafts Foundation	55
10.	Talagang Mud Pottery	Chakwal	NRSP	
11.	Behley Pilot Project	Bahawalpur	BRDP	20
12.	Hasilpur Pilot Project	Bahawalpur	BRDP	20

Stories Depicting Potential for Growth

Hand Embroidery Project – Bahawalpur

The objective of the project was to enhance the capacity of rural women to produce quality products and to link them with the market particularly with the leading fashion designers such as Rizwan Beg who usually believe in quality products rather than quantity of the products. This was done through conducting a Skill Enhancement Training for 50 females at Basti Januwala and Basti Sandhan, Bahawalpur from 3–24 June 2007.

According to Rizwan Beg, "the products made during the training have improved to a great extent". He also added that if the trainees continue to work in the same manner their products can be sold in the international markets. However, he also mentioned that this pilot project is a drop in the ocean and a lot needs to be done. Therefore, he suggested replicating similar projects in the vicinity and other districts of the country in order to take such cottage industries to an improved state of production. Many designers have shown interest in working with these trained women for production of high quality embroidery work. More training are being arranged to enhance the skill of the cluster, which would subsequently lead to increase in their income level.

Silver Jewellery - Bahawalpur

Bahawalpur is known for its silver jewellery since in the past. It has been famous for its peculiar designs and the craftsmanship. The major clusters have been concentrated in Cholistan, Hasilpur and Ahmedpur. Still the hub of this activity is Bahawalpur City. A Skill Enhancement Training was organized for 12 craftsmen from Bahawalpur at Beacon House National University (BNU) from 31 July to 11 August 2006. They were exposed to new trends in jewellery and were taught on new techniques for preparing their products in sterling silver, which is recognized worldwide as standard silver jewellery and silverware.

Mr. Junaid Ahmad Qureshi is a Kundan specialist and also the owner of the Junaid Goldsmith in Bahawalpur. He has been in the silver jewelry business for years and this training has helped him and his colleagues to a great extent in terms of improving quality of the products. As a result of this trail, this cluster is now getting orders from famous designers including Amna Sharif, Amir Adnan and Zeenat Hassan (Taneez).

AHAN - Sindh Region

Sindh, often referred to as "Bab-ul-Islam" (Gate of Islam), has been and still very rich in handicrafts. The textiles are very famous especially block printing and tie-and-dye. It also enjoys a reputation for textiles, pottery, clay, lacquer furniture, leatherwork and carpets. Each color, style, design and motif carries with it a unique symbol portraying the culture of that particular area and builds on people's indigenous skills.

There are different kinds of handicrafts produced in Sindh, the main ones are: leather goods, tableware, wood crafts, textile and block printing fabrics, embroideries, glass beads, traditional and contemporary jewellery, pottery, hand knotted carpets and rugs, shell and bone products etc. Karachi, apart from being the hub of economic activity in the country', is also a home of famous designers. Therefore, it has also become a centre of activity for AHAN to link rural artisans with these high profile designers for capacity building of crafts persons.

Project Profiles

The following is the list of initiatives under taken/under consideration by AHAN in Sindh:

Sl.No.	Product Name	District	Partner Organization	Direct Beneficiaries
1.	Embroidery, Textile and Accessories	Dadu	Amir Adnan	8
2.	Embroidery, Ralli	Shikarpur	Deepak Pervani	6
3.	Ralli	Karachi	Indus Valley School of Art and Architecture	275
4.	Silverware and Jewellery	Hyderabad	Taneez and jewelry associations	112
5.	Khes	Khairpur	Deepak Pervani, District Govt Khairpur and LEAD Pakistan	46
6.	Farasi Carpets	Badin	Development Consultants(DEVCON)	30
7.	Embroidery Fabric	All Pakistan	Amir Adnan	208
8.	Thari hand woven carpets	Umerkot	Thardeep Rural Development Programme (TRDP)	400
9.	Lacquer Work	Matyari	SHED, Centre of Excellence in Arts and Design (CEAD)	60
10.	Blue Pottery	Hala	SPO,CEAD, Mehran Forum-MSCF	45
11.	Embroidery	NawabShah	Nawabshah Social Development Council (NSDC), CEAD	75
12.	Family Resource Centre	Malir	DOTZ technologies	232
13.	Ceramics Products	Matyari	Sindh Small Industries Corporation (SSIC), CEAD	180

Stories Depicting Potential for Growth

Farasi Carpets

The pilot project is located in village Baghli, Taluka Golarchi, District Badin, Sindh, Pakistan. Farasi rugs come under the category of hand-woven carpets/rugs made of camel and sheep yarn. A six month pilot project of "Farasi Carpets/Rugs" is focusing on products diversification and skill base enhancement inclusive of new designs, color combination, use of raw material and processing interventions.

A visit of Japanese exporter – Mr. Khalid ul Hassan (Karizua Gallery) facilitated under AHAN who was highly impressed with the products and purchased samples of more than Rs. 1,000,000. A German consultant - Mr. Beck also visited the villages

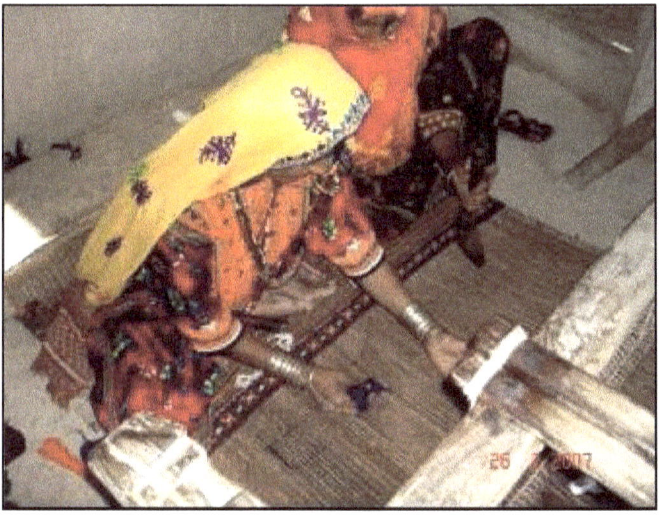

and examined the products and found them excellent. He suggested interventions in design, color and size marketable for international market. Finally but not the least, at this stage of beginning of AHAN interventions, craftswomen are receiving the placement of international orders. The order has been placed by Karuizawa Japan for preparing first consignment of camel wool rugs.

Textiles – Amir Adnan

FNKASIA – Amir Adnan is a six month old pilot project of AHAN through which craftswomen were trained as the master craftswomen at Amir Adnan. Two groups of eight craftswomen from Dadu and Johi in support with Social Welfare Department got trained in Karachi at Amir Adnan factory and produced the products such as necklaces, bangles, earrings, kurtis, purses and etc. Followed by the training of master craftswomen, in the back end some sixty girls were further trained. The craftswomen received an order by Amir Adnan amounting Rs. 108,000 to produce the products consisting of more than 1,600 necklaces, 620 sets of necklaces and earrings, 750 mobile pouches, 200 wallets and 300 bags.

Ms. Farzana Soomro - a craftswoman from Johi after completing the training from Amir Adnan got an opportunity to train more craftswomen of Choondko and near by areas of Taluka Nara, District Khairpur. She was working with Women Skill Development Organization, Dadu and the vocational training centre at Johi. Her monthly income was Rs. 5,000. On basis of her experience and training of master craftswomen from AHAN, she was considered eligible by Indus Resource Centre (IRC) to work at the capacity of 'Trainer' at Vocational Centre Choondko established by OMV Pakistan. She is appointed as a 'Trainer' with effect from 5th December, 2006 with monthly salary of Rs. 20,000.

AHAN – Balochistan Region

The province of Balochistan is unique in its geo-political significance. It forms 44 per cent of Pakistan's landmass and has a 770-km long coast line. It consists of arid basins and various hill ranges, sharply marked off from the Indus plain by the kirther and sulaiman ramparts. It exhibits a great variety of physical features, consisting of vast rocky desert with extremes of climate and very low rainfall.

Balochistan is famous for handicrafts skills, where a variety of colorful handicrafts in embroidery and needlework is designed in different areas. The art of handicrafts in Balochistan has survived since old ages without undergoing any scientific and industrial advancement. The handicraft work is common in the rural areas where females having learnt the handicraft skills, are actively engaged in sewing them to raise livelihood for their families.

The needlework in Balochistan contains different designs and techniques. It varies from place to place and tribe to tribe. The handicrafts are famous for their beautiful and sophisticated designs. A variety of needlework designing, sold at expensive boutiques and crafts shops, shows sale value in national and international markets

Project Profiles

The following is the list of initiatives under taken/under consideration by Balochistan:

Sl.No.	*Product Namc*	*District*	*Partner Organization*	*Direct Beneficiaries*
1.	Participatory Small Scale Enterprise Development	Kharan	Islamic Relief (IR)	132
2.	Killim	Quetta	Catholic Relief Services (CRS)-Pakistan	50
3.	Stone Cutting and Polishing	Chamman	Save the Children	500
4.	Hazargi Cross Stitch	Quetta	Sardar Bahadur Khan Women University (SBKWU) and Behbud Association	80
5.	Churma/Khamak Embroidery	Quetta	Catholic Relief Services-Pakistan	60
6.	Niarie-Ta-Hunar	Mastung	Balochistan Rural Support Programme (BRSP)	121
7.	Galabtoon Dozi	Gulabatoon & Zanjiri Stitch	American Refugee Committee	200
8.	Poost Chaka Kundi kar	Leather Embroidery	Balochistan Small Industries Directorate (BSID)	200
9.	Dano kar	Bead Work	Institute of Development Studies and Practice	
10.	Balochi Chakkan	Leather Embroidery	Bright Star Development Organization	100
11.	Women Empower-ment through Marble Mosaic	Marble Mosaic	PASDEC, GPIG and Taraqee Foundation (TF)	250
12.	Zartoo kar	Zartoo Stitch	APWA	80
13.	Killim	Killim	Institute of Development studies and Practice	80

Stories Depicting Potential for Growth

Story of Saima Kakar

Saima was living in village "Killi Sawan Khan"of Union Council Dasht, was a member of a poor family who had no concept of income generation by a female. She initiated a skill development center for the girls to provide training in different trades. This gave idea to an NGO for initiating a project for income generation when Aik Huner Aik Nagar (AHAN) intervened. In June 2007, AHAN and the NGO signed a memorandum of understanding (MOU) to initiate a project of collaboration in the same area to create income generation activities. In July 2007, Saima was selected as a master trainer for the project and was supposed to attend a one-year training programme in the factory of the famous designer Amir Adnan in Karachi. The mode of training was decided that she would receive training sessions of 15 days each month in the factory of Mr. Amir Adnan and will come back to the village to transfer her new skill/learning to other women involved in the same skill. She is happy now that she is in a position to train her community in new skills, which will open up new avenues of income for the rural community.

Story of Bibi Shazroo

Bibi Shazaroo lives in a Pushtoon village Tajik Abad of Union Council Kechi bieg. Bibi Shazaroo supports her family in generating income through killim making. As initial support, an NGO in collaboration with AHAN, established a training center for the community. The purpose of the center is product development and redesigning. During the last 3 months, Bibi Shazaroo has not only earned income from sale of her killims but has also participated in all activities regarding the project.

She attended a product review exhibition organized by Social Welfare Department and AHAN in Quetta and displayed her products on the stall. She also attended meeting of Steering Committee of AHAN at Karachi, which was an alien concept in the community.

Looking at the devotion towards the work she has showed, it is clearly visible that she will become a successful entrepreneur of killim very soon. This skill may become a regular source of income for her and her whole family. Her zeal shows how rural women belonging to far flung areas can be empowered to become entrepreneurs and become productive members of the society.

(4) AHAN – Khyber Pukhtoon Khwa Region

North West Frontier Province is rich, not only in terms of its cultural heritage but also in terms of biodiversity and human resources of arts and artistry. Almost each district of Khyber Pukhtoon Khwa has its own special products being produced by local rural industrious artisans. These artisans comprise both men and women. Phulkari stitch and Justi embroidery of Haripur, Swati embroidery of districts Malakand, Swat and Dir, Embroidery of Kaka Sahib Nowshera and Cross Stitch of DI Khan hold their distinction in terms of its stitch.

In addition, the lacquer and Gillit work of DI Khan, weaved products of Islampur (Swat) and Matta Mughal Khel (Shabqadar), Chitralli Patti Products and Chitrali rugs and Swati Rugs, all hand made items, are famous world over because of their beautiful designs and patterns. The leather products particularly hand made footwear items of Charsadda are popular not only within the province but all over the country.

Project Profiles

The following is the list of initiatives under taken/under consideration by Khyber Pukhtoon Khwa:

Sl.No.	*Product Name*	*District*	*Partner Organization*
1.	Weaving, Khaddar	Charsadda	Weavers Association
2.	Embroidery	Dir	Development Empowerment Women Association (DEWA)
3.	Swati Embroidery	Malakand	Bint-e-Malakand
4.	Shawls, Clothing	Swat	Islampur Weavers Association
5.	Justi Embroidery	Haripur	Alfalah Foundation
6.	Beads Jewellery, Dress Accessories and Home Décor Items	Abbottabad	Mashal (skill development center)
7.	Mosaic	Peshawar	Small Industries Development Board (SIDB) and Pakistan Stone Development Company (PASDEC)
8.	Nowshera Embroidery	Nowshera	SIDB and APWA
9.	Furniture Business Centre	Peshawar	Furniture Manufacture and Dealer Association Peshawar
10.	Crotia and Cross Stitch	Haripur	Pind Hashim Khan Network
11.	Business Centre for Women	Malakand	Bint-e-Malakand

Stories Depicting Potential for Growth

Dewa – Development Initiative

A project "Enhancing efficiency of rural women in district Lower Dir through improvement in quality skills" with a minimal cost of 81,000 was able to produce 15 entrepreneurs who are now managing their own production activities. The

intervention has infused enthusiasm for development and income generation in this conservative area. Four groups of females totaling to eighty three individuals are ready for training on the same pattern as carried out during AHAN intervention. Order of 50,000 and 10,000 each has been received from a member of Women Chamber of Commerce, Peshawar. Products of Rs. 15,000 were sold during an exhibition at Abbottabad arranged by Sungi Foundation.

Bint-e-Malakand Entrepreneurs

Binte Malakand is a non-government organization based in Batkhela, Malakand working for the economic uplift of the women entrepreneurs of District Malakand and Swat. AHAN Khyber Pukhtoon Khwa initiated a project on value addition to Swati embroidery products with the said organization. The intervention involved capacity building of 30 women of village Alla Dhand in developing new products of Swati Embroidery with vibrant colors and innovative designs. Moreover, it was also directed at developing entrepreneurship skills of the participants. AHAN sponsored their visit to Islamabad market.

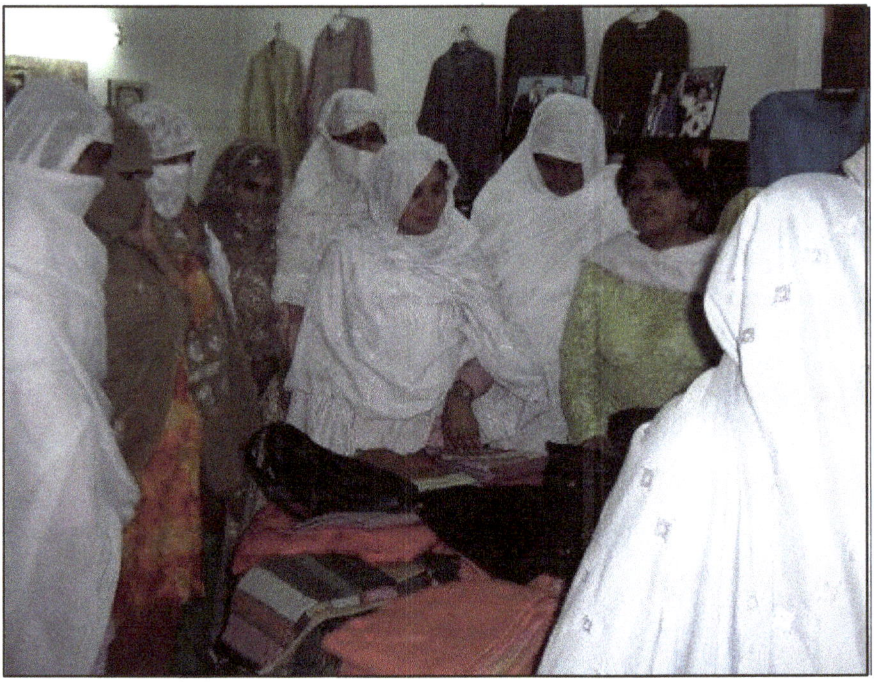

The project, though bearing a short duration and minimal cost, achieved tremendous success immediately after its conclusion. Not only the women involved have been able to obtain orders for the production of new products developed, but they have also been able to grow and mature as successful independent entrepreneurs.

AHAN – Marketing Department

The Marketing Department has been set up to provide marketing support both to AHAN as a corporate entity and to AHAN's target clients *i.e.* Crafts workers and

micro and small enterprises (MSEs) in the country. MSEs in Pakistan have always faced some kinds of marketing constraints due to lack of support services or their relatively higher unit cost. Marketing department is responsible for providing comprehensive marketing support programmes, which would improve MSEs' competitiveness through better market planning. Support programmes are classified into internal and external support programmes.

Some of the activities of the department are:

Marketing Input in Strategy Formulation

The Marketing Department, at this stage, proactively assists the pilot teams whilst ensuring that the strategy is properly conceptualized and is ready for implementation from a marketing standpoint.

Marketing – AHAN

Marketing AHAN as a corporate entity clarifies its role and helps create a conducive environment for working. This is done through AHAN website (www.ahan.org.pk) and printed material related to overall project. Linkages are created with stakeholders including the Chambers of Commerce and Industry, microfinance organizations (e.g. Tameer bank, Pak-Oman) etc.

Marketing of Pilot Projects

AHAN has participated in Women EXPO 2007, Hassle Free Business Environment and its products were also displayed at OIC Economic Forum.

Craft Bazaars

AHAN Steering Committee has recently decided to test a model of craft bazaar at one place to provide market access to rural artisans and helped them sell their products in existing form. Planning is being done to implement the decision.

Retail survey of handicrafts has been initiated to analyze demand for handicrafts in national market.

AHAN – Shell Tameer Work Shop on Entrepreneurship

Shell Pakistan launched Shell Live WIRE in 2003 with the local name Tameer with the aim of helping youth across the country to achieve their socio-economic development. Shell Tameer encourages young people (aged between 18-32 years) by enhancing their capacity building to start a business. A one-day workshop on entrepreneurship orientation programme was conducted in Balochistan and Sindh, involving beneficiaries of the pilot projects and participants from partner organizations of AHAN. The work shop provided awareness among people on entrepreneurship, which includes; setting up and managing a business (micro enterprise activity), knowing about banking procedures in terms of availing credit for micro enterprises, etc.

Development of Renewable Energy Technologies for Rural Areas/Villages

More recently, concern over global warming and acid rain have resulted in call for increased reliance on renewable sources of energy such as solar energy, biomass, wind energy, hydropower, etc. The abundance of sunshine makes it an obvious source of energy. To be practical source, however, the energy derive from sun light must be converted into a form that can be transported and stored.

Many different techniques can be used to convert sunlight into useful forms of energy. Active and passive solar energy technologies are generally used for space conditioning (heating and cooling) while solar electric technologies such as photovoltaic cells convert sunlight into electricity. Although the distinction between active and passive solar is blurry, the use of integral building components to capture the sun's energy is considered passive solar. Active solar technologies are generally add-on features, which utilize mechanical means to distribute captured solar energy. An example of active solar energy is a solar hot water heater, while passive solar features may be as simple as south facing windows.

Conventional energy usage has some negative impact on the environment. Burning fossil fuels such as coal and oil produces emissions of greenhouse and acid gases, which results in global warming and acid rain respectively. Fossil fuels are also responsible for air pollution and its associated health hazards. Nuclear power plants expose the environment to low levels of radiation during many stages of the nuclear fuel cycle, and also impose the risk of a major accident such as Chernobyl. Even renewable technologies using energy from the sun have some negative impacts on the environment. Hydro-electric dams, for example, can flood vast areas of land and damage aquatic eco-systems. In general, however, renewable such as solar and wind energy, small hydro-powers, bio-fuels have a smaller impact on the environment than fossil fuels and nuclear power.

Solar Energy

Pakistan lies in favorable latitudes as far as solar radiation is concerned and has long sunshine hours and high insulation levels. In the recent years, significant work has been done to map the solar radiations in Pakistan and a set of solid data on the solar intensity, the hours and the angles of the sun at various times of the day through out the year is available now. The solar intensity increases from north to south, and can be utilized for power generation through photovoltaic systems.

Photovoltaic (PV)

Photovoltaic devices convert the energy contained in sunlight directly into electricity. Photovoltaic power generation is a mature technology, for small power applications. The world's PV market is still very small but market growth has been impressive internationally. It however has not been able to meet its projection made in early 90s. PV systems are currently cost-effective in some consumer products. The costs are expected to continue to decline in the future allowing PV devices to compete with large-scale conventional power generation.

Achievements

Pakistan has developed the know-how and processing technologies in the field of solar cells, modules and systems. Its research laboratories are equipped with facilities of growing silicon mono-crystalline ingots, slicing these ingots into wafers, fabrication of solar cells and devices, and lamination of cells into PV modules. As a result, a number of products are being fabricated, albeit on a limited scale. These include silicon wafers, solar cells, PV modules, PV systems such as solar lantern, home light systems, and garden and streetlights etc. A number of such systems have been designed and installed for applications of lighting, fencing, water pumping, and telecommunications.

Number of houses, mosques, community center, schools, dispensaries, and parks electrified through PV systems are given below:

Number of houses electrified	=	132
Number of mosques electrified	=	02
Dispensaries	=	02
Schools	=	02
Parks electrified	=	02 (170 lights)

(For detail see Annex-I)

Solar Thermal

The activities here include modeling, designing and fabrication of low cost efficient solar thermal appliances such as Solar Cookers, Solar Water Heaters, Solar Room Heaters, Solar Stills, and Solar Dryers etc.

Solar Water Heaters

According to the HESS survey (UNDP/EW/WB, 1993a) conducted in 1991, water heating in the domestic sector in Pakistan, consumes about 10 per cent of the total residential supply of natural gas. HESS estimated that in 1991, water heating in the domestic sector consumed 145,600 TOE of natural gas. This figure is expected to rise to 1,000,000 TOE by the year 2020, owing to urbanization and expansion of the gas network into rural areas. (At present, natural gas supply and distribution is mainly in the urban areas). According to the HESS demand survey report, the number of gas water heaters in the domestic sector was 174,000 in 1991. The annual demand is estimated to be 80,000 units of water heaters and is growing at a rate of about 7 per cent annually. This need can at least partially,

be met by the use of solar water heaters. There is no technological barrier in using this technology in the country, especially since it has already been introduced, albeit on a limited scale.

Several models of solar thermal appliances like solar cookers, solar ovens, solar dryers, solar water heaters, and solar stills have been developed and are being disseminated. Recently solar hybrid dryer for de-hydration of apricots on commercial basis was designed and developed. Solar thermal devices developed and disseminated so far are given below:

Solar cookers (Box type)	=	585
Solar cooker (Concentrator)	=	103
Solar dryers (Ping pong, slanting, chamber type)	=	150
Solar stills	=	15
Solar water heater	=	10
Solar de-hydrators with electric backup system	=	02

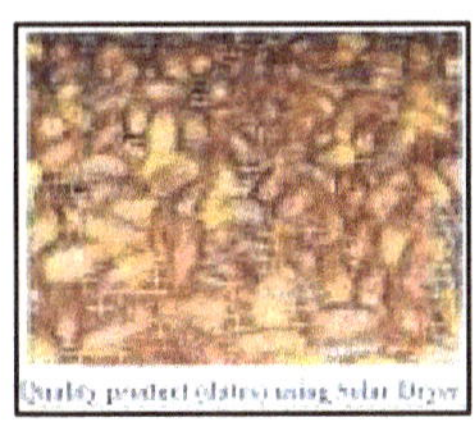

Other Renewable Sources of Energy

Cook Stoves

The improved cook stoves for village use have received much attention; both research into the technology and dissemination efforts and they offer many advantages such as decrease fuel consumption, reducing deforestation and time spent looking for fire wood, reduction in smoke that leads to better health, faster cooking that saves more time of women for other activities such as child care.

An appreciable amount of energy consumption goes into firewood, and hence, an increase in cook-stove efficiency is worthwhile. The improved cook stoves however, have not spread as quickly as one might have expected. The Government of Pakistan has been engaging in the development and dissemination of efficient cook stoves for the last so many years.

Salient features of different efficient cook stoves developed are given below:

Cook Stove (MA-1)

It is a round shaped smoke free metallic stove with a chimney and a baffle to control the fuel-burning rate. It is suitable for room heating as well as cooking/baking simultaneously. A metal baffle, which slows down the hot gases, surrounds the firebox of the stove and thus the hot flue gases transfer maximum their heat to the stove body before leaving through chimney. The improved design consumes less fuels, yet radiates as much heat as any traditional stove of same volume in a given time. It is operated at high power though operation of traditional stove at low power is very difficult. The improved stove however, performs well at low as well as high power. However, it needs care and regular cleaning than the traditional stove because due to reduce speed of combustion gases, more suite particles settle around the baffle, in elbow and pipes. It has primarily been developed for the cold hilly areas of the country. A baffle has been provided in the chimney to control air draft; besides the sliding door at the combustion chamber automatically adjusts the volume of air to the volume of fuel load, which contributes to a well regulated air circulation. As for field performance, on the average, 25 per cent fuel saving was possible as against the traditional heating cum-cooking stoves and material such as wood, dung, bushes, coal cakes etc., can be used as fuel.

Cook Stove (MA-II)

It is an improved version of MA-I. The principles of design parameters are the same, but a water container of 12L capacity has been added. The chimney passes through the water container so that the heat of exhaust gases is utilized for heating water. It thus performs three functions *i.e.* cooking/baking, water heating and room heating simultaneously. In the water container the temperature rises from 10°C to 50°C during a single operation while cooking meals. On the average, 35 per cent fuel saving can be done compared to the MA-I and wood, dung, bushes, coal cakes etc., can be used as fuel.

Cook Stove (MA-III)

It is also metallic and smoke free stove. It contains two potholes to facilitate baking and cooking simultaneously. Besides, it has been provided with an oven to keep the meal warm. The flue gases are used to operate this oven. As such this unique cook stove performs four functions at a time *i.e.*, cooking, baking, meal/bread warming and room heating. This model has, on the average, 40 per cent fuel saving as against traditional heating cum-cooking stoves and it works better on fuel wood rather than on other fuels like dung, bushes, and coal cakes.

Cook Stove (MA-IV)

It is an improved version of mud stove already being disseminated, which require 2 to 3 days for construction. It was noted that the critical parameters of the mud stove contributing to its efficiency were not being strictly observed by the villagers and as a result, the required efficiency was not achieved, which affected adversely its acceptance/ propagation.

In order to observing the critical design parameters and also to reduce construction time, the prefab baked stove has been designed. It comprises of four prefab components, three of clay and one of metal (top). Chimney has also been provided with baffle to control the burning rate. It is easy to assemble in the field. One person can assemble about 50 stoves in a day. This improved stove has been designed for plane areas of the country where metallic stove are not popularized. As compared to the traditional stoves, it saves 30 - 40 per cent fuel. Wood, dung, and bushes can be used as fuel.

Cook Stove (MA-V)

It is prefab baked clay stove with two potholes. The 1st pothole works directly on heat of combustion, the heat of flue gases are utilized in the 2nd pothole. Like MA-IV, it is easy to assemble. It comprises of three components; two of baked clay and one metallic (top). Chimney has also been provided with baffle to control the burning rate. It has also been designed for the plain area of the country. As compared to the traditional stoves, it saves 30 - 45 per cent fuel wood. Other fuel sources can be dung and bushes.

Achievements

1. Six models of efficient cook stoves have been developed (mud stoves, backed clay stoves and metallic stoves) for application in plain areas as well as cold hilly areas of the country.
2. 70,000 efficient cook stoves have been disseminated.
3. 20 cook stoves manufacturing shops established
4. 200 sales centers opened

Benefits

Financial Benefits

1. Number of households provided with improved cook stoves	=	70,000
2. Saving of fuel wood per household	=	30 kg per month
3. Saving of Fuel wood for 70,000 household	=	2.1 million kg/ month or 25.2 million kg/yr (18.4 million M^3/yr)
4. Financial saving @Rs.3/- per kg fuel wood	=	Rs.75.6 million/ year

Environmental and other Benefits

1. No emission of additional CO_2 in the atmosphere
2. Safe disposal of smoke through chimney
3. Smoke free kitchen, clean utensils
4. Healthy environment
5. Protection from eye cataracts and respiratory diseases of house wives and children
6. Reduction in drudgery of repeated blowing to keep the fire burning
7. Protection of forests and ecosystem
8. Saving on FEC which otherwise is spent on import of Kerosene oil.

Potential of Cook Stoves in the Country

✰ Total number of households in the country 18.0M households (about 80 per cent) use biomass (fuel wood etc.) to meet domestic fuel needs.	=	23.41 Million

- ☆ Annual fuel wood consumption in the country (Source: HESS 1991 estimate for household sector) = 17882000 TOE
- ☆ Expected saving of fuel wood (25 per cent) through efficient stoves. = 4470500 TOE per year

Biogas Technology

The biogas technology makes optimal utilization of the valuable natural resource of dung. It provides (soot-free) clean gas for meeting the cooking and energy requirements as well as enriched bio-fertilizer for improvement of fertility/

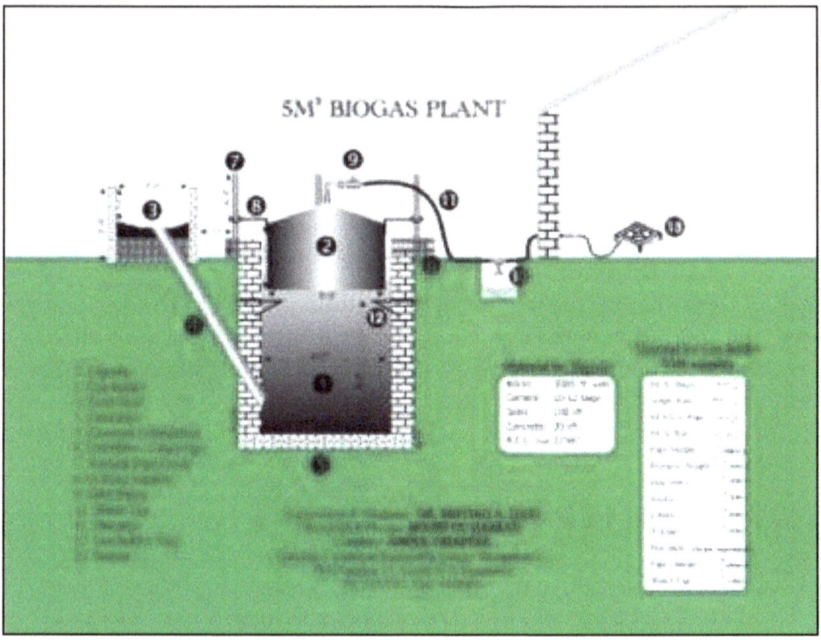

productivity of agricultural lands. Promotion of the biogas technology, seems to be one of the best options, which cannot only partially offset the fossil fuel from wood consumption but also facilitates recycling of agro-animal residues as a bio-fertilizer. Moreover, being clean and renewable, it will also contribute towards environment protection, sustenance of ecosystem and conservation of biodiversity.

Benefits of Biogas

The biogas technology provides nearly three times more energy than useful when dung is directly burnt, and also produces nutrient-rich manure, as depicted below:

As a cooking fuel, biogas is cheap and extremely convenient. Based on the effective heat produced, a 2M^3 biogas plant could replace, in a month, fuel equivalent of 26 kg LPG or 37 litres of kerosene, or 88kg of charcoal, or 210 kg of fuel wood, or 740 kg of animal dung.

To the housewife, a biogas is easy to use and saves time in the kitchen. Biogas stove has an efficiency of about 55 per cent. Cooking on biogas is free from smoke and soot, and can substantially reduce the health problems, which are otherwise quite common in most rural areas where biomass is the chief source of fuel. The use of biogas is helpful to improve the quality of life in household.It can also be used, to run IC (internal combustion) engines as well as dual fuel engine (80 per cent biogas and 20 per cent diesel) for electricity generation, water pumping; small industries like flour mill, saw mill, oil mill etc. This would not only reduce dependence on diesel, but also help in reducing carbon pollutants, which adversely affect the atmosphere.

While biogas has multiple benefits at the individual family level, it also has several qualitative and quantitative benefits at the societal level. Firstly, a shift to biogas from traditional biomass fuels results in less dependence on natural resources such as forests, checking their indiscriminate and unsustainable exploitation. Since dung is collected systematically, when used in biogas, environment can be kept clean and hygienic. The other advantage is that, unlike centralized systems such as thermal power plants and fertilizer factories, which entail huge capital investments and need to elaborate distribution networks, biogas plants are decentralized systems which can be installed even in remote areas with very low investments.

Biogas Potential in Pakistan

As per livestock census 2002-03, there are 48 million of animals (Buffaloes, Cows, Bullocks) in Pakistan. The average (daily dung dropping) of these animals @ 15Kg comes to be 690 million kg. Assuming 50 per cent collect-ability; the availability of fresh dung comes to be 345 million kg/per day. Thus, 17.25 Million M^3biogas per day can be produced by anaerobic fermentation of dung through installation of about 5.0 million family size biogas plants, which could meet the cooking need of 50 million people.

The total population of Pakistan is about 140 million, out of which 70 per cent reside in the rural areas, which comes to be 98 million. Therefore, we can meet about 50 per cent cooking requirements of the rural masses from this source of energy (biogas) alone. Besides, producing 96.6 million Kg of bio-fertilizer per day or 35.04 million tons of bio-fertilizer per year is an essential requirement for sustaining the fertility of agricultural lands.

Achievements

1500 biogas plants have been installed during the last three years, which meet the domestic needs of 1500 households (6-15 each family members).

Financial Benefits

Annual biogas production from 1500 biogas plants and the financial benefits accrued from it from are summarized below:

1. Production of biogas	=	2.5 million M^3/year
Kerosene oil substitute	=	1.5 million litres
Worth (@Rs.27/-litre	=	Rs.41.0 million/year
2. Production of bio-fertilizer	=	8.75 million Kg/Year
Worth	=	Rs.3.0 million
(based on 2% nitrogen contents only)		
3. Total Gain (41.0 + 3.0)	=	Rs. 44.0 million/year

Allied Benefits

1. Replacement/substituting 1.5 million litre Kerosene oil per year (worth Rs. 41 million FEC) Or saving of 8.0 million kg wood per year
2. Provision of easy, clean and soot free fuel
3. Reduction in eye cataract and respiratory diseases
4. Healthier and cleaner environment
5. No additional CO_2 emission in the atmosphere
6. Checking of forests
7. Conversion of biodiversity and eco-system
8. Substituting hazardous chemical fertilizer
9. Creation of employment opportunities.

Micro Hydropower Plants

The northern rural areas have not been electrified because the local population is sparse and scattered across mountainous terrain. The national grid extension is an economically non-viable option in these areas and is likely to remain so during the next ten to twenty years. Small streams and tributaries running through the area can provide sufficient water head and flow to support micro, mini and small-hydropower plants. These plants could generate 5 to 2,000 kW of power, thus providing electricity for lighting, other household uses as well as for running small industries.

The concept of micro hydropower is adhered to non-grid strategy in implementing rural micro hydraulic schemes and adopted a flexible approach that caters for socio-economic conditions and skill level of the target areas. A cross flow turbine was adopted after examining different designs, as it is easy to fabricate locally and suites the head and flow conditions most frequently available in the hilly areas.

A new approach has also been adopted for promotion of MHP technology which envisages the under mentioned parameters.

- ☆ Non-conventional use of readily available materials.
- ☆ Design suited to the ground realities and communities
- ☆ Community involvement in the initiation, implementation, management, operation and maintenance of hydropower scheme.

Wind Energy Technology

The wind energy is the most dynamic source of renewable power for many years. Being clean and green electricity, this relatively young industry continues to expand rapidly, generating a new way of meeting the world's generating energy needs. As one of the most advanced and commercially available renewable energy technologies, the wind is now becoming more price – competitive. Given an increasing awareness of the need to reduce green-house-gas emission and the equally important need for new sources of electricity, wind has a significant role to play in the future.

Pakistan has a considerable potential of wind energy in the coastal belt of Balochistan, Sindh and as well as in the desert areas of Punjab and Sindh, which if utilized effectively, the electricity so produced can not only electrify all remote coastal villages but also can be fed to the national grid. This renewable source of energy has however, not so far been utilized significantly.

Achievements

Two hundred (200) houses in the remote coastal villages have been electrified through installation of 50 small wind turbines (stand alone) and a survey of other 400 houses has been completed. These will be electrified by June 2005? The local beneficiaries provide land for installation of wind turbines, room for batteries and invertors and also bear the installation cost of wind turbine. The beneficiaries are also responsible for post installation operation and maintenance of wind turbine. A proper training in operation and maintenance of complete system of wind turbine is imparted to the beneficiaries during installation and before handing over the wind turbines to them.

Outcome of the Project (on completion)

Electrification of 600 houses (in remote coastal villages), which are not feasible to electrifying through national grid.

Allied Benefits

1. Provision of clean and green (emission free) electricity
2. Replacement of kerosene oil has resulted in saving of foreign exchange
3. Protection of health
4. Improvement in the quality of life
5. Increase in literacy rate and capacity building
6. Promotion of small scale agricultural activities
7. Creation of job opportunities.

Success Story

Pakistan started using renewable energy in 1960s, when 18 villages were electrified using photovoltaic, and solar stills were installed to desalinate the seawater at Gawadar, however the progress remained slow. The net installed capacity of different renewable energy systems are as under:

1. Pakistan Installed 538 micro hydraulic power plants (5-50 KW capacity) with total capacity of 7.8 MW, for electrifying 70,000 houses in the hilly terrain of Phkhtoon Kwa province and northern areas on cost sharing basis. These plants meet energy demands of the local people, which otherwise cannot be met from the national grid for long time. The beneficiaries (Cottage Industries) have established by themselves small industrial units like flourmills, rice sellers, saw machine, lathe machine etc. at 100 sites. It is estimated that 20MW electricity will be generated through installation of MHP plants by the year 2015.
2. Installed 155 small wind turbines (0.5 KW to 10 KW) with total capacity of 61KW in Sindh and Balochistan, electrifying 1560 houses and 9-coast guard check posts,
3. Installed 300 Solar PV systems with total capacity of 100 KW electrifying 500 houses, mosques, schools, and 265-street/garden lights,
4. Installed 4000 biogas plants (size 3&5M3/day: producing 18000 M3/day), 5. Developed 6-models of efficient smokeless cook stoves for cooking and space heating,
6. 100,000 mud stoves have been built in rural houses; saving 36500 tons of fuel wood per year,
7. Installed 21 solar dryers with total capacity of 5230 kg/day fruit.

Photovoltaic (PV) is economical source of electric power for remote areas. The cost comparison shows that the initial investment in the case of PV is higher. But, the integrated cost due to extremely low maintenance of PV becomes lower after five years, as compared to wind and ten years as compared to diesel. The extension of grid for small load in rural areas is also expensive as compared to PV. The extension of the gridline is not economical in the target areas that are more than 10 km away from the gridline. A total of about 650 KW of PV has been installed in the country. More recently, three villages of Balochistan have been electrified using wind/PV/ diesel hybrid and solar home light systems. Another four villages are in the pipeline for electrification through PV. A number of parks have been illuminated through solar garden and streetlights. Photovoltaic are also being used to power the telecommunication systems.

In the early 1980s, eighteen PV stations were set up by the government in different parts of the country for village electrification, with an installed capacity of nearly 440 kW. However, because of lack of technical know-how and follow up, these systems have not performed as required. Currently solar technology is being used in Pakistan for stand-alone rural telephone exchanges, highway emergency telephones, cathodes protection, refrigeration for vaccine and medicines in hospitals etc. The Public Health Department has installed about 20 solar water pumps for drinking purposes in different parts of Balochistan.

Empower consultants of New Zealand, a Wellington-based renewable energy NGO, is working to bring affordable electricity to poor villagers in the harsh deserts of Gawadar area in Balochistan province. The project commenced in February 2001 and by June 2002 the team had completed a feasibility study and detailed design and equipment purchasing is now underway. The 'project has a 6 kW solar PV array, with a 10 kW inverter, large capacity (90 kWh) battery bank, 100 Amp 120 VDC solar charge controller and 7 kW backup diesel generator. The system is expected to provide approximately 30 kWh per day more depending on diesel backup system usage.

Two plants each consisting of 250 stills with capacity to clean water up to 6000 gallons per day, were installed in Gawadar in sixties. This was followed by other small plants of capacity 250 gallons per day each in rural areas to convert brackish water into potable water. A number of other solar thermal appliances such as solar cooker, solar water heater, and solar stills have been designed and introduced in the country. Solar drying is another application that is effectively being employed for drying fruits in the Northern mountain region.

The biogas activity was initiated in the country in the early seventies when more than 4000 plants of capacity varying from 5-15 cubic meters per day were installed in the country. Recently 1200 biogas plants have been installed throughout Pakistan on cost sharing basis, where 50 per cent cost is to be borne by the beneficiary.

Fuel wood is still the cheapest of all energy sources in the country and is the main source of energy in rural areas. In 1995, Fuel Saving Technology (FST) project was launched to conserve energy at a domestic level. It reduces the family's consumption by 30-40 per cent and eliminates smokes. So far more than 60,000 energy conserving, improved stoves devices have been installed/disseminated all over the country and training is provided in construction and use of such stoves to the workers of various NGOs and local population.

A project was launched in 1976-77 for design, development and the installation of micro hydro power plants. One of the features of MHP programme is the participation of local communities that are involved in every phase of the project *i.e.* from initiation to implementation and subsequent operation and maintenance. Up till now over 300 micro hydro-power plants with generation capacity exceeding 3 MW has been installed. Another project that will utilize the hydropower of the canals to produce the electricity on micro scale is at its development stage.

For the purpose of harnessing wind energy, a project, now in its final stage, was initiated to produce wind maps for the coastal areas of Balochistan and Sindh. About 30 wind water-pumping machines have been installed during 1976-88. Manufacturing capability of such windmills also exists in the country. At the same time, another project 134 small wind turbines for power generation is being initiated.

Solar drying is largely and effectively being used in the Northern Mountains of Pakistan. The technology is used to dry apricots and other fruits that are grown in the area. The dry product is packed and sold in the market at higher prices. Four community size solar dryers have been planned to be installed in date growing areas of Punjab, Sindh and Balochistan

Outcome

1. Villages electrified (through installation of 300 MHP plants) = 300 (25000 houses)

Financial Benefits

1. Installed capacity = 4.0MW
2. Electricity generation/year (300 days) = 28800MWh
3. Financial gain @ Rs.3/- per kWh = Rs.86.4 million/year

Allied Benefits

1. Provision of cheap and green electricity
2. Promotion of small scale industrial activities
3. Increase in agricultural produce
4. Improvement in quality of life
5. Enhancement in productivity/per capita income
6. Socio-economic development
7. Education and awareness.

Future Plan – Upto 2020

Development and Dissemination of Renewable Energy Technologies

1. Installation of 100,000 biogas plants for meeting the domestic cooking and lighting needs of 1.5 million inhabitant of rural/semi urban areas, as well as producing 1000 million kg bio-fertilizer.

2. Design, development and installation of community size biogas plants based on agro industrial and municipal organic wastes for industrial applications.
3. Gasification of biomass (cotton stalk, rice husk, corn cobs, and organic waste) for meeting domestic fuel needs as well as industrial energy needs particularly for agro processing industries.
4. Electrification of 50,000 houses through solar system.
5. Development and local manufacture of Kaplan turbine for harnessing the power of small canal falls.
6. Generation of 250MW electricity through installation of MHP plants (50-500kW capacity each) at canal falls.
7. Generation of 20MW electricity through installation of MHP Plants in remote hilly areas.
8. Generation of 20MW electricity using wind energy through installation of small wind turbines (500Watts – 10,000Watts capacity each) stand-alone system.
9. Development and mass dissemination of solar thermal devices like solar water heater, solar cookers, solar dehydrators, solar desalination plants etc.

Capacity Building for Development, Promotion and Dissemination of Rets?

1. Establishing the facilities for pilot production of solar cells and module to prove the commercial use of the developed industrial processes and generate industrial and commercial activities via indigenous self-reliance in the area.
2. Establishing the "Solar System and Component Testing" facility to test the solar systems, their components to meet the demand of local market and industry and to standardize and improve the products according to international standards
3. Establishing technology promotion and marketing facility for fast technology dissemination of the know-how and transfer of the technology from laboratory to industry; to get necessary feed back on products for their improvements and to collect data regarding the market requirements and development.
4. Establishing of Training Wings to meet the need of human resource development in the sector at all different levels.

References

The paper is based on the personal knowledge experience and information available under different Government Departments and no specific journal/book is consulted in writing the paper.

Annexure-I
The Sites/Villages Electrified through PV System are Given Below

Name of Site	*System Installed*	*Capacity*	*Number of Houses Electrified*
Zardalo, Sibi, Balochistan	Wind-Solar	Wind-7kW Solar-500Watts	40-houses
Gawadar (Chibkalmatay) Ballochistan	Diesel-Solar	Diesel-10kW Solar-4kW	35-houses
Loralai (Bughat), Balochistan	Solar, Stand Alone	40kW	55-houses
Dera Bugti (Bekkar) Balochistan	Solar-	5-kW Stand Alone	01-School 01-Mosque 01-Dispensary 02-Houses
Jalalabad Park, Muzafarabad, Azad Kashmir	Solar	2-kW	165-lights
Community Center, Thana Bola Khan, Sindh	Solar	300-Watts	For Community Center and Street Light
Sukkher, Sindh	Solar	300-Watts	For Water pumping
Haroonabad, Bhawalnagar	Solar	250-Watts	For Lighting and Sound system for Mosque
Edhi Emergency Center, PIMS, Islamabad	Solar	150-Watts	For Lighting and Communication system
PSF, Building, Islamabad	Solar	150-Watts	For Lighting
Natural History Museum, PSF, Faislalabad	Solar	150-Watts	For Lighting and Fountain
NESCOM, Texila	Solar	150-Watts	For Lighting
Zaida Village, Sawabi, Khyber Pukhtoon Khwa	Solar	-	Solar Fence
PAEC, Islamabad	Solar Cells		3000-Solar Cells supplied for Laser Land Leveling System

Chapter 5

Strategies of Nepal in Making Use of Science and Technology to Develop Village Occupation

Dilip Subba

Secretary, Nepal Academy of Science and Technology,
Khumaltar, G.P.O. Box 3323 Kathmandu, Nepal
E-mail: dilipsubba2009@yahoo.com

ABSTRACT

Nepal is an agricultural country, with over 65 per cent of the population living in villages. Agriculture and agri-business are the two main village occupations. These occupations constitute the cultivation and production of various agriculture products such as cereals and grains, legumes and oilseeds, fruits and vegetables, spices and herbs, tea, coffee, sugarcane, jute, milk, meat, poultry, egg etc. The farming of sheep, goat, pig, poultry and fish and the production of meat, poultry and egg, honey, silk and wool are other important village occupations. A large number of cottage, small and medium scale industries for the processing of various agricultural products have been established. Besides, collection and drying of medicinal and aromatic plants, saw mill, plantation of oil bearing plants and production of bio-diesel, biogas production, biomass briquetting, brick making, hand made paper making, handicraft production from local materials such as wood, bamboo, paper, textile and rice straw and weaving are other village based occupations. The micro-hydro and pico-hydro power generation are also village based operations. The present situation of these occupations is not very satisfactory. The development of village based occupations can be achieved by commercialization of agriculture with application of new and modern agriculture knowledge and practices, setting up of industries based on agriculture, farm, forestry products and other natural resources and the domestic and the export market. Needless to mention that technology has strength to significantly prosper village occupations. The areas where technology can play important roles are the introduction of new technology; modernization of equipment and process mechanization; improvement of product and service quality; new product development and value addition; power generation etc. Even though some positive cases have been witnessed in this direction; a lot more should to be done. The conclusion is that the full use of science and technology can develop the village occupations and the rural economy.

Keywords: *village occupational development, Village occupations, Agriculture and agro-business, Forestry products, Miscellaneous but important occupations, Science, Technology and Innovation, Nepal.*

Introduction

Nepal is a small mountain country; situated in South Asia, having a total area of 147,181 sq. km, of which mountains, hills and plains occupy 35 per cent, 42 per cent and 23 per cent area respectively. It has fragile and steep topography (about 200 m in the south to 8,848 m in the north). Seventy one per cent of the Nepalese land has green coverage (29 per cent forest land, 10.6 per cent shrub land, 20 per cent agricultural land, 12 per cent rangeland). The Plains has the major agricultural land and the hill belt has the potential for horticultural development while the mountain zone is considered suitable for livestock farming. Nepal is a country of villages where over 65 per cent of 28.5 million people are living. There are 4000 villages in Nepal.

Village Occupations

The main occupations of Nepalese people are agriculture and agro-business; about 71 per cent of the population earns their living directly from agricultural occupation. Agriculture contributes to 33 per cent of the gross GDP. There is quite a great range of agriculture related occupations and include cultivation and production of cereals and grains such as rice, maize, wheat, millet, buckwheat, production of varieties of legumes and oilseeds, production of varieties of fruits and vegetables, spices and herbs, production of plantation crops *viz.*, tea, coffee, coconut, beetle nuts and sugarcane, production of milk, meat, poultry and egg, bee keeping, sericulture, production of jute, wool and so on. Farming integrated with agriculture is part of rural life. Single family farming and group farming of sheep, goat, pig and poultry and the production of meat, poultry and egg are the important village occupations. The occupation of fish farming is growing popular rapidly these days. People living in the vicinity of rivers also have a profession of catching and selling of fishes in the nearby market. A large number of cottage, small and medium scale industries for the processing of these agricultural products have been established that include rice mills, flour mills, oil pressing mills, vegetable processing industries producing pickles, dried vegetables etc, fruit processing industries for the preparation of juice, squash, jam, candied, crystallized and dried fruits, tea and coffee industries, milk processing industries to prepare curd, *ghee* etc, sugar industries, wine industries, industries for production and packaging of honey and slaughterhouses for poultry birds, buffalo and pig. Fish drying is a traditional practice.

A large number of villages, especially those in the hills and mountains find themselves situated close to the forest. The village people depend on forest products for their livelihood. They cut wood from forest and make wooden products such as leg-pound, mortar and pestle for pounding and grinding of cereals and grains. Wood is also used for making bullock-carts, arm for axe etc. Leaves are used for making plates and containers of various shape and size. Forests are natural resources for varieties of herbs, which are collected and dried. Bamboo has multiple uses, from fabricating handicraft items to *Doko, Dalo, Dhakar* (bamboo woven sacks to carry at back), sieve and stool. Woolen carpets are made manually in villages for which state of the art wooden weaving machines are used. Many people know tailoring and they make textile and readymade garments. Handicraft making is a very popular occupation of village people. Varieties of handicraft products made from wood, bamboo and handmade paper are sold in the local market and also exported. Among them *Thanka* is a special one. There is good business of manufacturing handmade paper (so called Nepali paper). Clay work and iron work are old traditional occupations. The pottery products are less expensive and they are environment friendly, unlike the plastic. The iron work products are mainly knife, axe and many agricultural tools. Floriculture and bio-briquette making have emerged as rural business in the recent past. Collection of precious stones is another business. Small to medium size brick kilns are operating in the plains. Hotel and restaurant business and tourism are also flourishing. Jungle safari, trekking, eco-tourism, recreational tourism and cultural tourism are some of the brilliant concepts in the rural tourism. There are hundreds of small high current rivers and rivulets in Nepal flowing across mountainous villages and a good number of micro-hydro power plants have been erected and operated by the local communities with the support of government.

Present Situation of Village Occupations

Being an agricultural country, the status of agriculture need to be discussed first. The present agriculture is at large, subsistence based. Commercial agriculture is at introducing stage. The new technologies in agriculture are being introduced in the recent past. Although about 71 per cent of population is directly engaged in agriculture, the economic output is low. The contribution from agriculture sector is merely 20 per cent to the total manufacturing output. The agro-business sector is facing many problems. The first and foremost problem is that the availability of sufficient funds to set up and run the enterprise. The inadequate roads and transport facility, large load shedding schedules and shortage of electric power are other problems. Also, there is a need of an adequate number of qualified and skilled manpower for running the enterprises. The industries can be classified into cottage, small and medium enterprises. Except rice mill, sugar mill, jute mill, cement and brick factories, the other enterprises are either cottage or small enterprises. The cottage and small enterprises are not modern and the operations are labour intensive. Knowledge of business is important for these enterprises, which is also lacking.

There is lack of knowledge about modern management systems such as Good Agriculture Practices (GAP), Good Veterinary Practices (GVP), Good Manufacturing Practices (GMP), Good Hygienic Practices (GHP) and Hazard Analysis and Critical Control Point (HACCP). The entrepreneurs are unaware of product standards, standards certificates, Intellectual Property Right (IPR) and relevant World Trade Organization (WTO) provisions which are important for the promotion of business.

The business support and facilitating services by the government and private companies can be very helpful in the establishment and growth of enterprises. There is no research culture in industries and R and D activities are completely lacking. As a result, product and service quality is not up to the required standard.

Key to Village Occupational Development

The keys to village occupational development can be categorized as follows.

- ☆ Commercialization of agriculture with application of new and modern agriculture knowledge and practices
- ☆ Set up of industries based on agriculture, farm, forestry products and other natural resources
- ☆ Domestic and export market.

Support Necessary for Village Occupational Development

The solution is easy, if one analyses the situation of village enterprises. First of all, there should be a flow of funds. The answer to question, where the funds come in is seemingly not easy, but it is not difficult either. The government support is unquestionable and it should be there from the registration of enterprises by making the right policies and legislative support. The entrepreneurial models of Public Private Partnership (PPP) and Rural Urban Partnership (RUP) could be very useful. There should be effective information system for market and services facilities. It is observed that without civil facilities in a village such as school, college, road, electricity, comfortable houses, market, hospital, bus-park, cinema and other means of entertainment it will be difficult to have high expectations as these facilities are required to retain the employees in the village. It is strongly advised to provide good knowledge of GAP, GVP, GMP, food safety, standards etc., to the owners and employees of the industries.

How Science and Technology can Intervene?

In above context, let us examine the intervening role of science and technology. As is accepted by everybody, science and technology are powerful tools applicable in almost all spheres of life and deed while having boundless scope in developing all kinds of occupations. In the present context, its main thrust should be: a) improving the existing technologies, b) innovation and c) building and enhancing the capacity of the required for starting and growing the enterprises.

The intervention area can be following and many more:

- ☆ Introduction of new technologies and mechanization in agriculture to increase production (yield) of crops by developing high yielding varieties, development of drought, disease resistant seeds and crop varieties suitable for different soil and climate conditions, development of non seasonal vegetable varieties and so on, by the application of genetic engineering, use of bio-pesticides and bio-fertilizers, water management, IPM etc.
- ☆ Pesticide residue monitoring to produce safe food

- ☆ Horticultural orchid management to get best benefit
- ☆ Development and modification of agricultural machinery
- ☆ Increasing the production of livestock (meat, milk), fish and poultry products through breeding techniques, artificial insemination, management of pasture and feed the livestock
- ☆ Reduction of the post-harvest losses through storage, post-harvest handling management, and development of appropriate food preservation technologies
- ☆ Grading and packing of fruits, value addition to horticultural products
- ☆ Modernization of slaughtering and meat production operations
- ☆ Modernization of traditional food processing equipment
- ☆ Improvement of cooking efficiency of briquette and cooking stoves, designing and fabrication of briquetting machinery
- ☆ Design improvement of water-mills
- ☆ Modification of brick kiln for reduced smoke emission
- ☆ Development of solar and hybrid solar dryer for drying cereals, vegetables, high value cash crops, spices, medicinal plants, meat, fish etc.
- ☆ Design improvement of solar cooker
- ☆ Scientific harvesting of medicinal plants
- ☆ Biogas production at cold climatic high altitude regions
- ☆ Design improvement and mechanization of knitting and weaving machines
- ☆ Improvement of product and service quality
- ☆ Product and process standardization
- ☆ Making the production process economic
- ☆ Carrying out R and D in natural dyes used for dyeing textiles and metal work
- ☆ Electricity generation from solar and wind energy
- ☆ Establishment and operation of wind mills at appropriate locations
- ☆ Grid connection of electricity generated by micro hydro-power generation system to highway grid
- ☆ Improvement of solar home system
- ☆ Technology of solar PV water pumping
- ☆ Pumping of river water in the crop field at river bank
- ☆ Scientific harnessing of different local resources
- ☆ Application of modern techniques of IT for enhancing information facilities required for market and service
- ☆ Establishment and operation of village technology business incubation centers
- ☆ Establishment and service of technology transfer centers

Cases of Positive Development

It is not entirely correct to say; that Nepal has done nothing in using science and technology to improve village technology and develop the village occupations. Following are some of the instances for the case.

- ☆ Myagasesse Prize winner Mahabir Pun, built an IT network in the Mygdi district for the sake of farmers for easy access to information on market price of food commodities, livestock and other information useful for agricultural occupation.
- ☆ Centre for Energy Studies, Tribhuvan University modified the design of traditional water mills and improved its grain milling performance.
- ☆ Government has launched an Improved Water Mills (IWM) Support Programme.
- ☆ Nepal Academy of Science and Technology with cooperation of a Japanese agency designed a natural force boat for water transport of goods in high current rivers, from one bank to the other.
- ☆ Nepal Academy of Science and Technology is doing research on new briquetting materials, briquette binders, briquette extruder and briquette stoves.
- ☆ Over 200,000 biogas plants have been manufactured and are being used in villages.
- ☆ Department of Cottage Industries of the Ministry of Industry has established the business incubation centers.
- ☆ Department of Food Technology and Quality Control of the Ministry of Agriculture and Cooperatives is providing technical knowhow to farmers and women entrepreneurs on processing and preservation of fruits and vegetables by means of training.
- ☆ Research institutes and research stations of Nepal Agriculture Research Council are involved in research for developing new rice, maize and wheat varieties and carrying out extension work and outreach activities for technology diffusion. Likewise, they have developed high meat yielding fowl cross-breed.
- ☆ Some private research organizations are engaged in research, in animal biotechnology with the aim of developing an improved goat breed.
- ☆ Government is in the process of establishing village information centers.
- ☆ Few banks have come up with investment schemes in energy sector.
- ☆ Communities of local people are being formed for conservation and sustainable management of forests.

Conclusion

In the case of Nepal, development of villages is the key to development of the country. For this, village occupations must be developed. There is a great opportunity to develop the village occupations, since there are many resources, namely agriculture, forests, natural beauty, rivers and rivulettes. Thus, the country has enormous potential of development through optimum mobilization of these natural resources. The maximum benefit can be obtained by the input of science and technology in the existing village occupations. Innovation in use of science and technology should be the motto. Less fortunately, Nepal belongs to the group of countries that are scientifically lagging behind. Inferentially, it can be said that the development of science and technology is prerequisite to the village occupational development.

References

Ale, B., Kaini, B., Dahal, J., Shrestha, J.N. and Pradhan, S. (2008). Performance analysis of natural force ferry boat at different river conditions. Abstracts of Fifth National Conference on Science and Technology, November 10-12, Kathmandu, pp 239.

Bajracharya, D. (2001). Science and Technology in Nepal. Royal Nepal Academy of Science and Technology, Kathamandu, Nepal.

Choudhari, R.P. (1998). Biodiversity in Nepal: Status and Conservation. Saharanpur Press, Kathmandu.

CRT/N (2005). "National Energy Situation Report Nepal-Focus on Renewable Energy Poverty Reduction". Centre for Rural Technology/Nepal, Kathmandu.

Gajurel, C.L. and Vaidya, K.K. (1983). Traditional Arts and Crafts of Nepal. S. Chand & Company, New Delhi.

Lekhak, H.D. and Lakhe, B. (2003). Natural Resource conservation and Sustainable Development in Nepal. Kshitiz Publication, Kathmandu.

NBPG (2007). "Biogas Sector in Nepal". Nepal Biogas Promotion Group, Kathmandu.

PACT (2010). http//www.pact.gov.np.

Sharma, U., Munankami, R. and Karki, D. (2003). Improved water mills (Ghatta): An energy centre for rural communities. In: Proceedings of International Conference on Renewablw energy Technology for Rural Development, 12-14 October 2003, Kathmandu, Nepal, pp 289-292.

Subba, D. (2008). SMEs and Technology Transfer in Nepal. Oral presentation in: International Conference on South-South Cooperation For Technology Transfer and Development of Small and Medium Enterprises (SME). Negombo, Sri Lanka, 18-22 August 2008, NAM S & T Centre, New Delhi and National Science Foundation, Sri Lanka (org).

Subba, D. (2010). Status and Initiatives of Biodiversity and Energy Resources Utilization in Nepal. In: Sustainable Utilization of Energy and Biodiversity Resources for Wealth Creation and Development (B.O. Solomon, ed.), Scitech Publications (India) Pvt. Ltd.

Chapter 6

Current Status of R&D Activities Conducted in Myanmar for Rural Community Development

Aye Aye Khai[1] *and Mya Mya Oo*[2]

[1]*Associate Professor, Department of Biotechnology, Technological University (Kyaukse), Myanmar*
E-mail: aakhai@gmail.com
[2]*Rectors, Mandalay Technological University, Myanmar*
E-mail: ytu.rector@gmail.com

ABSTRACT

The development of rural areas is one of the main objectives of Myanmar, and hence there are some research programmes conducted under the purview of the Ministry of Science and Technology. These are the biogas plants, the Gasifier and the production of biofertilizer, bamboo vinegar; chitosan and natural liquid fertilizer. In addition, there are some on-going projects such as the home-made fermented fish sausage, fruit juice, anti-malarial drug using medicinal plant extracts in Myanmar, and the cultivation of *Musa* spp. through plant tissue culture. This paper discusses the above mentioned activities in detail.

Keywords: *Rural development, Biogas, Gasifier, Bamboo vinegar, Chitosan, Fermented fish Sausage, Anti-malarial drug.*

Introduction

Myanmar has a population about 60 million population and the majority is the Buddhist, while the others are Christian and Islamic. It is situated in the Southeast

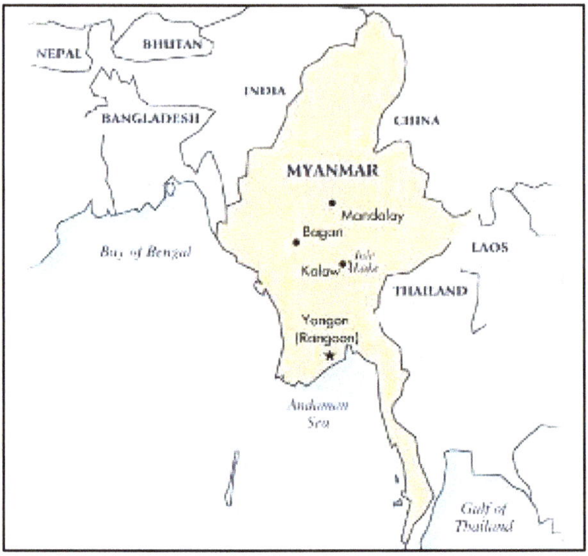

Figure 6.1: The Map of Myanmar

Asia and bordered on the north and northeast by China; on the east and southeast by Laos and Thailand; on the south by the Andaman Sea and the Bay of Bengal; and on the west by Bangladesh and India. It is located between latitudes 09°32′N and 28°31′N and longitudes 92°10′E and 101°11′E. The total area of Myanmar is 678,500 sq km where 657,740 sq km occupies the land and 20,760 sq km occupies the water. It has three main seasons namely, the hot season, the rainy season and the cold season. Myanmar is rich in natural resources such as petroleum, timber, tin, antimony, zinc, copper, tungsten, lead, coal, some marble, limestone, precious stones, natural gas, and hydro-power.

Excellent Performance in Rural Development

Since the food security and rural livelihoods contribute to increase income of the rural people, increasing agricultural productivity is the core policy for Myanmar. In order to increase agricultural productivity, promoting access to irrigation water and its sustainable uses are the main necessity. Myanmar has heavily invested in the construction of 228 irrigation dams, 322 river water-pumping projects, construction and repair of small irrigation tanks throughout dry-zone, and enhancement of undergroundwater utilization through more than 7000 tube wells. In addition to promoting access to irrigation water for increasing crop production, the rural electrification is also being promoted by installation of mini-hydro-power along with the canal drop structures of irrigation net works.

Knowledge Development Programme in Village

The government is setting up self-reliant libraries in all villages to widen the horizons of the people, with calling for follow-up programmes for sustainable progress of the already-established libraries.

Apart from that, there are other rural area development programmes such as the supply of safe drinking water, health care and education,

Human Resources Development Programme

A huge number of scientists and engineers would need to be trained to cater to the requirements of research, application, production and process development, extension activities, and also for the awareness creation and educational programmes. Therefore, the Ministry of Science and Technology (MOST) has been implementing the human resources development programmes for science and technology since its establishment on 2nd October 1996 and has been making the following efforts:

- ☆ To carry out research and development programmes for the national economic development;
- ☆ To strengthen the national economy utilizing the national resources and to raise the living standard of the people;
- ☆ To enhance production in agricultural and industrial sectors through technology transfer and distribution of knowledge gained from the research and development work;
- ☆ To plan and carry out human resources development so as to get specialists and professionals in science and technology;
- ☆ To carry out tasks such as laboratory analysis, quality control and standardization of industrial raw materials and finished products.

Education Programme

For the future growth with the recognition of indigenous knowledge system, the advancement of modern science and knowledge transfer is also further encouraged. In order to achieve this objective, the vocational schools such as 23 Universities of Computer Science and 26 Technological Universities that cover the different States and Divisions, also function at their full strength under MOST. Meanwhile, the students from villages have been trained at the Government Technological High Schools covering their preferred areas with the aim of providing their own villages requirements effectively.

Research Programmes

Under the close guidance of the Hon. Minister, Ministry of Science and Technology, the following research works for rural area development are being conducted.

Biogas

Myanmar is an agro-based country and there are many resources of biomass. Due to more effective and efficient ways being applied in biotechnology, biogas can be produced by using biomasses such as animal wastes, agricultural wastes, and human wastes. For the development of rural area, economy biogas digesters need to be constructed in rural area, (*i*) to get biogas for cooking and generating electricity, (*ii*) to get a wood- substituted fuel, and (*iii*) to improve the rural environment in the agricultural sector and in the livestock sector.

Since the year 2002 to 2010, one hundred and fifty six (156) numbers of fixed dome type electricity producing biogas plants with 50 cubic meters in size, which use anaerobic fermentation process in which cow dung is used as a raw material, had been constructed and utilized in Mandalay, Sagaing, Magway Division and in the Northern Shan State. They can drive a 25 Hp gas engine and 15 KVA generator, using biogas produced from the cow dung. All houses, monasteries and streets in these villages have been provided light with biogas for two hours in the early morning and four hours in the evenings.

Figure 6.2: G.T.H.S Students who are Working in the Led Lamp Preparation

Figure 6.3: Led lamp Lighting in Monastery

In 2008, a fixed dome type biogas plant with 100 cubic meters in size was constructed at a diary cow breeding farm in Mandalay Division. It can produce biogas for cooking, chopping the forage for cows, and producing the electricity to provide light to the whole farm. At the same time, a fixed dome type biogas plant, twenty five cubic meter in size, which use human wastes as raw material was constructed at the Sin min cement factory. Biogas from the said plant can be utilized for cooking and the effluent from the biogas plant is used as an organic fertilizer for plant growth.[5]

In 2009, a project to construct six cubic meters to twelve cubic meters in size (family-sized), fixed dome type biogas digester has been started, by the biogas research group functioning under Ministry of Science and Technology, to increase the public awareness for multi-utilization of biogas, which provide beneficial support for the rural area development. The biogas research department in Myanmar cooperates with people in the rural areas to succeed the plan for biogas production and utilization in cooking, gas lamp lighting, and production of engine driven electricity. Using one family-sized biogas digester, effluent slurry of 30 - 50 gals per day can be obtained as organic fertilizer and it can be used in soaking seeds for germination, spraying on the plant for growth, feeding pigs, keeping fishes and also as a value added organic fertilizer.

Figure 6.4: Outlet Tank of Biogas Plant in Thu Nge Taw Village

Figure 6.5: Digester of Biogas Plant

Biogas from Gasification

The Down draft gasification system is another type of biogas digester, which is based on the woodchips and rice husk. It is designed and constructed to produce electricity of 30 KW and 50 KW. A number of Chemical, Mechanical engineers from Technological Universities attended training programmes held in the Kyauk Se Research Centre of Mandalay Division (approximately 70 engineers attended to the Design and Construction of Gasifier course in the year of 2009). The hands on experience helped them to understand the techniques used in design and fabrication of gasification system and they received a good working knowledge of troubleshooting and repairing these systems.[5]

Figure 6.6: Artificial Inseminated cow and her Child

Figure 6.7: Female Friesian

Animal Breeding

Most of the people in rural areas are usually poor. To overcome this problem or to increase their income otherwise, the selective breeding for milk production in local

Friesian crossbreed and Zebu cows had been done under MOST in Myanmar. Under good housing, feeding and management systems, the cows were used for breeding with both the breeding of bulls and the artificial insemination to improve the performance of the breed through selection procedures.[7]

Research on Production of Natural Fertilizer, Bio-fertilizer, Bio-pesticide, Natural Liquid Fertilizer [2,3,8]

The biofertilizer with seven kinds of effective microbes is produced, packed into bags with 50 kilograms per bag and distributed at the cheap or reasonable prices. By using the bio-fertilizer and the chemical fertilzers alternatively, it has been possible to reduce the use of chemical fertilizer to 35 per cent. The bio-fertilizer can widely replace the expensive chemical fertilizer and hence, the farmers were able to reduce cultivation cost dramatically, up to 30 per cent. This is an approach where soil will be made fertile while promoting the agricultural products that are hazard free.

Figure 6.8: Shwe Zi Wa Biofertilizer Product

Figure 6.9: Compost Used in Biofertilizer

1. The production of natural fertilizer extracted from unmarketable fish and fish waste, which consists of various nutrient elements, enforces the plant growth and it is a solution for replacement of costly chemical fertilizer. Research on this product, especially needed for plants, is continuously conducted.
2. The bamboo vinegar extracted from bamboo is applied as a type of pesticide and this product, as a liquid pesticide, can control pests and the diseases caused by either fungi or bacteria. In addition, it contains various different nutrient compounds, which help root growth and chitosan is also combined with this bamboo vinegar for plant gowth. Meanwhile, the demonstration of how to use the pesticide was already conducted at the rural areas under knowledge sharing and extension programmes.

Figure 6.10: Shwe Poe Hnin Bamboo Vinegar Product

Figure 6.11: Shwe Sein Chitosan Product

3. A natural plant growth stimulant is produced using chitosan that is extracted scientifically from the waste of shrimp-shells is manufactured as a liquid foliar. This product helps strengthening all kinds of plants to grow rapidly and to yield well. [6]
4. Another natural liquid fertilizer containing organic liquid fertilizer extracted from fish, chitosan [6] and bamboo vinegar makes plants yield well. This product also prevents plants from pests and helps them grow rapidly as produced and applied when the plants is at the bud stage.

Figure 6.12: Shwe Se Ye Organic Liquid Fertilizer Product

Other On-going Projects which Approach as Solutions to the Rural Development

Mutation Induction of Rice Varieties through Gamma Radiation [4]

To obtain the high yielding varieties, the mutation induction of rice through r-ray from Co60 source is being carried out since 2008.These Collaborative projects are IAEA/RAS/5/045 and IAEA/TCA/MYA/017. The latter has been started in the rainy season of this year.

Fruits Juice Production

In this fruit juice production project, the country has intended to produce mango and guava juices by using local fruits. As a result of this project, the rural community can fetch an extra income while distributing the high quality fruit juice product to the general consumers. In addition, the research groups can share knowledge in food juice production and the appropriate technology to the public. Consequently, the

Figure 6.13: High Yielding Rice var. Sin Thwe Latt

Figure 6.14: Field View of Rice var. Sin Thwe Latt in M_5 Generation

village women can use the technology by themselves and they can get extra money. [7]

Fermented Fish Production

Fermented fish sausage is one of the most popular Myanmar traditional foods that are mainly and widely produced in lower parts of Myanmar. Lactic Acid Bacteria and other health beneficial microbes are involved in this project, which is good to human health. Usually, this product is a little expensive in the Central Regions of the Country due to the difficulty in get the fresh fish because of high transportation charges and the time involved for transportation. Now, this food item is being produced using the fish caught from Swan Ye' Lake near Kyauk-se District in Myanmar solving the above problem partly.[7]

Figure 6.15: Fermented Fish Sausage Product

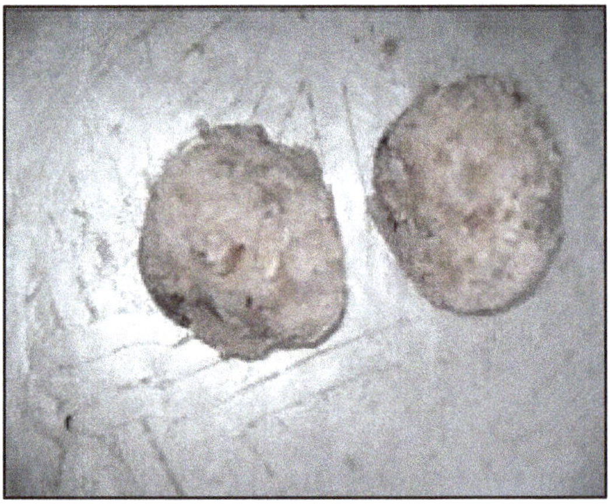

Figure 6.16: Grinding Fish Ball along with Rice

It is aimed at distributing this fish food product in fresh and safe manner and also at cheap price to the general consumers as a traditional home-made item and this has helped in improving the living standards of people in the Swan Ye' village. Meanwhile, it is planned to support the national income by producing health beneficial fermented fish sausage product.

Small Scale Production of Antimalarial Drugs Using some Myanmar Medicinal Plant Extracts [1]

Malaria is one of the widely distributed severe infectious diseases, that can cause mortality tremendously. And most of the anti-malarial drugs are now not resistant to malaria parasites especially to *P. falciparum*. Hence, research is now aiming at obtaining the anti-malarial agents of plant origin in Myanmar with low toxicity and high efficiency. At present, an anti-malarial medicine has been produced in combination of three kinds of the Myanmar medicinal plant extracts, under investigation.

Figure 6.17: Shwe Pin Swan Anti-malarial Medicine

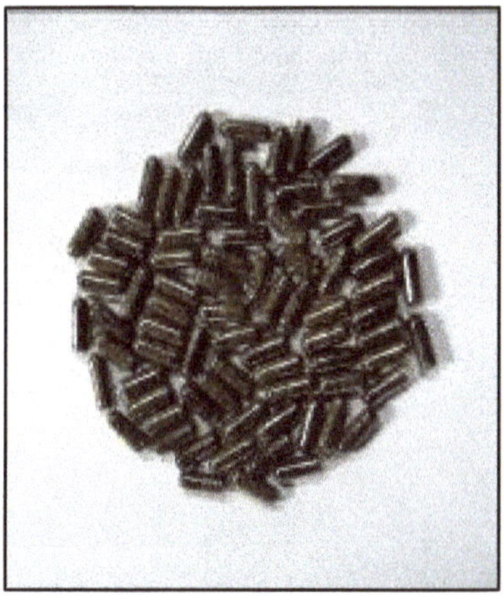

Figure 6.18: Anti-malarial Medicine (Capsules containing the Myanmar medicinal plant extracts)

Production of *Musa* spp. through Plant Tissue Culture Technique

Musa spp var. Khunwar and Khunsein have been already produced and handed over to the farmers since 2008. At present, this research technique is being carried out totally free of contamination and the banana plants are multiplied in large scale.[9,10]

Figure 6.19: Multiplication of Shoot and Root of *Musa* spp. through PTC Technique

Conclusion

The application of Science and Technology will be more beneficial to the rural development compared to the

Figure 6.20: Hardening *Musa* spp. Plants

conventional methods. The researches under MOST could improve not only the farmers' income but also their standard of living. From 2002 to 2010, one hundred and fifty six numbers of biogas plants have been constructed. In addition, a large number of bio-fertilizer bags and bamboo vinegar and chitosan natural liquid fertilizer have been delivered to farmers with the lowest price. There are other ongoing projects, directing at the rural development are being carried out in the country.

Acknowledgements

The author would like to express her deepest gratitude to the honorable Minister, His Excellency U Thaung, Ministry of Science and Technology, for the close guidance extended in the rural research programmes. Then, the author is deeply grateful to Prof. Dr.Mya Mya Oo, Rector of Mandalay Technological University and Yangon Technological University, for her kind encouragement and valuable suggestions. Finally, thanks are due to all colleagues at the Department of Biotechnology covering different States and Divisions under MOST.

References

1. Hlaing, A.K.S, Oo, Z.K.and Mon, H.M. (2008).Evaluation of anti-malarial activity of selected Myanmar medicinal plants.The third GMSARN international conference 2008 on sustainable development: Issues and prospects for the greater Mekong Region.Uchoice Hotel, Kuming, China, 12-14 November, 2008.
2. Latt, Z.K., Yu, S.S. and *et al.* (2008).Screening and characterization of mutant strains of some cellulolytic nitrogen fixing bacteria for production of N rich biofertilizers. Annual meeting and international conference of the Thai Society for Biotechnology TSB 2008: "Biotechnology for global care"Taksila Hotel, Mahasarakham Octobacter 14th-17th 2008.
3. Linn, T.M.,Yu,.S.S. and *et al.*, Developmentof mutants from wild type phosphate solubilizing bacterial isolates for effective bio-fertilizer production. Annual meeting and international conference of the Thai Society for Biotechnology TSB 2008: "Biotechnology for global care" Taksila Hotel, Mahasarakham October 14th-17th, 2008.
4. Minn, M., Khai, A.A. and Lwin, K.M.(2008).Study on the effect of gamma radiation on rice var.Sin Thwe Latt (IR53936). The third GMSARN international conference 2008 on sustainable development: Issues and prospects for the greater Mekong Region.Uchoice Hotel, Kuming, China, 12-14 November, 2008.
5. Ohn, S.S. and Oo,M.M.(2008).Comparative study on biogas production and soil fertility of various effluents.Annual meeting and international conference of the Thai Society for Biotechnology TSB 2008: "Biotechnology for global care"Taksila Hotel, Mahasarakham Octobacter 14th-17th 2008.
6. Pranee, L., Ng, C.H., Suwalee, C. and Willem, F.S., 2002. Effect of chemical treatment on the characteristics of shrimp chitosan. Journal of metals, materials and minerals. Vol 12 No. 1, pp. 11-18.
7. Rural development programme under MoST in Myanmar (not written in any paper until date).
8. Than, W.M., Khai, A.A. and *et al.* (2008).Management of common plant pathogenic fungi via biofungicidal function of beneficial microbes. The third GMSARN international conference 2008 on sustainable development: Issues and prospects for the greater Mekong Region.Uchoice Hotel, Kuming, China, 12-14 November, 2008.
9. Thandar, S.and Tun,O.M.(2011).Isolation,identification and prevention of contaminated bacteria during in vitro culture of red banana (*Musa acuminate* Colla).The fourth National conference on Science and Technology, NCSE 2011.
10. Thein, E.E., Aye T. H. and Tun, O.M. (2011).Investigation of the antibacterial agents to control the bacterial contamination occurred in culture of *Musa sapientum* (Khun War). The fourth National conference on Science and Technology, NCSE 2011.

— *Section II* —

Case Studies of Varied Approaches to Rural Industrialization

Chapter 7

Design and Development of a Village GIS for the Rajasthan State:
A Complementary Approach Towards Rural Development in India

V. Madhava Rao, R.R. Hermon, P. Kesava Rao, Ravi Shankar S.S. and T. Phanidra Kumar*

Centre on Geoinformatics Application in Rural Development (CGARD), National Institute of Rural Development, Ministry of Rural Development, Government of India, Hyderabad, India
**E-mail: madhava@nird.gov.in, madhavaraov@gmail.com*

ABSTRACT

The use of GIS for a micro administrative unit for planning, implementation, monitoring and evaluation has immensely helped Panchayat(Village) level Planning process to promote decentralized Governance for greater socio-economic development in a participatory manner. Geographic Information System (GIS) has the capacity to analyze both spatial and temporal data on a cost effective manner for effective planning and management, in an interactive and participatory manner, ensuring judicious spatial growth, reduce the cleavage between the rural and the urban areas through appropriate development planning for the villages, faster economic development and offer rational, efficient and effective solutions. Further, the rapid advances in the hardware and software technologies coupled with a growing competition among the related vendors have brought down the cost of Geomatics/GIS technology by manifold, making it affordable for deployment on a large scale for use in decentralized planning. The Village GIS is a Template which is a customized GIS Package, simple, user friendly, customized, local need based, interactive and can be handled by the local youth and Gram Panchayat functionaries. It could be called Panchayat GIS, which has all inbuilt features of various decisions at the Gram Panchayat level integrating various layers of information.

The successes and failures of micro-level development planning over the last 60 years have made it clear that participation and ownership is the key to success and any technological intervention and knowledge level decision support system, at micro level solutions must address to the pressing problem of the area in terms of their own perceptions of needs and priorities. This is where GIS can be of direct use, in visualizing for the villages, the consequence of alternative priorities and is therefore is accepted by the villagers and is proving as a potential and as a powerful decision making tool for the people in Indian villages.

GIS is best suited to capturing data from diverse sources and is an excellent tool for management of large bodies of spatially extensive data with precision, consistency, and absence of computational errors. A GIS enables one to zoom a map instantly to any scale to view every feature observed in the field.

The GIS technology is geared to prove most appropriate in every aspect of planning, decision making, modeling, disaster management, alternative scenario generation and as a decentralized governance tool in India. Mobile and ubiquitous devices has made the information sharing, accessing, decision making and processing so simple that once which was considered impossible is now, a happening and a possibility.

Keywords: *Geographical information system, Rural development, Science and technology, Small and medium scale enterprises, Gram panchayat, Decision support system, Knowldege management.*

Introduction

The Gram Panchayat concept of India is emerging as a hub of activities at the micro level and proving as a decision making process of the community and the people for developing their livelihood and the area they live, for their sustenance. Gram panchayats are local governments at the village or small town level in India. There are about 265,000 gram panchayats in India. Panchayati Raj Institutions – the grass-roots units of self-government – have been proclaimed as the vehicles of socio-economic transformation in rural India. Effective and meaningful functioning of these bodies would depend on active involvement, contribution and participation of its citizens both male and female. The aim of every village being a republic and Panchayats having powers has been translated into reality with the introduction of the three-tier Panchayati Raj system to enlist people's participation in rural reconstruction.

The Ministry of Rural Development has embarked upon numerous development programmes in the country to assist the local people to develop themselves while providing a safety net, for their empowerment and economic development. A modern technology such as the Geographical Information System (GIS) proves to be a decision supporting tool at the local level for judicious and rational decisions, by integrating all knowledge bases at the village level. This system has the capabilities of a) modeling, b) scenario generation and c) strong visualization for making the communities and the people who participate in the process of development programmes in the country. This paper presents an attempt of applying the GIS concept at the village level in the Bhilwara district in the Rajasthan State of India.

What is a Village GIS?

A village GIS is thought to comprise all relevant information of the individual village, which is lacking in the present scenario where all the researchers and the data agencies provide the outputs mostly up to the level of Block boundaries only. It is an attempt to reach the actual village parcel boundaries and settlements to provide data on their land and settlements. It will work as a village wise data bank and an information system that could be used by the progressive villagers, various agencies at the government level and the non-government level, workers in the village or at higher levels and by the researchers.

In the first instance, a prototype was tested to comprise all the related information as much as possible for three villages namely Surajpura, Bera and Yaswantpura in the Bhilwara district in the Rajastan State. However, due to time shortage, only the Suarjpura village had been concentrated for inclusion of all related available information on actual basis. The result can be expressed in different categories as given below.

Cadastre

A parcel map of all three villages has been digitized and geo-coded with the help of GPS points and Google reference points. The maps were further spatially adjusted for mosaic and matching the boundaries. A rough settlement map has been digitized from the Google village map and further spatially adjusted to fit properly in the spatially corrected map of cadastre. It is planned to go to the village in the next phase with Google map as the reference map and onsite mapping of the settlement boundaries, are being carried on with due consultation of Patwari and Sarpanch and the house owners and the proper boundary adjustments of houses are being rectified and corrected. The Patwari is the lowest state functionary in the Revenue Collection system, whose job encompasses visiting agricultural lands and maintaining record of ownership, where as the Sarpanch is a democratically elected head of a village level statutory institution of local self-government called the gram panchayat (village government) in India, a focal point of contact between government officers and the village community.

In the cadastral map, all the Khasara Jamabandi and other details present in the Record of Rights (RoR) for each parcel are incurred like ownership with photo of owner, bank loan, revenue, land category, cropping pattern details etc. A khasra is a legal agricultural document used in India that specifies land and crop details often used in conjunction with a reference map of the village that administers the land described all the fields and their areas, measurement, who owns and what cultivators he employs, what crops, what sort of soil, what trees are on the land.

Settlement

In the settlement maps all the details of the houses such as ownership and family details with photographs, all the house-wise details as per the Livelihood survey conducted by an NGO (Bishwas) for the District under the guidance and instructions of the district officials.

Others

Other than the cadastre and settlement details, all the details of the related department *viz.*, school locations and other details attributed; health center locations and details; PHED water quality; well locations and details; village-wise soil details; agriculture input status details; weather; livestock details etc.

Assets

It is planned to incorporate all the structures and assets constructed for the village under Mahatma Gandhi National Rural Employment Guarantee Scheme (MGNREGS), Watershed or other Departments working in the village with their location and other details. In the next stage, the facility will be given to the user to update the data (locational and attribute) on their own into the system, so that the timely departmental changes can be incorporated into the system.

Change Detection

The Indian Latest Satellite images of IRS P6(Liss IV) for Rabi season are being associated for two years 2005 and 2008, for the ready reference to check the parcel wise actual change in the vegetation status and the village wise changes in the water bodies and vegetative cover. Many other readily available location based data have been collected from the State Remote Sensing Agency, Jodhpur such as geomorphology, lithology, soil series, soil capability, water bodies, drainage line, and groundwater.

As the output other than the direct view of the layers and the interactive information on the system, we may get many comparative charts and graphs along with the reports. Presently the interface is a mix of Hindi and English as per availability of data. Later, it will be separately made in Hindi and English both. A Google attachment is also there to visualize the map and parcels overlaid on Google, if Net connectivity is present. After fully testifying and updating the system it will be useful for villagers, patwaries and revenue officials, relevant departmental officials for planning and monitoring of the village programmes and assets.

Objectives of the Village GIS

The following are the broad objectives of the Village GIS that are being proposed for the hundred and one (101) villages of the Bhilwara district of the Rajasthan State:

1. Design and development of a digital Village GIS for each of the 101 villages integrating each household and cadastre, as well as all the spatial and attribute databases of the villages;
2. Integration of other schemes such as Watershed, MGNREGS etc, as a part of village GIS;

3. All the village level details of other departments such as demography, resources, LULC, livestock, soil, rainfall, water quality, high resolution satellite data, GPS/DGPS Coordinates, household photograph of the family members along with the house and the authentic identification number etc., with automatic report generation, capturing graphic *viz.*, audio and videos, and language interface.
4. Training and capacity building for managing the Village GIS by Gram Panchayat locally.

Process Methodology

The process methodology for the design and development of Village GIS for 101 villages in the Bhilwara district will follow the following steps:

1. Collection of Cadastral maps, Khasara Girdavari copies, household photographs, land marks, DGPS reference points of selected points, relational databases from all available sources, satellite imageries etc.
2. Collection of other departments' data and field location of Watershed and NREGS works
3. Digitization of cadastral maps and geo-referencing using DGPS points;
4. Procurement of the best available high resolution satellite imageries;
5. Seamless DEM and 3D view generation using Cartosat-1 data and LISS-IV data
6. Procurement of the basic spatial value added digital data available with other departments such as SRSAC etc;
7. Mapping of settlements and houses in the field using high resolution satellite data, DGPS. Surveys, digital conversion of hard copy data and the data input/feeding of cadastral data (Khasara Girdawari) and other departments' hard copy data;
8. Integration of spatial and attribute data base;
9. Analysis and mapping;
10. On Screen window data inputting menus and icons.
11. Training and capacity building of the users.

Design and develop basic GIS layer for the Gram Panchayats with available GP databases, map layers, satellite imageries, GPS/DGPS Coordinates, PRA, etc., and the subsequent development of all GIS functionalities for making the system fully operational and for wide scale training and capacity building for familiarization, skill development, enhancing knowledge level and managing the Village GIS, are key milestones of geospatial technology intervention at village level.

Deliverables and Output

The following will be the deliverables and the output for the Village GIS for 101 villages in the Bhilwara District of the Rajasthan State:

1. GIS based Decision Support System with information up to household and cadastral levels;
2. Timely up-gradable database and DSS for Watershed and MGNREGS;
3. Strong capabilities of mapping and analysis;
4. Automatic summary reports both in Hindi and English;
5. Facility to generate 3 D analysis, geo-visualization and virtualization;
6. Facility to generate statistical graphic analysis;
7. Facility to capturing and displaying audio and videos;
8. Both the Web and the Standalone versions for wider use;
9. Template mode and with a customized GIS software, for micro level application.

The following figures show different types of maps that could be obtained through village GIS systems.

Displaying Settlement of Surajpura Village based on Family Code

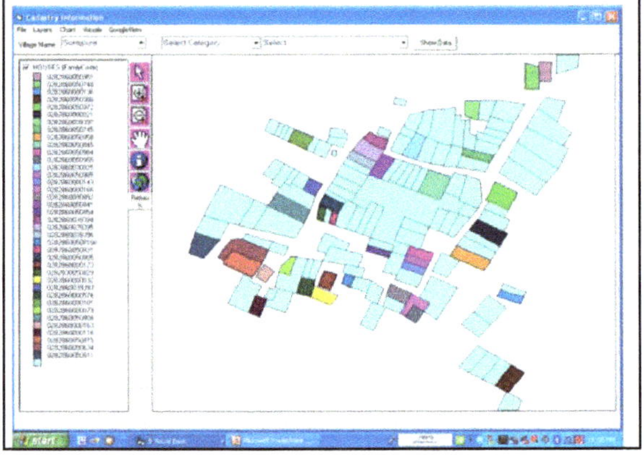

Identifying Family Codes on the map

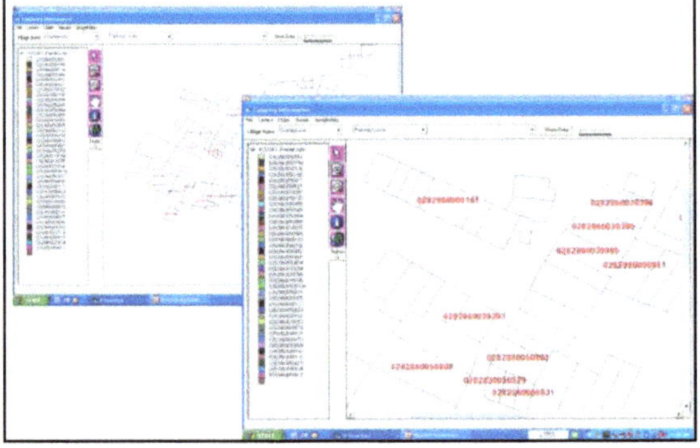

Identifying Family Information along with their photograph

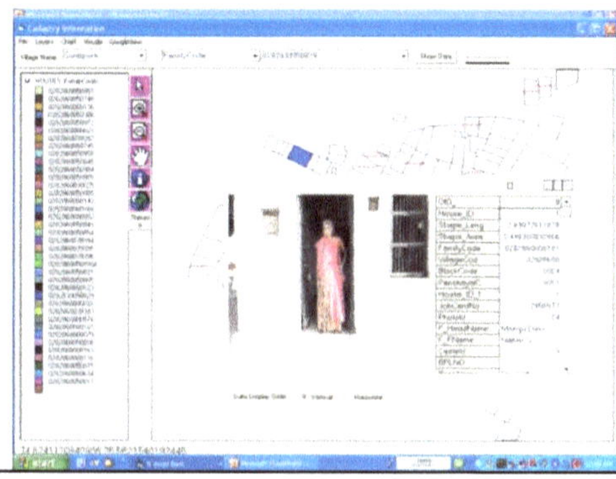

Based on Account Holder

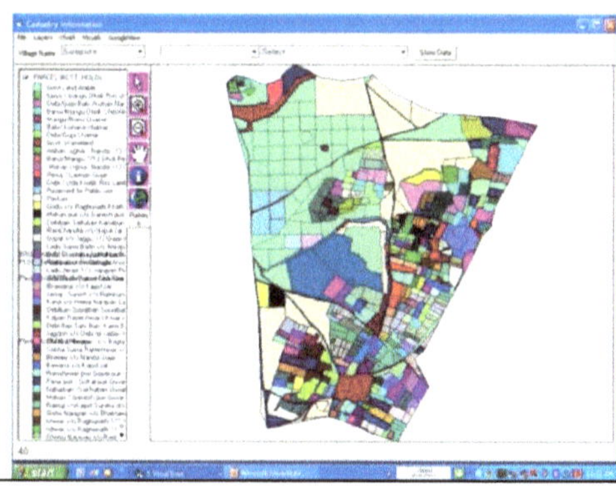

Displaying Cadastral Map based on Survey Number

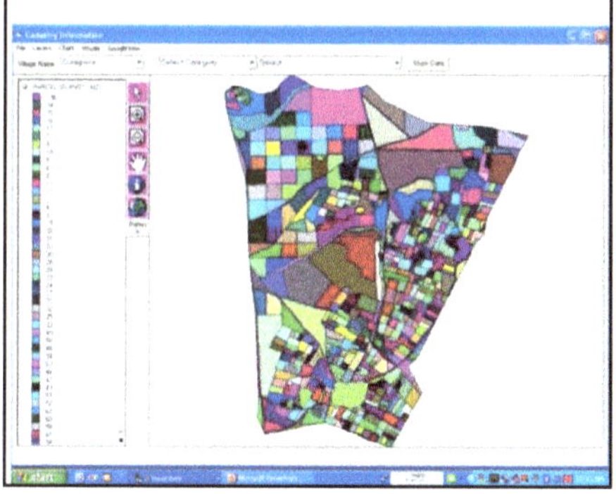

Identifying NREGS Structures on Satellite Imagery

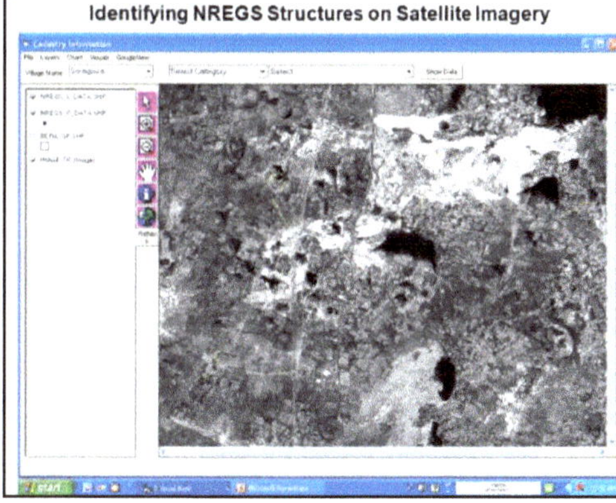

Observations and Findings

The striking revelations in this study is the fact that local people are greatly enthusiastic in adapting to new technology and new way of working together and takes lot of interest in knowing new ways of doing things. Secondly, the people at the local level raise above political, economic and social affiliations, and express their solidarity in effecting the new tools and technology in their day to day work.

The women participation was quite encouraging and they come forward for propagating the use and adoptability to new way of doing things. The younger mass feel social upheaval and increase in their social status in exposure and use of new IT Tools at the local level.

The people's representatives give all encouragement and support and stand by the people in introduction of new technology and in new ways of doing things, irrespective of their political affiliations.

The district and local administration takes great interest in introducing new IT Tools and request development of these tools with inbuilt functionality of day to day administrative use at all level, in improving their performance and effectiveness and express their support and co-operation in any attempt in this regards.

Replicability

The Interactive GIS developed at the Gram Panchayat can be replicated. There are also possibilities and potential available for adding additional features and in making it wide spectrum software.

The database required may vary depending on the type of problem faced in a particular area. The level of social mobilization needed to ensure participation in data collection and analysis can also vary from place to place, which can be handled by the software with necessary modifications.

Conclusion

The user friendly interactive GIS for a micro administrative unit has helped in ensuring better participation of the stakeholders in decision making. The variation in the level of education of the community was bridged by pictorial representation of data. The project helped in empowerment of the community and generated land literacy.

User friendly interactive GIS data base generated at a micro administrative unit (panchayat) can improve the efficiency of administration, improve resource mobilization and help in informed decision making.

The software being simple and customized and open for modifications hold lot of promise for local level applications. The strong aspect of this software is the capabilities of liking wide graphic aspects, like photographs, audios, videos, imageries and analyzed maps etc, which project field realities and help taking appropriate decision making, by people themselves.

Chapter 8

Renewable Energy Technologies and Rural Development in Tanzania

Omar Jumanne Bakari

Commission for Science and Technology (COSTECH)
4302, Dar es Salaam, Tanzania
E-mail: obakari@costech.or.tz

ABSTRACT

Tanzania is one of the poorest countries in the world with a population of almost 40million with 6 per cent GDP in 2009. 80 per cent of its population lives in rural areas where less than 2 per cent have access to electricity. Regardless of low supply capacity, Tanzania Electric Supplies Company (TANESCO) is the only major company, which supplies electricity in the country. Lack of access to modern energy services create a vicious cycle of poverty to rural communities due to continued limited improved production opportunities and social facilities.

This paper shows the impacts of renewable energy technologies to rural development as presented in two case studies. It shows how the poverty status of a rural community can be changed with the application of appropriate technology. It also indicates how the energy technology transfer appears to be a major challenge due to the price factor in relation to purchase power in the rural community.

This paper recommends the needs of government intervention on creating awareness, capacity building and involvement of rural communities in planning and implementation of different energy technology development projects in the rural areas. Such a strategy will facilitate sustainable technology transfer and increase the technology impact on the rural development. Also, it recommends creation of a credit system or synergy system (e.g. sisal farmer for sisal biogas), which will enable the rural community to acquire the said technologies.

It concludes that the availability of appropriate renewable energy technologies is the key element for all the key stakeholders to focus on. This will create jobs, increase production and protect the environment, while reducing soil erosion and climate change. Without appropriate technologies the impact on rural development will be insignificant and the country will not attain Millennium Development Goals (MDGs). Further, consistency in Government financing, planning, implementing and monitoring processes in renewable energy programmes will bring sustainability to all renewable energy programmes in the country.

Keywords: *Biogas, Solar energy, COSTECH, Tanzania, Rural development, Poverty, Renewable energy.*

Introduction

The United Republic of Tanzania is a result of the union of Zanzibar (Unguja and Pemba) and Tanganyika in 1964. It is located in the East Africa with a population of almost 40 million. About 80 per cent of the total population lives in rural areas. Geographically this country lies between latitude1° and 12° South and Longitude 29°

Figure 8.1: Map of Tanzania

and 41° East with an estimated area of about 945,087 square kilometers. According to the Government of Tanzania (2007), the economy of this country mainly depends on agriculture with a total contribution of 25.8 per cent to the GDP in 2007.

Tanzania is blessed with abundant energy resources in different forms, *viz.*, biomass, solar, wind, hydro and geothermal. Energy consumption is dominated by biomass (90 per cent), petroleum (8 per cent), hydro electricity (1 per cent) and others (1 per cent) (World Bank, 2009). Average national wood fuel consumption per capita is around 1.0 to 1.5 m^3 of solid wood. Main consumers of wood fuels include households, both in rural and urban areas, small-scale industries and processing of agricultural crops mainly tobacco curing and tea drying. The estimated national annual wood fuel consumption is about 36 million m^3, which is already over and above the estimated annual sustainable yield of 24 million m^3 from the forest standing stock.

Regardless of all these available energy resources, there is a higher probability of not achieving the Millennium Development Goals (MDGs) by 2015 because of the lack of modern energy. Energy is an essential service and the availability and affordability of which will determine the success of development endeavors. Modern Energy is a prerequisite for the proper functioning of all sectors of the economy and the Macro-economy development. From the 80 per cent of Tanzanian population who lives in rural areas less than 2 per cent have access to electricity. This is a very challenging problem to the economic development of the country. The next section gives some details on the energy sector and its policy.

Energy Sector and Its Policy

Ministry of Energy and Mineral is responsible for policy formulation and implementation. It is also responsible for the development of the energy sector in Tanzania. The vision of the energy sector is to contribute to the growth of national economy and improve the standard of living of the population in a sustainable and environmentally sound manner. The mission of the energy sector is to create situation for provision of safe, reliable, efficient, cost effective and environmentally compatible energy services on a sustainable basis to the widest cross section of the population. The revised National Policy came out in 2003 after reviewing the 1992 National Energy Policy. This was influenced by the changes, which were taking place in all sectors of the economy. The policy takes into consideration the need to:

- Have affordable and accessible energy supplies countrywide.
- Reform the market for energy services.
- Form an Institutional framework, which facilitates investment, expansion of services, efficient pricing mechanisms and other financial incentives.
- Adequately take into account environmental considerations for all energy activities
- Promote energy efficiency and conservation in all sectors; and increase energy education and build gender-balanced capacity in energy planning, implementation and monitoring

The main objectives of a national energy policy are: to ensure availability of reliable and affordable energy supplies and their use in a rational and sustainable manner, in order to support national development goals. In other words, the energy policy strives to ensure sustainable energy supply, security, reliability and environment protection.

The specific objectives of the energy policy are: to develop and utilize natural gas and coal resources to step up petroleum exploration activities, to promote development and utilization of appropriate new and renewable sources of energy, to reduce deforestation through efficient wood fuel conversion and end-use technologies and techniques, and to promote energy conservation and efficiency. To achieve these objectives, the focus is on increasing electricity supply and distribution, sustainability and increasing oil and gas exploration, regional interconnection, rural electrification and reaching rural households. The last two focuses are very important for rural development.

To succeed in the development of the modern energy, some of the factors to be considered include market economy, independent regulatory regime, national interest versus market forces, regional cooperation and trade, energy conservation and efficiency, environmental management, gender issues (social role of women and men) and appropriate technologies. In this paper, the focus has been on the appropriate energy technology and its impact on rural development. The next section discusses some key organizations responsible for energy development in Tanzania.

Key Organizations for Rural Energy Development

In the facilitation of the energy in rural development, the Ministry of Energy and Mineral is collaborating with different organizations to facilitate the implementation of different strategies for rural development. The following are some of those organizations who played the roles in the energy technology development and commercialization.

Rural Energy Agency

This is an autonomous body established under the Rural Energy Act No. 8 of 2005 governed by the Representative Board of Directors. The members of this board are appointed from the key institutions and ministries. The Rural Energy Fund (REF) was established by this Act and is responsible for grant payment and financial disbursement through the Trust Agent. The main objectives of this agency are to promote, facilitate, and improve modern energy access for the social and productive use in rural areas to facilitate the activities of key stakeholders with interest in rural energy and also to promote the rational and efficient production and use of energy. The main sources of income for this Agency include up to 5 per cent levies from both commercial generations of electricity from the national grid and specified isolation systems. Also government annual budget allocation, service fees, interests on investment and donation from the development partner.

Tanzania Commission for Science and Technology (COSTECH)

The Tanzania Commission for Science and Technology (COSTECH) Act No: 7 of 1986 (Revised) was established in order to repeal and replace the Tanzania National

Scientific Research Council Act, 1968. This Act extended to Zanzibar as well as to Mainland Tanzania. COSTECH is the main principal advisory organ of Government on all matters relating to scientific research and technology development. The Commission operates under the Ministry of Communication, Science and Technology but there is a very close link with other Ministries.

COSTECH organization structure includes the Commission, the R&D Advisory Committees and the Secretariat. The Commission is the governing authority of COSTECH; it gives directives and guidelines to the Secretariat and R&D Advisory Committee. The Chairman is the Presidential appointee for a term of three years. Members of the Commission are heading the affiliated institutions, Professors from Public and Private Universities, decision makers from ministries both Zanzibar and Mainland and representatives from key sectors of the economy. The secretariat is governed by Director General who is also a presidential appointee. The Secretariat is the main organ for day to day activities of the COSTECH.

There are twelve Research and Development Committees,which act as the "Think Tank" on various science and technology issues. All matters related to energy are dealt with either the Industry and Energy Committee or the Development and Technology Transfer Committee or both committees.

COSTECH vision is to be *a prime driver of science, technology and innovation for sustainable development.* The commission's mission is *to foster knowledge based economy through promotion, coordination of research, technology development and innovation for sustainable development in Tanzania.*

The roles of the COSTECH are as follows:

- ✰ To advise the Government on all matters relating to Science, Technology and Innovation (STI) including, but not limited to the formulation of S&T policy, priority setting for Research and Development (R&D), allocation and utilization of resources.
- ✰ To promote, coordinate, monitor and evaluate scientific research and technology development and technology transfer activities in the country.
- ✰ To facilitate national, regional and international cooperation in scientific research and technology development and transfer.
- ✰ To acquire, store, and disseminate scientific and technological information and popularize S&T.

COSTECH has got Regional and International collaboration with WIPO, UNDP, World Bank, EU, SIDA to mention a few. In the area of technology development, institutions, which are highly collaborating with the energy technologies development and transfer include Centre for Agricultural Mechanization and Rural Technology (CARMATEC), Tanzania Engineering and Manufacturing Design Organization (TEMDO), College of Engineering and Technology (CoET), Mbeya Institute of Technology(MIST), Tanzania Industrial Research and Development Organization (TIRDO), Small Industry Development Organization (SIDO), National Environmental Management Council (NEMC), Appropriate Rural Technology Institute –Tanzania and TaTEDO as discussed below.

Tanzania Technology Development Organization (TaTEDO)

TaTEDO is a sustainable energy development national NGO based in Dar es Salaam, Tanzania with more than 14 years of active involvement in sustainable energy development projects and programmes in the rural and urban areas. TaTEDO's vision is to facilitate communities with better access to modern sustainable energy services for poverty reduction, economic growth and social development while conserving the environment. The mission of TaTEDO is to develop and promote greater uptake and rational use of Sustainable Energy Technologies (SETs) and environmentally sound practices through provision of relevant information, knowledge and skills geared towards sustainable development of communities. The goals of TaTEDO are: to improve quality of life of the poor majority in Tanzania by contributing to availability of improved and sustainable energy services, employment and income generating opportunities, which are essential for poverty reduction and sustainable development. Also to reduce environmental degradation resulting from increased use of wood and fossil fuels and to contribute in reducing the country's dependence on imported energy (fuels and electricity).

To achieve these goals, TaTEDO undertakes the following activities:

- Promote the use of modern sustainable energy services for productive uses (industrial, agriculture and commerce etc.), consumptive uses and for basic needs (households and communities) and an institutional and social need uses (education, health, water etc.).
- Implement sustainable energy programmes and projects.
- Provide sustainable energy related consultancy services.
- Develop networking and partnership with local and international organizations,
- Manage and disseminate energy information to stakeholders.
- Undertake lobbying and advocacy to influence energy related policies, strategies regulations, legislations and pricing.
- Support sustainable energy enterprises development.
- Conduct energy related applied and adaptive research on technologies.

Other Key Organizations

Other key organizations include Energy and Water Utility Regulatory Agency (EWURA), which has the responsibility of promoting the availability of energy regulated services to all consumers including low income, rural and disadvantaged consumers; Tanzania Bureau of Standards (TBS), which has the role of formulating standards for energy equipments. Tanzania Engineering and Manufacturing Design Organization (TEMDO) has a mission of promoting engineering design, technology development and enhancement of the competitiveness of local manufacturing enterprises through provision of quality technical support services. Center for Agricultural Mechanization and Rural Technology (CARMATEC) is another organization, which has a mission to develop and disseminate improved technologies suitable for agricultural and rural development. The activities are aimed at boosting agricultural production and improving the quality of life and alleviation of rural poverty and finally the Tanzania Industrial Research and Development Organization (TIRDO) whose mission is to undertake applied research that leads to Industrial utilization of local materials and support the industries in technological and technical services so as to become locally and internationally competitive in industrial production.

Energy Needs and Millennium Development Goals

In the rural areas, poverty and lack of income activities are major problems which this has led to an increase in deforestation. This is because charcoal has become one of the major sources of the income and employment in urban and rural areas. This is mainly due to demand of the charcoal both in rural and urban areas. According to World Bank (2010), an estimated 90 per cent of Tanzania's energy needs are satisfied through the use of wood fuels. Charcoal is the single largest source of household energy in urban areas, as it is considered cheap and easy to transport, distribute, and store. Between 2001 and 2007, the proportion of households in Dar es Salaam using charcoal climbed from 47 per cent to 71 per cent. Approximately half of Tanzania's annual consumption of charcoal takes place in Dar es Salaam amounting to approximately 500,000 tons.

At present, the contribution of Tanzania's charcoal sector to employment, rural livelihoods, and the wider economy is estimated to be in the region of US$650 million per year, providing income to several hundred thousand people in both urban and rural areas. These people tend to be members of poorer households, who work as the small-scale producers or traders, and who often have limited alternatives for earning a living. Despite the important role charcoal plays in local economic development, its contribution to government revenues and the broader tax base are limited due to widespread evasion of licensing fees and transport levies. National and local governments are estimated to lose about US$100 million per year due to their failure to effectively regulate the charcoal sector (World Bank, 2009).

The increase in deforestation has highly impact in change climate due to irregular rains and soil erosion. In Tanzania climate change has started to affect the pastoralists due to lack of animal feeds, which has resulted in an increased number of animals' deaths and conflict between pastoralists and agriculturalists. According to Mohamed (2008), Urbanization has been mainly influenced by the poverty status in rural community. People from rural areas believe that they can reduce poverty if they will migrate to the urban areas. Today the urbanization has started to show very high impact. One of the impacts is uncontrollable increase in population in urban areas. Another impact is about infrastructure sustainability. In a period of short time there will be a need of repairing or putting new infrastructure.

According to (Reddy 2000), the energy poverty has been defined as the lack of sufficient choice that would give access to adequate, affordable, effective and environmentally sustainable energy services that could support economic and human development. There is a direct relationship between energy and poverty reduction. Energy technology will reduce respiratory disease in women, reduce burden to women who always have to collect fuel and finally, will increase income in the community.

In customer's perspective, the services which the technology is going to offer will be the one which has impact on their social and economic development. Access to modern services can help in increasing incomes by improving productivity, creating employment, and providing access to the markets (Figure 8.2).

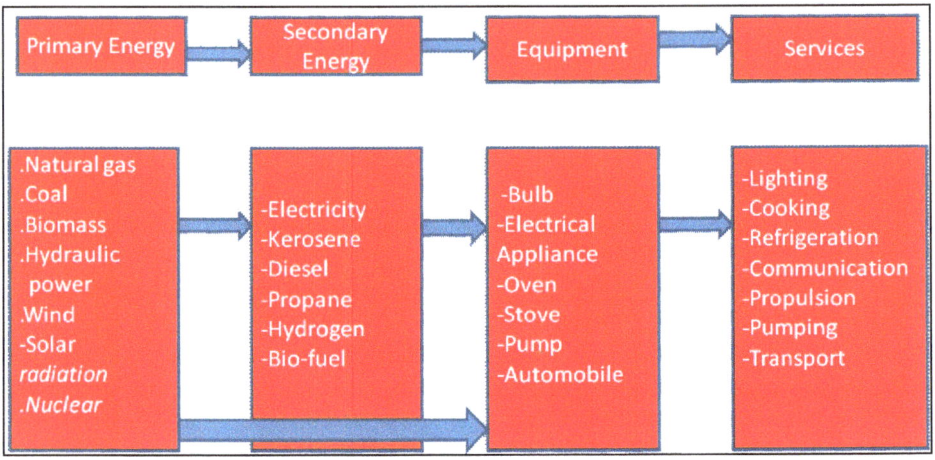

Figure 8.2: The Need for Energy Services

***Source*: Adapted from East African Community, 2006**

According to UNDP (2006), there is a direct link between energy and Millennium Development Goals. Facilitating the access of energy is one of the major strategies in order to achieve millennium development goals. Table 8.1 shows the linkages between energy and achieving millennium development goals:

Table 8.1: Linkages between Energy and Millennium Development Goals

Sl.No.	*MDG*	*Energy Linkages*
1.	Eradicate extreme poverty and hunger	Energy inputs such as electricity and fuels are essential to generate jobs, industrial activities, transportation, commerce, micro enterprises, and agriculture outputs. Most staple foods must be processed, conserved, and cooked, requiring energy from various fuels.
2.	Achieve universal primary education	To attract teachers to rural areas, electricity is needed for homes and schools. After desk-study requires illumination. Many children, especially girls, do not attend primary schools in order to find wood and carry water to meet subsistence needs
3.	Promote gender equality and empower women	Lack of access to modern fuels and electricity contributes to gender inequality. Women are responsible for most of the household cooking and water boiling activities. This takes time away from other productive activities as well as from educational and social participation. Access to modern fuels eases women's domestic burden and allows them to pursue educational, economic, and other opportunities.
4.	Reduce child mortality	Disease caused by un-boiled water, and respiratory illness caused by the effects of indoor air pollution from traditional fuels and stoves, directly contribute to infant and child diseases and mortality.

Contd...

Table 8.1–Contd...

Sl.No.	MDG	Energy Linkages
5.	Improve maternal health	Women are disproportionately affected by indoor air pollution plus water and food-borne illnesses. Lack of electricity in health clinics, lack of illumination during night time deliveries, the daily drudgery and physical burden of fuel collection and transport, all contribute to poor maternal health conditions, especially in rural areas.
6.	Combat HIV/AIDS, Malaria, and other diseases	Electricity for communication such as radio and television can spread important public health information to combat deadly disease. Health care facilities, doctors and nurses, all require electricity and services (illumination, refrigeration, sterilization etc) to deliver effective health services.
7.	Ensure environmental sustainability	Energy production, distribution and consumption have many adverse effects on the local, regional, and global environment. These effects include indoor, local, and regional air pollution; local particulates; land degradation; acidification of land and water; and climate change. Cleaner energy systems are needed to address all of these effects and to contribute to environmental sustainability.
8.	Develop a global partnership for development	The World Summit for Sustainable Development called for partnerships between public entities, development agencies, civil society, and the private sector to support sustainable development, including the delivery of affordable, reliable, and environmentally sustainable energy services.

Source: UNDP 2005.

Current Status of Energy Technologies

Biomass Technologies

Average national wood fuel consumption per capita is around 1.0 to 1.5 cubic meters (m^3) of solid wood. Main consumers of wood fuels include households both in rural and urban areas, small-scale industries and processing of agricultural crops mainly tobacco curing and tea drying. The estimated national annual wood fuel consumption is about 36 millions m^3, which is already over and above the estimated annual sustainable yield of 24 millions m^3 from the forest standing stock. Different energy saving technologies have been inverted or adopted to minimize unnecessary waste of energy. These technologies include improved cook stove technology and charcoal briquettes technologies while biogas technology has been adopted to provide alternative source of energy. These technologies are discussed below:

Charcoal Briquettes Technology

There are different types of charcoal briquettes' technology in Tanzania. Most of them need electricity for operation and different raw materials are used. The promising technology is the one, which is adopted by ARTI (Appropriate Rural Technology Institute – Tanzania). ARTI is an NGO that deals with the development of rural areas through renewable energy. Unfortunately until now, less that 0.1 per cent uses this technology. This NGO was registered in 2007 and one of the technologies adopted (under technology transfer agreement with ARTI – India) was charcoal briquettes technology using the waste as a raw material to make briquettes and doesn't need electricity to operate. The process to make that charcoal briquettes is as follows:

Figure 8.3: The Same Biomass is Filled into all the 4 Inner Drums. The filled inner drums are closed with their lids and placed inside the kiln on the layer of biomass.

Figure 8.4: Additional Biomass is then Stuffed in between, Around and Above the Drums

Figure 8.5: The Lid of the Kiln and the Chimney are Placed Over the Filled Kiln

Figure 8.6: The Area around the Kiln is Cleared of all Dry Biomass and the Kiln is Fired Using the 4 Small Fire Windows Provided

Figure 8.7: The Biomass inside the Kiln will Burn for 1.5 hours

Figure 8.8: Lid of the Kiln is Removed. Any small fires remaining inside the kiln are put off using water. The biomass that was in the Kiln but outside the inner drums has been converted to ash. This ash can be sprinkled on the farms to increase soil fertility.

The char powder is bagged into woven sacks. The kiln and the inner drums are filled once again to start the next cycle of pyrolising the dry biomass into char powder. The time between the cycles can be utilized to manufacture briquettes using the manual extruder. The bags of char powder are weighed, paid for and collected by Joint Environmental Techniques Limited (JET) every 2 weeks. The char powder is transported to JET's briquette manufacturing site located in Dar es Salaam.

Cooking Stove Technologies

The main objective of this technology was to reduce the consumption of charcoal. Different design of improved cook stove came up and the performance of this

Figure 8.9: The Kiln is Lifted to Expose the Inner Drums

Figure 8.10: The Inner Drums are Opened. The biomass inside the drums has been converted into charcoal powder.

Figure 8.11: Motor Driven Screw Extruders with a Capacity of Producing 300–500 kg/day of Briquettes are Used to Manufacture Briquettes

Figure 8.12: The Briquettes are then Laid Out in the Sun to Dry for 2–3 Days, before they are Packed for Sale in the Market

technology has been very well. Some of the improved cook stove includes KUUTE from COSTECH, Jiko Bora from (TaTEDO), Heavy Duty Energy Saver and Rocket Energy from Appropriate Technology.

According to Salvatory Mushi, "KUUTE cook stove reduce by 50 per cent the quantity of charcoal to be used compared to the traditional type of cook stove. It is affordable and most of the artisans have been trained to make this type of cook stove. According to TaTEDO (2008), a total number of households using their improved stove in rural, urban and in the institutions is 1428, 20,040 and 35 respectively. This shows some positive trends in urban areas and to some extent in rural areas.

Biogas Technology

In Tanzania, bio gas technology was started in 1975. SIDO (Small Industry Development Organisation) built 120 floating-drum plants up to 1984. Until 2007 there is an estimate of almost 3000 reported plants installed in Tanzania. According to GTZ (2007), there is no specific statement addressing bio gas in the country except the general policy statement in the energy policy of 2003, which is "Promote energy efficient buildings and facilitate wider application of alternative sources of energy for cooking, heating, cooling, lighting and other applications". Some of the organizations, which deal with bio gas technology include:

- ✰ *CAMARTEC* (Centre for Agricultural Mechanization and Rural Technology) is a government institution located in Arusha under the Ministry of Industry, Trade and Marketing with a vision of developing and disseminating improved technologies suitable for agricultural and rural development, Beside engagements in biogas issues, CAMARTEC researches and carries out training with respect to other energy and agricultural technology.
- ✰ *Katani Ltd.* is a private Company which has managed to build the first Sisal Biogas in the World (150KW) with the support from UNIDO.
- ✰ *MIGESADO* is the Swahili acronym for 'Gas from Cow Dung in Dodoma Region'. It is an NGO for developing alternative energy sources in Dodoma and was started in 1994 and registered in 1999. The head office is based in Dodoma Municipality. The objective of MIGESADO is to conserve the environment by the use of biogas as an alternative source of energy in rural areas.
- ✰ *SURUDE* (Sustainable Rural Development) was registered in December, 1994 in Morogoro. SURUDE is closely associated with Sokoine University of Agriculture and has implemented projects together with other international organizations, such as FAO, DANIDA and SIDA. SURUDE's biogas expertise is mainly focused on tubular plastic digesters, which was later found to be not sustainable in the Tanzania context. In collaboration with Superflex (Denmark), a new plastic biogas system was, for example, tested in Tanzania.
- ✰ *KAKUTE Ltd.* (Kampuni ya Kusambaza Teknolojia) is a Tanzanian-based company, which was established and registered in 1995 with a social justice agenda to reduce poverty through real business opportunities. Currently,

research is undertaken to see if cow dung can be replaced for the residues of Jatropha tree. KAKUTE Ltd. is very active in activities concerning products of the Jatropha tree.

- *ELCT* (Evangelical Lutheran Church in Tanzania) seems to be a very powerful organization, which has a long history in undertaking development projects within Tanzania, in particular with energy projects. As a consequence, ELCT claims to have built 200 biogas digesters themselves and 300 units with masons trained by ELCT.
- *SUDERETA* (Sustainable Development and Renewable Energies in Tanzania) was founded out of ELCT in 2001. Currently, GTZ has a project in Tanzania together with SUDERETA called PRET (Promotion of Renewable Energy Technologies) focusing on the dissemination of solar home systems and improved stoves. SUDERETA's staff consists primarily of one person, however, who seems to have access to ELCT staff. Main activities of the NGO are aimed at solar, wind, mini-hydro, biomass, and biogas energy.
- *FIDE* (Friends in Development Trust Fund) is an organization based in Babati, Manyara and received funding mainly from Belgium and Austria. FIDE used to be active in promotion of biogas in 1992. In 2006, FIDE returned to foster biogas in Manyara region and has currently undertaken projects. This organization is also involved in training the "fundis" to construct biogas plants and house hold members to operate them correctly. As a consequence, FIDE built altogether about 70 plants, 20 each in 2005 and 2006 and already 10 units this year and are planning to build about 20 digesters annually in the future. FIDE follows the approach in financing biogas digesters with a revolving fund. According to FIDE, a 16 m^3 digester cost about 800,000 TZS. 300,000 TZS are financed by a loan from SACCOs with a payback period of 3 years, and the remaining 500,000 TZS are subsidies provided by FIDE.
- *ARTI* (Appropriate Rural Technology Institute) – NGO was registered in 2007. It is focusing on compact biogas technology. The main focus is the urban area whereby until May 2010 there were 43 plants installed.

According to GTZ (2007), bio gas technologies existing in Tanzania are:

CAMARTEC Design

It is fixed and concrete dome design (originally based on Chinese technology, modified in to the Tanzanian context by CAMARTEC). This type has been built over 1000 times in Tanzania and experiences show long lifetimes and relatively high costs. This type has not only been adopted by the Tanzanian private sector, but also other African countries installed this design successfully. Therefore, this design demonstrated technical suitability for African context, but costs remains a barrier to wide-spread dissemination.

MIGESADO Design

It is fixed dome design (originally based on Indian technology, modified to the Tanzanian context by MIGESADO using to a limited extent experiences from CAMARTEC). This type has been built about 850 times in central Tanzania and experiences show long lifetimes and relatively high costs.

Tubular Plastic Digesters

This type mainly built by SURUDE and Kakute Ltd. up to 300 - 500 units exist within Tanzania so far. Price per digester is about $US 150. The advantage of cheap costs is confronted with limited lifetime of maximum of five to seven years. However, the average lifetime is far below. Additionally, the low gas pressure does not allow the use of gas lamps; consequently, biogas can only be used with this technology for cooking. Due to the limited lifetime, trust of customers in tubular plastic digesters is limited. As a consequence, SURUDE, which built about 200 plants of this kind, does not consider this technology as sustainable and, therefore, does not follow this technology any more.

Super Flex Design

For this type very few have been reported to be installed in Tanzania. This design also uses plastic as the main material with modified plastic water tanks accommodating biogas digestion and gas storage. According to SURUDE, the cost is around US$600 with the latest design. However, SURUDE expects limited lifetime due to the use of plastic and this type cannot be produced within Tanzania. Super flex claims that several plastic water tank manufacturers in Dar es Salaam could easily modify their designs to be convertible to biogas. ZALWEDA follows this type of digesters currently building several demonstration digesters in Zanzibar that utilize an automated mixing technique to maximize biogas production with minimized digester volume.

Floating Drum Designs

In early days of biogas dissemination in Tanzania, *floating drum designs* have been installed. Due to the limited lifetime of the metal drum and relatively high costs, this technology was not continued. Consequently, this technology cannot be considered as appropriate or suitable for the Tanzanian environment.

In general, the total number of biogas plants installed in Tanzania falls between 3000-5000. The following chart shows the reported number of biogas installed by respective organizations. Due to poor record keeping some more bio-gas plants might be available. The ARTI data is for 2010 but other information are based on the feasibility study done by GTZ in 2007 (Figure 8.13).

Wind Technology

The wind technology in Tanzania has been adopted mainly for the purpose of rural development. This technology has been in use since the early 1950s with the purpose water pumping for agriculture, domestic and livestock development activities. Most of the wind mills have been mainly installed in Dodoma (25), Singida (36) and Iringa(16) due to the prevailing water problems.

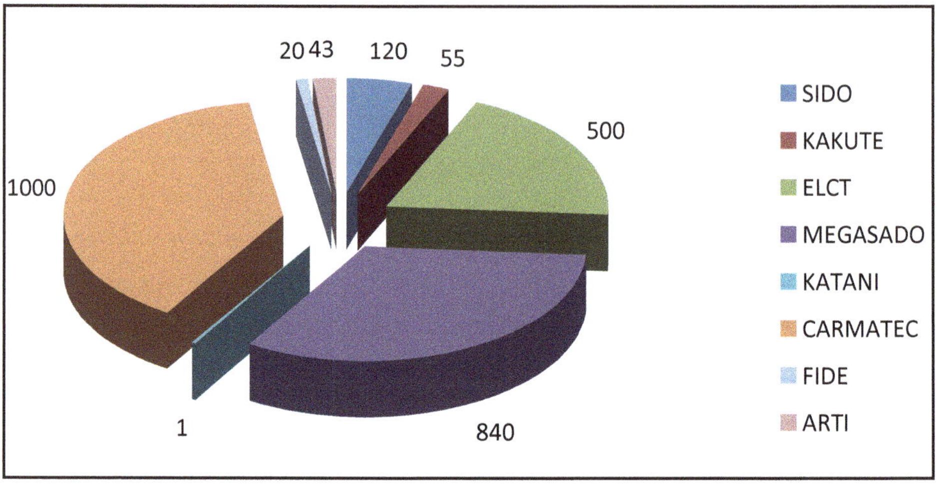

Figure 8.13: Reported Biogas Plants Installed in Tanzania
***Source*: GTZ (2007)**

According to Nzali *et al.* (2003), there are 106 windmills in twelve regions (Arusha, Singida, Dodoma, Tabora, Musoma, Shinyana, Kilimanjaro,Tanga, Lindi, Iringa, Mbeya and Rukwa).Most of them are multi - bladed horizontal axis windmills. In terms of working and not working, 56 per cent of them are not working. This is due to different reasons, including lack of spare parts, lack of maintenance, vandalism, poor wind regimes in places where installed.

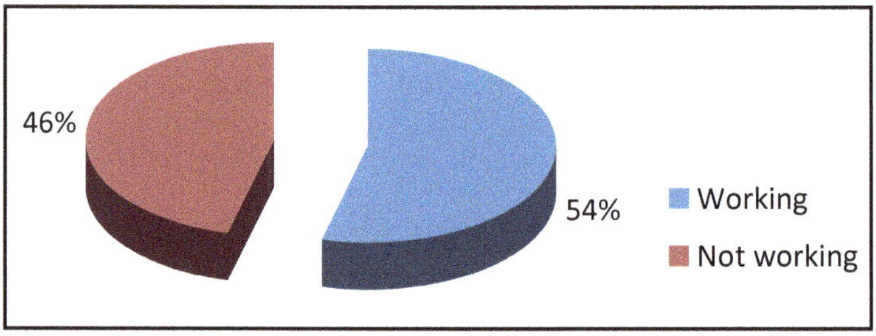

Figure 8.14: Wind Technology in Terms of Working or Not Working
***Source*: Nzali *et al.*, 2006**

In terms of ownership, 8 per cent own by the government, 7 per cent Private, 9 per cent Church organizations and 76 per cent Communities.

Solar Technology

The solar technology has been getting attention for the last three decade in Tanzania due to little or unavailability of energy in rural areas. For Tanzania rural development, this technology is the best due to the fact that the country is in '*Solar Belt*' (2800-3500 hours per year). It can be used for: water pumps, hospitals, schools and other educational institutions and for the tourist hotels in the national parks *e.g.*

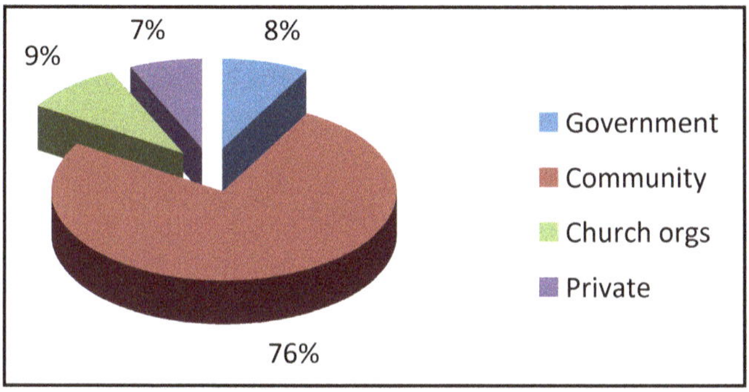

Figure 8.5: Ownership of Wind Technology
***Source*: Nzali *et al.*, 2003**

Mary Leakey's Camp Site (it will be discussed as a case study). Until now, sun drying is the most used technology in the rural areas. In promoting technology transfer, different strategies are in place to make sure the intention of lighting the rural area is successful. Among the strategies mode were;

- ✰ The tax reforms,
- ✰ Increased the involvement of the number of private sectors organizations
- ✰ Tanzania Solar Energy Association (TASEA) was registered in May 2001 with the following objectives:
    - Solar energy market development;
    - Encourage research and education;
    - Lobbying and advocacy;
    - Information and knowledge dissemination;
    - Networking and information sharing;
    - Encourage research and education
- ✰ Establishment of Zanzibar Solar Energy Association (ZASEA) pioneered by TASEA
- ✰ Increase on the NGOs dealing with solar energy such as TATEDO,

The estimated average sales between 2000 and 2005 were about 500-600 PV systems per annum. The system components are imported through various private sector initiatives. The common PV applications in Tanzania are for telecommunications, lighting, vaccine refrigeration, water pumping, powering electronic accessories *e.g.* radios, computers, TVs, etc. The schools, health centers, rural dispensaries and missionaries are the major users. The current price for 360 Watts Solar System (one can use ten tube lights, fridge and Iron for 30 minutes daily) is 4500 USD. This is very high compared to the purchasing power of the rural community.

Technology Impact in Rural Development

The following case studies enlighten the impact of the energy technologies adopted or innovated in Tanzania and the need to facilitate technology transfer to rural area.

Case Example One: Makange Village

Commander is a person who lives in Makange village, he always go to the bush to cut trees and make charcoal. He spends two to three weeks to make charcoal and earn about 40,000TZS per month. He has inherited 40acres, which he cannot develop due to the financial crisis. Commander was lucky to meet ARTI management in his village, and he was lucky to be the pilot of this technology in Makange village. According to ARTI, Commander was advised to start clearing the bush from his farm slowly and use the waste as raw material for making charcoal powder. The charcoal powder is transported to Dar es Salaam for final processing to make charcoal briquettes. Commander is now making an average income of 150,000TZS per month, he is not going to the bush to cut trees anymore, and he has started to plant some crops in his farm. He is expecting to earn more money through his farm too; after harvesting. He is planning to buy a home solar system.

Case Example two- Mary Leakey's Campsite

This camp is one of the famous places in the world. This is the place where the late Dr Leakey found the origins of human being. This place is located within Ngorongoro National Park at Olduvai Gorge. There are a good number of tourists, researchers, academicians who always visit the camp. This camp uses wind generator and solar system. According to the camp attendant "*wind and solar technologies installed has helped us to help our tourists even in late hours. It helps our tourists and us to be able to charge phones and communicate with our friends, relatives within and outside the country. We can watch TV (i.e. Football World Cup 2011) with our tourists and know what is going on in this world. The sales of the kiosk have increased due to the fact that most of the tourists prefer to sleep in this camp. This technology is very good and many of us would like to install in our houses but we cannot afford. It is also very difficult to do maintenance ourselves because we did not get any training after installation.*"

Discussion

In Commander's case study, it was shown that appropriate technology can bring changes to the life of the poor. Commander is not going to the bush any more to cut the trees; this means technology has contributed to saving of our environment. He used to earn a little amount of money while spending a lot of time, and the situation is different now. He is earning almost four times of money he earned in the past while spending less time. He spends rest of the money to develop his farm. Now he is thinking of different ideas of business after he has started to get some incomes.

For Mary Leakey's case study, technology has contributed to the increase in sales to of the Kiosk. Number of evening events has been increased due to the availability of light. Researchers from all over the world who visit the Centre can continue working in late hours to accomplish their plan. Provision of communication and information is also an added value to the availability of these technologies.

Conclusion and Recommendations

Energy technologies are very important for the development of rural area as has been seen in the above case studies. There might be a lot of people like Commander who would benefit from this type of technology. Hence, it is of much importance to focus on the commercialization strategies for these energy technologies and this will only be done by creating efficient and effective network among the government, private sector, technology innovators (R&D) and other stakeholders. There is a need to adopt an Indian Strategy on commercialization; focusing on the credit sales of the technology. This will result in an increase in the participation of the rural community in maintaining these technologies. Creating awareness and involvement of the rural community in planning and implementation of technology transfer process are vital elements towards successful technology adoption and diffusion. Training of a group of villagers in maintenance and repairs is also important because it will not only maintain the technology but also will be a source of income to some villagers. Pricing of the technology is a challenge to the private sector due to the fact that purchasing power of the rural communities is very low.

The encouraging trend is that, for some time now, the political commitment to rural energy problems has been growing. There have been many initiatives and willingness at the international, national and local levels to support the rural energy development. There are also good policies in relation to rural development. Despite this encouraging and supportive atmosphere, there have been proportionately very little attempts actually happening on the ground to bring about the desired impacts in the rural areas. Planning, implementation and monitoring is essential not only for energy development but also for other rural development projects that are in the priority areas in order to bring the changes one wants to make in the rural community. This should go hand in hand with the government support in order to achieve sustainability of the development activities.

Acknowledgement

I would like to thank the Commission for Science and Technology (COSTECH) for giving me this opportunity to represent my Country at this workshop. Also I would like to thank ARTI –Tanzania for their valuable inputs in this paper. I really appreciate the contribution of individual members who were consulted by me from different organizations such as CAMARTEC, TaTEDO, REA and TEMDO. Last but not least I would like to thank everyone who is not mentioned by his/her name in this acknowledgment.

References

1. Bariki, K.,2007, "Monitoring and Evaluation of rural energy development in Kigoma Region" UNDP/GVEP, Dar es Salaam.
2. East African Community, 2006. Regional Strategy for scaling up access to modern energy services in the East Africa.
3. Government of Tanzania, 2007. The Economic Survey, United Republic of Tanzania.

4. GTZ, 2007. Feasibility study for National Domestic Biogas Programme in Tanzania
5. http://ec.europa.eu/development/geographical/regionscountries/countries/country_profile.cfm?cid=tz&type=short&lng=en (accessed 5/7/2010)
6. Kimambo, C., 2006. *Solar Energy: Utilization Status and Constraints in Tanzania* in Energy Resource Book Volume 1, COSTECH, Tanzania.
7. Ministry of Energy and Minerals., 2003. Energy Policy of Tanzania" Dar es Salaam
8. Mohamed, F., 2008. Does the Money Matter? Micro-credit and Women Empowerment in Tanzania, A PhD Thesis at London School of Economics.
9. Nzali, A and Mushi, S., 2006. Wind Energy application in Tanzania in Energy Resources Book Volume 1, COSTECH, Tanzania.
10. Reddy, A., 2000. *Energy and Social Issues,* in "World Energy Assessment", UNPD, New York, ISBN 92-1-126-126-0.
11. TaTeDO.,2008. Annual report for the year ended 2008
12. World Bank., 2009. Environmental Crisis or Sustainable Development Opportunity? Transforming the Charcoal Sector in Tanzania.
13. World Bank., 2010. Enabling Reforms: A Stakeholder-Based Analysis of the Political Economy of Tanzania's Charcoal Sector and the Poverty and Social Impacts of Proposed Reforms in the rural areas.
14. www.gtz.ze (accessed on 26/6/2010)
15. www.mem.go.tz (accessed on 25/06/2010)
16. www.tanzania.go.tz/economicsurveyf.html (accessed on 15/6/2010)
17. www.unhabitat.org/../7803_74079_EAC) per cent 20Regional per cent 20Strategy per cent 20for per cent 20SCALING per cent 20U (accessed on 22nd June 2010)

Chapter 9

Efforts of Nigeria in Application of Science and Technology towards Occupational Villages Development: The RMRDC Approach

Angela C. Nwanya, I. Jolaoso, Moyo O., Obasi Seb C. and Onwualu, Azikiwe P.*

**Chief Scientific Officer (Food Division), Raw Materials Research and Development Council (HQ), Abuja, Nigeria*

**E-mail: angyeke@yahoo.com, aekezie@rmrdc.gov.ng*

ABSTRACT

Nigeria is one of the most blessed nations in the world in terms of the abundance of agricultural and mineral resources. However, it is yet to fully harness these potentials for its development. This situation can be correctly attributed to the slow pace at which the governments accepted the pivotal role that science and technology can play in transforming its largely agrarian economy to an industrialized one, particularly, the rural areas. In recent years, the role of S&T in other sectors of the economy has received tremendous attention. For it to make a meaningful impact on the nation, research and development results need to be effectively transmitted to the rural dwellers. The Co-operative Societies of Nigeria, equivalent to the occupational village development concept, have been the most effective channel of disseminating innovations in S&T to rural dwellers. They are also used as agents of economic emancipation of their members through establishment of productive ventures.

The Raw Materials Research and Development Council (RMRDC), a parastatal under the supervision of FMST has continued to lead in developing new ways of using S&T to address the rural industrialization drive of the government. Some of its programmes such as the resource based processing clusters, upgrading of indigenous technology, capacity building programmes and technology expositions have been used to publicize and promote investment potentials in the country.

Keywords: *Science and technology, Occupational villages, Rural development, RMRDC, Shea butter.*

Introduction

Nigeria is a country in West Africa, which shares boundary with the Republic of Benin to the West; Chad and Cameroun to the East; and Niger Republic to the North. The Atlantic Ocean borders the country to the south indented by lagoons and the immense Niger River Delta. It has a population of more than 140million[15, 18].

The country has nine distinct ecological zones, which can be streamlined into five namely- sahel/sudan savanna, guinea savanna, derived savanna, lowland rainforest/mountain forest and freshwater swamp forest and coastal vegetation[18].

Nigeria has a land area of 98.3 million hectares (ha), out of which, 74 million ha is good for farming, but less than half has been explored. Understandably, agriculture is the dominant economic activity in terms of employment and linkages with the rest of the economy. The major agricultural produce of the country are roots and tubers, cereals, legumes, livestock and fishery, spices and economic tree crops. Its agriculture which suffic food and fibre needs of the population[12] and provide raw materials for agro-industries. It is equally endowed with mineral resources such as crude oil, natural gas, tin ore, lead- zinc ore, gem stones, granites, limestone etc. Table 9.1 shows the major agricultural and mineral resources in the six zones of the country.

More than 70 per cent of the country's population of 140 million resides in rural areas[11] and derives their income from agriculture and agro-allied activities. The sector accounts for 5 per cent of total export, provides 88 per cent of non-oil earnings and contributes to about 41 per cent of GDP and 85 per cent of which is from the crops sub-sector.

About 94 per cent of agricultural output is accounted by the small scale, resource poor, subsistent farmers cultivating less than 1 hectare farm units characterized by the use of hoe-cutlass technology, rain dependent and use of minimum inputs. The soils are also fragile and of low fertility. Consequently, the productivity of agriculture in Nigeria is relatively low considering the hectares of land it cultivates annually.

Nigeria operated an import substitution industrial policy in the 1960s up to the early 1980s. The country exported mostly agricultural commodities with little value addition due to the dearth of a solid technological base to transform these raw materials into industrial inputs. The continuous over dependence on foreign technology as well as industrial inputs in critical sectors of the economy such as oil and gas, solid minerals, and chemical industries, posed a serious security threat to the existence of the country.

Table 9.1: Some Natural Resources in the Six Geopolitical Zones in Nigeria

Sl.No.	*Zones*	*Major Agricultural Resources*	*Major Mineral Resources*
1.	**SOUTH EAST:** Anambra, Enugu, Ebonyi, Abia and Imo states	Non-timber forest products, Roots and Tubers, Oil palm produce, Rice, Fruits and vegetables, Indigenous Spices, Rubber	Lead – Zinc ore, Granite, Crude oil, Bentonitic clay, Laterite, Kaolin, Ileminite, Gypsum, Salt, Coal
2.	**SOUTH SOUTH:** Akwa-Ibom, Cross River, Rivers, Edo, Bayelsa states and Delta states	Sea foods, Coconut, Oil palm produce, Indigenous Spices, Rubber	Crude Oil, Limestone, Baryte, Kaolin, Salt, Bauxite
3.	**SOUTH WEST:** Lagos, Ogun, Oyo, Ondo, Ekiti, Osun states	Cocoa, Kolanut, Timber, Non-timber forest products, coconut	Bitumen, Gold, Limestone, Talc, Phosphate, Bauxite, Granite
4.	**NORTH CENTRAL:** Benue, Kogi, Niger, Plateau, Kwara, Nassarawa states	Yam, Beniseed, Vegetables, Potatoes, Rice, Fruits, Sheanut	Limestone, Ileminite, Gem-stones, Columbite, Marble, Granite, Gold, Kaolin, Talc, Baryte, Salt, Iron ore, Tin ore, Uranium
5.	**NORTH EAST:** Bornu, Gombe, Yobe, Katsina, Adamawa, Bauchi, Taraba states	Rice, Cereals, Sugarcane, Tea, Livestock,	Lead-zinc ore, Soda ash, Diatomite, Coal, Borno
6.	**NORTH WEST:** Kaduna, Kano, Sokoto, Kebbi, Jigawa, Zamfara states	Tomatoes, Pepper, Ginger, Sorghum, Rice, Wheat, Livestock	Gold, Phosphate, Gypsum

Source: Self compiled from various sources.

This uncomfortable situation prompted the country on the need to develop a local science and technology base in the 1970s. Furthermore, in the mid 1980s the need to reduce the country's dependence on imported industrial inputs and technology, led to the establishment of the Raw Materials Research and Development Council (RMRDC) in 1989.

The Council was mandated to facilitate the industrial development of Nigeria through maximum utilization of local raw materials as industrial inputs. This policy thrust of government was in realization of the critical role that S&T had to play in the industrialization drive of the country, which was to be transmitted to the rural areas through the activities of extension workers in the Ministry of Agriculture. Naturally, the initial impact of S&T was on the agricultural sector, of which the rural populace constitutes the major driving force. Consequently, S&T research findings were to be transferred to rural based agro- industrial enterprises. To effectively manage these enterprises, the pivotal role of co-operative societies came into function. These co-operative societies are equivalent to the occupational village development concept in Nigeria.

Many initiatives are currently been tried, which are targeted at micro-enterprise innovations and improvement of management skills through science and technology.

Government agencies in the ministries of agriculture, science and technology and that of commerce and industry are working hard on these initiatives to ensure they succeed.

Poverty and Unemployment Situation in Nigeria

Poverty is a major problem faced by Nigeria. Thus, the reduction of poverty is one of the most difficult challenges that the country is facing at present. The evidence from the 1999 National Bureau of Statistics further buttresses the fact that in Nigeria the number of those in the poverty bracket has continued to increase. It increased from 1.27 per cent in 1980, 45 per cent in 1985, 42 per cent in 1992, 67 per cent in 1996, to more than 70 per cent in 1999[3].

The view of poverty in the rural sector is not new as the nature of rural economy (which is absolute subsistence) has over a long time been a cause of poverty. Furthermore, the poor in Nigeria are not a homogenous group. They can be found among several social/occupational groups[10]. The poverty alleviation measures implemented so far have focused more on specific issues such as provision of food, shelter, potable water, roads and electrification programmes. This integrated approach towards rural development no doubt provided the necessary basic infrastructure that can stimulate the agro-allied small-scale enterprises in the rural areas[3]. Unemployment is also a very serious issue in rural Nigeria and closely linked with the poverty situation. According to a study, many of the chronically poor people only emerge briefly from poverty because of seasonal employment but lack the impetus to contribute to technological change and economic growth[3]. Thus, a pre-requisite for creating decent employment is the transformation of low producing traditional agriculture to labour – intensive high value agriculture and processing and to the growing industrial and service sectors, taking advantage of globalization opportunities.

Agri-business and agro- industries can create wealth and new jobs in Nigeria and Africa at large, and shrink poverty. According to a UNIDO report, agricultural advancement can help producers and companies go beyond local markets and secure a niche in the globalized world and sustainable growth. It is the key to reducing hunger and achieving the Millennium Development Goals (MDGs), seriously threatened by the global financial crisis and soaring food prices.

Rural Industrialization in Nigeria

Although the country is endowed with sufficient natural resources, it is yet to exploit these resources for industrial growth. The problem can be linked to its undeveloped local technology base. The country lacked simple process technology to transform its natural resources to value added products. Consequently, it continued to export its raw agricultural and solid mineral resources and in turn imported the value added equivalents as industrial raw materials at very high foreign exchange. A critical look at the situation in the country will help us appreciate it better.

Current Status

There is little evidence to show that rural industrialization has been pursued as a national strategy in the development of Nigeria[9]. Over the years different governments have initiated various programmes to develop the rural areas through the provision of some physical and social infrastructures. There have also been attempts to encourage industrialization of the rural areas by provision of incentives such as pioneer status, tax holidays etc., by the federal and state governments. However, these have not had the desired impact on the rural communities, because of obvious problems. Rural Nigeria is agrarian in nature. Agriculture, which employs more than 70 per cent of the 140 million Nigerian population is characterized by low productivity growth and thus has not provided enough real employment and income security to the rural population.

Furthermore, the inter-relationship between agriculture and industry has not been recognized until recently. All along there has been an advanced urban economy based on rapid industrialization propelled largely by foreign direct investments in crude oil, manufacturing and service delivery, existing besides a backward rural economy, based largely on agriculture[9]. The critical role S&T had to play in linking agriculture to industry was not fully appreciated. Research and Development (R and D) results from research institutes and tertiary institutions were not translated to commercializable products. However, in recent years the realization of the crucial need for development of S&T propelled agro- allied industries, has resulted in the emergence of many cottage and small scale industries across the country. Notable among them are food processing industries, small scale mining, agricultural produce marketing, metal works etc.

Common Rural Based Enterprises

Given the vast natural resources of Nigeria, the occupations of the villages where these resources are located are largely determined by what they are naturally endowed with. Consequently, agrarian communities concentrate on food processing, while those blessed with minerals are into small scale mining/quarry or craft making. A list of the most common enterprises found in Nigerian villages based on its natural resources is as follows.

1. *Oil Palm*: Production of palm oil, crude palm kernel oil, palm kernel cracking, broom making, thatch roof making, mat making, black soap making etc.
2. *Rapphia palm*: Production of palm wine, "Ogogoro"/local gin.
3. *Cassava*: Production of cassava chips, garri, fufu/akpu, lafun, starch etc.
4. *Cereals and Grains (rice, corn, sorghum, millet, etc)*: Alcoholic and non-alcoholic drinks *e.g.* pito, burukutu, kunu), pap, "agidi", "massa" etc.
5. *Beans (bambara, soya, local beans)*: Okpa, "akara"/fried beans meal, "moi moi" etc.
6. *Oil seeds (sheanut, locust bean, melon, African oil bean seed)*: Production of local seasoning *e.g.* dawadawa, Ogiri; snacks; cooking oils etc,
7. *Salt water*: Local salt processing,
8. *Livestock*: Meat processing into suya, kilishi, leather products *e.g.* shoes, slippers, bags etc; dairy products *e.g.* wara/local cheese, fura/yoghurt etc,

9. *Cocoa*: Cocoa pod juice
10. *Natural sweeteners (T. daniellii)*: Harvesting and sale of leaves for food wrapping.
11. *Non-timber forest products*: Harvesting and sale of raw or processed snails, mushroom, grass cutter, antelope, squirrel, giant maggot etc.
12. *Granite and Stones*: Crushing, sizing and stone picking for sale to construction workers mainly.
13. *Gem stones and other metallic and non-metallic minerals*: Quarry/small scale mining of minerals for export.
14. *Handicrafts*: Besides the main occupations of farming and mining, villagers in their spare time also use some of the agricultural "wastes" and solid minerals to make household items or arty-crafty for the rich and the tourists. Some of these handicrafts centres[8] are:
    (*a*) *Benin*: World renowned ancient bronze arts, iron implements, ivory, ebony and wood carvings
    (*b*) *Oyo*: carved calabashes (large hollowed gourds)
    (*c*) *Kano*: Tie-dyed textiles, leather wears
    (*d*) *Ihiala*: Basket making
    (*e*) *Abia*: Akwete woven material
    (*f*) *Abeokuta*: Adire and Aso-Oke textile designs
    (*g*) *Oyo/Osun*: Stone works

Other handicrafts practiced in the Nigerian villages include- braided and sewed materials, musical instruments, local mats and brooms, weaving, pottery[8,20], ceramic wares, traditional ware embroidery, local beads for dancers, ceremonies and tourists.

Challenges of Rural Industries

It goes without saying that the establishment of an industrial outfit is usually froth with challenges, more so when it is to be located in a rural setting. Some of the problems rural industries in Nigeria face can be summarized as:

1. *Infrastructural problems*: Nigerian industries have to contend with some infrastructural problems. Top on the list of this problem is electricity supply. However, with the assurances and concrete actions being taken by the present government at the state and federal levels, it is hoped that the problem of infrastructure will soon be adequately addressed.
2. *Financial limitations*: Access to credit is another major factor militating against rural industrialization. Rural borrowers still encounter difficulties in accessing credit from formal financial institutions, because the institutions were not designed to function as "true" financial intermediaries that mobilize deposits to make loans available for their economic activities. Traditional savings and credit associations are more prevalent in the rural areas. However, these financial intermediaries are not accommodated in the country's money market development policies[7].

3. *Labour problems*: Rural – urban migration, the movement from rural to urban areas is another problem militating against rural industries. Vibrant youths within the employable age group, have left and are still leaving the villages en mass for greener pastures in the cities. Migration of young adults from rural areas thus places a great burden on the aged farmers[1].
4. *Managerial problems*: It is a known fact that most operators of rural enterprises lack the requisite managerial skills to successfully run the business operations. Most times, cognate experience is not considered in recruiting managerial staff.
5. *Quality*: Products from rural enterprises often have crude finishing unattractive to urban consumers. Due to lack of quality control practices, some food products from villages cannot attract prime prices, limiting the income power of the villagers. Furthermore, they often lack resources to meet the requirements of regulatory bodies such as the National Food and Drug Administration and Control (NAFDAC).
6. *Access to markets*: Due to limited capital, illiteracy, inadequate market information, villagers are often at the mercy of the middle men in order to get their products to urban market or for export. They do not readily get the relevant information or technical assistance needed to improve their products.
7. *Culture*: Due to stiff resistance to outside influence on local cultures, villagers often refuse to change their process technologies or designs to meet the demand in markets outside their domain. Resistance to innovation means inability to meet up with changing sophisticated tastes and susceptibility to enterprise death.

Other challenges of rural industrialization (for sustainable development) are:

- High population density
- Vicious cycle of poverty[12]
- High level illiteracy
- Low social interaction
- Strict regulatory requirements
- Superficial implementation of government programmes (poor supervision, vague objective)
- Armed communal conflicts
- Inconsistency in government policies
- Corruption by public officers supposed to implement programmes etc.

Rural Industrial Policy of Nigeria

Various policies of the Nigerian government to improve the living condition of the rural areas show the zeal of different governments and non-governmental organizations, which led to the proliferation of development agencies[12]. Huge funds and material resources have been employed; however the peoples' expectations are not being fully met. The fact being that policy stability was required to achieve sustainable development of raw materials in the country. Policy instability has been one of the limiting factors to national development in Nigeria. Therefore, the rural industrialization policy is not an exception.

Over time, successive governments have embarked on several programmes targeted at rural development. Other approaches have been by Non- Governmental Organizations (NGOs), co-operatives, individuals through private initiatives, corporate bodies as well as International Organizations. The various programmes[12,13] initiated and chiefly targeted at the rural sector by government include the following:

- ☆ National Accelerated Food Production Programme (NAFPP)
- ☆ River-Basin Development Authority (RBDA)
- ☆ Agricultural Development Programme (ADP)
- ☆ Operation Feed the Nation (OFN)
- ☆ The Green Revolution
- ☆ Agricultural Credit Guarantee Scheme (ACGS)
- ☆ Directorate for Food, Road and Rural Infrastructure (DFFRI)
- ☆ Better Life for Rural Dwellers (BETTER LIFE)
- ☆ National Agricultural Insurance Corporation (NAIC)
- ☆ National Directorate of Employment (NDE)
- ☆ National Agricultural Land Development Authority (NALDA)
- ☆ National Poverty Eradication Programme (NAPEP)
- ☆ National Rural Roads Development (NRRDF)
- ☆ Rural Banking Scheme (RBS), including the Peoples Bank of Nigeria as well as Community Banks and Micro-Credit Banks
- ☆ Family Support Programme (FSP)
- ☆ Universal Basic Education (UBE)
- ☆ Rural Infrastructural Development Scheme (RIDS)

'Better Life for Rural Women' Programme

This programme was initiated by the wife of the military head of state, the late Dr. (Mrs) Maryam Babangida in 1986. One of the main objectives of the programme was to stimulate and motivate rural women towards achieving better living standards and sensitize the rest of Nigeria[18]. It was also to inculcate the spirit of self –help development, particularly of education, business, art and agriculture and small scale enterprises. As a matter of fact this was the first national programme that brought rural Nigeria to public limelight and attention. The programme massively mobilized rural women into co-operative societies. These co-operative societies were involved in processing activities such as fish smoking, cassava processing, handicrafts, soap making, pottery[18,] training and development programmes including seminars on science and technology for the girl child; supply of process equipment; guarantee of financial assistance for members; supply of farm inputs among other forms of help.

Agricultural Research and Rural Industrialization

The national agricultural policy for Nigeria approved in 1998 was premised on the expected role of the economy. The goal of the policy is the structural transformation necessary for the socio-economic development of the rural areas. The specific objectives of the policy were the attainment of self sufficiency in basic food commodities, increased production of agricultural raw materials, integrated agricultural and rural development and combating environmental threats to food security. The Nigerian Agricultural policy also focuses on food production, agro-industry, environmental conservation, agricultural credit and insurance, agricultural mechanization, poverty alleviation and agricultural information and management.

The place of agricultural research in Nigeria's quest to reduce poverty and industrialization has never been under estimated. To that effect, National Agricultural Research Institutes, Colleges and Universities of Agriculture were established in the bid to develop modern practices in agriculture for productivity as to furnish the agricultural sector with new techniques/technologies and manpower. World Bank assisted Agricultural Development Programmes (ADPs) have been pivotal in communicating these new technological findings to the villagers through their extension officers. However, the full realizations of these laudable objectives have been hampered due to the absence of articulated extension policy in the country.

It should be noted that researches which produce new models and methods in agriculture are not sufficient on their own without a chance of translating them to physical and practical reality. This is the point where agricultural extension gains prominence. It is the link between research and application as it informs, educates, communicates and teaches new methods, models and techniques to end users (farmers).

In order to effectively transfer these new knowledge and technologies, farmers have been encouraged to form farmers associations. To further empower these farmers, they were taught other profitable venture such as agro-processing techniques. However, due to inability of individuals to mobilize all the necessary resources to actualize these projects, they were encouraged to form co-operative societies. These community based co-operative societies have been instrumental to the uplift of the living standards of many rural dwellers. The membership of the societies cuts across families and gender.

It is worthy of note that Nigerian women form the majority in the rural areas and are involved in all stages of agricultural enterprise. In a study carried out in Oyo State in Western Nigeria, it was discovered that 50 per cent of the women were involved in planting, 97 per cent in weeding, 85 per cent in harvesting, 91 per cent in transportation, 92 per cent in processing and 79 per cent in marketing[5]. In order for adoption of innovations and improved varieties to the development of agriculture, the adoption behavior of women should be taken to cognizance in any rural based initiative. It is suggested that membership of co-operative societies could be one of the strategies to improve their adoption of innovations.

Status of Co-Operative Societies in Nigeria

In Nigeria, a co-operative society forms the hub of the nation's concept of occupational village development. Members of the same family can form a co-operative while inter- family co-operatives are possible and are in the majority. Co-operatives have existed in Nigeria for long. They were borne out of the need to meet the socio-economic needs of members. While some parts of the country have well developed and sophisticated co-operatives, others are still faced with developmental challenges. The state Ministries of Commerce and Industry and Agriculture have the statutory responsibility to co-operate in Nigeria.

The co-operatives are businesses owned and controlled by the people who use its service. They finance and operate the business or service for their mutual benefit. Co-operatives are seen as voluntary organizations established for the pursuance of economic, social and political interest of members. The International Labour Organization (ILO) in its studies and report series, No. 57 of 1960 on co-operative management and administration viewed the co-operatives as an association of persons usually of limited means, who have voluntarily joined together to achieve a common economic end through the formation of a democratically controlled business organization, making equitable contribution and accepting a fair share of the risk and benefits of the undertaking[17].

The major emphasis on the cooperative system is the self-help, thus people cooperate because they realize that it is extremely difficult to achieve some goals by working alone. Undoubtedly, the best way of pushing back the limit of economic problem of scarcity is by working together. The co-operative business is to afford members particularly the poor, to come together to promote their socio-economic well-being; a feat, which a single individual working alone cannot achieve if it is possible at all for him to do. Co-operative societies are based on the value of self-help, self-responsibility, democracy, equality, equity and solidarity. Each member of a co-operative organization is an equal owner. The rules and regulations formulated and applied by Rockdale equitable pioneers has become the co-operative principles worldwide[17] including Nigeria under which successful and aspiring societies must logically follow the facts given below. These are identified as:

- Open membership
- Democratic control
- Limited interest on capital
- Patronage refund
- Political and religious neutrality
- Case trading at market prices
- Education of members
- Sales of pure and unadulterated goods

The International Co-operative Alliance (ICA) principles (1995)[17] were modified inter alia.

- ☆ Voluntary and open membership
- ☆ Democratic members control
- ☆ Autonomy and independence
- ☆ Education, training and information
- ☆ Cooperation among cooperatives
- ☆ Concern for community

At the rural level, those who introduced formal co-operatives realized that individual farmers were too small in terms of farm holdings, total production and volume supplied. Thus, they encouraged co-operatives which facilitated quick transfer of technical knowledge of major commodities to groups of farmers. They demonstrated the efficacy of the co-operative approach to problem solution. They helped to nurture small groups with little financial means into larger groups by encouraging the creation of unions out of the societies.

Co-operative societies are potentially important instruments of social transformation, especially in the rural areas. Their methods have proved to be useful in achieving increased domestic production of food, industrial raw materials, manufactured products and equitable distribution of farm inputs, farm products and other commodities. Rural co-operative societies in the country engage in entrepreneurial activities in economic development such as "garri" processing, fish processing, palm oil processing, mat weaving, boat making, fashion designing, blacksmithing, pottery and craft making[3] to mention a few.

Statistics are not available, but is a general fact that the cooperative is second only to government in providing employment, particularly in the Southwest of Nigeria. It is also worthy of note that most rural co-operative societies in Nigeria are women dominated. Consequently, any plan to improve on the operations of rural co-operatives should give priority to the peculiar problems of women.

Problems of Co-operative Societies in Nigeria

Several factors have been identified as being responsible for poor performance, declining and death of many cooperatives. These problems faced by the cooperative movement in Nigeria are generally due to political and socio-economic factors17. Notable among them are:

1. Lackof adequate working capital
2. Bad leadership and succession problems characterized by mismanagement
3. Lack of modern business techniques and liability to compete
4. Bad record keeping and shortage of supervisory staff
5. Inadequate capital base to cope with the need of SMEs
6. Proliferation of weak cooperative societies, which results from lack of coordination of activities.
7. Inconsistency on the part of government in supervision and channeling fund through the cooperative societies

8. Widespread illiteracy and mass ignorance of the societies
9. Out-dated cooperative laws
10. Government interference and manipulation of cooperative programmes, and inconsistency in government policies
11. Corruption and embezzlement of the cooperative fund by leaders and paid workers

The land ownership situation in Nigeria is still a problem for success of rural enterprises. Rural Nigerians do not have proper records and titles that can be used as collateral to raise capital. Though rural dwellers may have large expanse of land bequeathed to them, such property is mortgage meaningless since they cannot the land to obtain loans for business ventures. The ambiguous role of government on cooperative societies is another major factor. The co-operative method remains the vibrant economic technique of poverty eradication, wealth creation, job creation and rural development and SMEs financing and should be assisted by every means; by any progressive government in Nigeria.

Nigerian Science and Technology Policy and Rural Industries

The inception of the Federal Ministry of Science and Technology (FMST) was as far back as 1966 with the promulgation of the first Science and Technology Policy. The first Science and Technology Policy led to the establishment of the Nigerian Council for Science and Technology (NCST) [16]. In the late 1970s, the NCST was transformed into the National Science and Technology Development Agency (NSTDA) [24]. Exhibiting the importance of the sector, in1979 the stature of Science and Technology was enhanced by the creation of the Ministry of Science and Technology. According to the 1986 National Policy of FMST, the Ministry was to "...co-ordinate and undertake scientific and technology research and development". This included both inward and outward looking efforts such as "...integrating foreign technology into the culture of the country; upgrading indigenous technology; and promotion of international co-operation in Science and Technology". Over two decades since its inception, the FMST has grown to be strengthened. In 1999, the Ministry witnessed more stability[16]. Today it is a ministry central to the present administration's economic development plans, which plays the lead role on issues related to scientific research and technological development in the country.

Parastatals of FMST

To facilitate actualization of the FMST mandate, it supervises the activities of its parastatals charged with aspects of S&T as it affects different sectors of the economy[21]. Some of these parastatals include the following:

1. Federal Institute for Industrial Research, Oshodi (FIIRO)
2. National Building and Road Research Institute (NBRRI)
3 National Agency for Biotechnological Development Agency (NABDA
4. National Centre for Genetic Research and Biotechnology (NCGRAB)
5. National Centre for Technology Management (NACETEM)

6 National Information Technology Development Agency (NITDA)
7. National Office for Technology Acquisition and Protection (NOTAP)
8. National Space Research and Development Agency (NSRDA)
9. Nigerian Institute for Trypanosomiasis Research Institute (NITRI)
10. Nigeria Atomic Energy Commission (NAEC)
11. Project Development Institute (PRODA)
12. National Agency for Science and Engineering Infrastructure (NAENI)
13. Raw Materials Research Development Council (RMRDC)

The RMRDC is mandated to facilitate the emergence of a strong industrial and technological base in Nigeria through the creation of a strong domestic source for necessary raw materials. Some of its activities are: sponsorship of R and D in projects, which eventually lead to establishment of process industries; upgrading of indigenous technologies; capacity building in micro, small and medium scale enterprises; boosting of improved varieties of industrial crops among farmers; and development of improved process equipment for local industries (*See* Figures 9.1–9.4 for some of these process equipment).

Support Mechanisms by S&T Agencies to Promote Rural Enterprises in Nigeria

Most agencies in S&T sector do not have programmes specifically targeted at the rural settings, although their impacts could be felt in the long run. That notwithstanding, the activities of some S&T agencies have been known to positively affect village enterprises. Two of such S&T agencies are the National Board for Incubation Centers and the Raw Materials Research and Development Council.

Technology Incubation Centers

The National Board for Technology Incubation Centers is the supervisory agency in charge of Technology Incubation Centers (TICs) in Nigeria. The National Technology Incubation Programme (NTIP) is a veritable institutional mechanism for commercialization of research and development (R&D) results. It is an integrated support programme designed to assist budding entrepreneurs in the development of new technology-based firms both start up and fledglings. It seeks to effectively link talent, technology, capital and know-how to leverage entrepreneurial talent in order to accelerate development of new companies and speedy commercialization of R&D and innovation. It also helps in value reorientation by creating an environment for changing the attitude towards personal initiatives, creativity innovation, risk-taking and entrepreneurship.

TICs develop demand driven and commercializable technology, science innovations and R and D results in Nigeria leading to the manufacturing industries of different scales for mass production. TICs train entreprenures[4]. There are more than twenty TICs across the country. These are located in Abuja, Aba, Kano, Calabar, Minna, Warri, Maiduguri, Gusau, Sokoto, Nnewi, Uyo, Birinin Kebbi, Igbotaka, Bauchi, Ibadan, Benin, Okwe Onuime, Abeokuta, Yola, Jos, and Yenegoa[21].

Figures 9.4–9.4: Pictures of Some Nigerian-Made Improved Equipment for Processing Food Materials and Solid Minerals

Figure 9.1: Palm Nut Digester with Fibre Separator

Figure 9.2: Melon sheller

Figure 9.3: Cassava Harvester

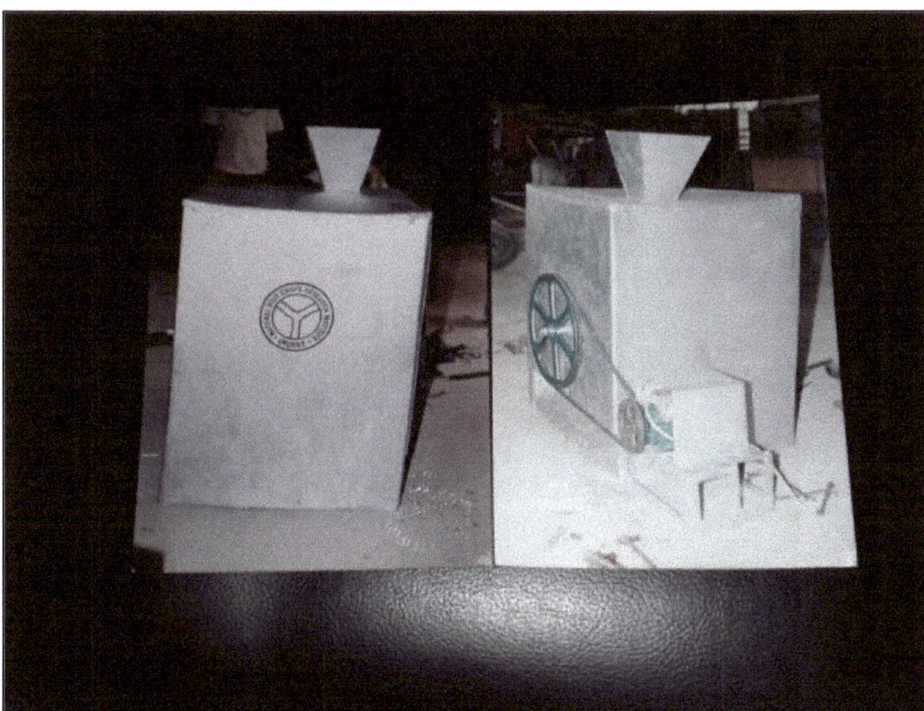

Figure 9.4: Ginger Splitting Machine

It is interesting to observe that all these TICs are located in state capitals or major towns in the states of the federation. However, there are plans on ground to replicate these centers in the rural areas of the country as soon as possible with the active involvement of the states' and local governments' administrations.

Raw Materials Research and Development Council Intervention Strategies

The Raw Materials Research and Development Council (RMRDC) undoubtedly is the major focal parastatal in the Ministry of Science and Technology in Nigeria. It propels the execution of many science based initiatives in the country's industrial development drive using local community based raw materials. It executes many projects most of which are based in rural areas where they serve as pivotal projects for others to emerge. Given below is a brief summary on some of these programmes.

Upgrading of Indigenous Technologies

In the early 90s, the Council carried out a nation-wide survey of indigenous technologies across the country. The essence of the survey was to have a graphic picture of indigenous technologies as practiced at the grassroots with a view to upgrading processes that make local technologies inefficient and ineffective. Characteristics of these traditional process technologies include – unhygienic, laborious, crude, low quality, time consuming, low efficiency, poor packaging, and heterogeneous product quality *i.e.* inconsistency in quality, among others.

The Council's interventions were to tackle the problems identified above. The findings resulted in sponsorship of the fabrications of numerous improved equipments, which were given to the educational institutions, rural co-operative, individuals etc. Some of these machines can be seen in Figures 9.5–9.6.

Among projects successfully executed include:

- Shea butter processing
- Spices processing
- Salt processing
- Mini-dairy processing
- "Kilishi" making machine
- Fish smoking
- Aso-oke (traditional hand-woven textile) Weaving Loom Development (on-going)

Boosting of Industrial Crops

Under this programme, the RMRDC has sponsored the relevant National Research Institutes to produce improved varieties of industrial crops, which are given to the rural based farmer groups. Some crops that have been boosted include cassava, sesame seed, ginger, cotton, castor seed, soya bean, pineapple, kenaf etc. The Council through its state offices encouraged the formation of co-operative societies where none existed.

Figure 9.5: Improved Spices Processing Hammer Mill being Handed Over by officials of RMRDC

Figure 9.6: Improved "Kilishi" (Dried Meat) Processing Machine

Figure 9.5–9.6: Results of Some R&D Projects on Upgrading Indigenous Technologies Sponsored by RMRDC

R&D Activities

The Council has sponsored many R&D research proposals, which have been commercialized and transformed to small scale projects that can be conveniently promoted in the rural areas. For example, soy fat and sesame seed oil projects handled by the Technology Incubation Centres, Agege, and Lagos. A few others are the production of traumatic, castor oil, alkyd resin from rubber seed oil.

Capacity Building

The Council organizes training workshops and seminars and entrepreneurial programmes for participants from villages such as farmer groups/co-operatives, minors and students, on improved technologies, products and investment opportunities.

Science and Technology Innovation Systems

The ability of a nation to achieve sustainable economic growth is to a large extent, depends on the level of its technological advancement and the ability of its science and technology expertise to initiate innovative mechanisms. Innovation capability consists of the skills, knowledge, resources, creativity and attitude that are needed to continuously generate and translate ideas into marketable new products, processes and services. A firm that intends to cope with the future must be innovative.

The four key elements of Nigeria's National Innovation System (NIS) are namely, education and research; industrial production; finance and public policy; and regulation[26]. There exist weak interactions among these elements, each one carrying out its programme without giving adequate information to the other elements. Furthermore, it is interesting to note that data on most of the standard S&T indicators are not available in Nigeria; and where they are, such data exist in no systematic manner. This has made it unnecessarily difficult to assess the country's performance in S&T compared to other countries. More difficult is its impact on rural occupants.

The issue of incorporation of indigenous knowledge in our innovation system has always been an issue. The policy makers and funding agencies often face the questions *viz.*, a) should indigenous knowledge systems (IKS) activities be supported in development projects?, and b) should investment be made in developing indigenous technologies or develop new technologies[2]?

According to an FAO report, the problem of commercialization of indigenous technology is that the development of indigenous technology by no means guarantees that it is going to be integrated into the main stream of industry and business in the respective developing country. Experiences of a number of developing countries such as India and Korea reflect a variable degree of utilization of results of national research and development (R&D) efforts within the business sector in general and within the small industry, in particular. The facts identified as crucial in the process of commercialization of indigenous technology include the R&D capabilities and the management of the R&D function; extend of multi disciplinary approach; the strength of the link between R&D efforts and applied industry problems; the expertise of the R&D personnel; the orientation of the R&D function towards the educational system;

and the needs of the country concerned etc. Some less encouraging factors could, however, restrain the process considerably. For example, the commonly observed absence of demand for technological innovation, the pre-occupation with technical excellence at the cost of commercial reality, conflict between the interest of industry and that of R&D institutions, technological gaps between R&D institutions and industry, the observed time lag in technology transfer, and lack of substantial follow up.

In Nigeria, the innovations are often not readily available to farmers, and where available, a few farmers, usually excluding women, have access to them. Users often find such innovations are difficult to maintain due to lack of maintenance and skill training[11]. It is also difficult to say that if explored innovatively, indigenous technologies can become an integral part of developmental projects. From the perspective of national competitiveness, systems of knowledge that are less accessible to others offer a potential competitive advantage.

In realization of this fact, the RMRDC embarked upon projects that geared towards value addition, by improving indigenous technology and building capacity for enhanced research and development of the industrial sector as already highlighted previously. Furthermore, it is worthy of note that the government of Nigeria with the support of Japan and technical assistance from UNSECO, is conducting a large-scale reforms of its S&T and Innovation system. The plan involves the following components[23]:

1. A thorough review of the functions of Nigeria S&T system, coupled with a preliminary evaluation of the performance of the country's government and academic S&T institutions,
2. In-depth international review of the programmes of the research institutes and universities,
3. Human resource development programmes designed to upgrade skills in the area of research management of technical change,
4. Development of a proposal for financing mechanism of S&T in Nigeria.

Impact of Science and Technology, Innovation Systems on Rural Industries

No matter how weak the link between Science and Technology and rural industrialization in Nigeria, it has nonetheless made considerable impact on entrepreneurial activities at the grassroots. Many hitherto tedious daily food processes have been made easier by the intervention of improved technologies. There are also improved non-food processes in use. Table 9.2 gives a picture of how Science and Technology is being applied by the rural communities of Nigeria to process cassava into many local foods. The improved technologies identified in the table are used mainly in commercial enterprises operated by co-operatives in the villages due to the high costs involved. The local people, to a large extent, still use some of the crude process tools mentioned for domestic food preparations.

In order to reach this stage, there have been the co-operative efforts of the national and international research institutes, government agencies, local fabricators, co-operative groups, NGOs, community and faith based groups. Let us review two processing technologies that have had a great impact on Science and Technology in Nigeria.

Cassava Processing

Cassava (*Manihot esculata*) can be decribed as the main staple crop in Nigeria, since virtually all communities in Nigeria cultivate and process it for food and other uses. Cassava production, processing and marketing to the domestic market is long standing. Both women and men are actively involved in various stages of the cassava development. At the rural level, some processors have demonstrated their ability to adopt low-tech, low cost improvements to processing of cassava such as the construction and use of water tanks and "double fermentation". A processing technology has also been developed for village level production of dried *fufu* flour using a simple drier that can operate in areas with or without electricity[25].

The fabrication of user-friendly equipment for cassava processing in Nigeria has also been witnessed a renewed interest. Local fabricators assisted by research institutes and tertiary institutions developed these equipment. Promotional and publicity activities were carried out by extension workers along with government agencies. RMRDC has promoted improved cassava equipment through its biennial TECHNOLOGY EXPOSITIONS since the early 1990s. A list of some of the traditional equipment alongside their improved counterparts in use across the country for processing cassava is indicated in Table 9.2 below.

Table 9.2: Traditional and Improved Cassava Processing Technologies/Equipment being Used in Rural Areas in Nigeria

Processing Stages	*Traditional Technology*	*Improved Technology*
A. GARRI		
1. Peeling	Knife made of bamboo, flint or metal	– Mechanical peeler – Motorised peeler – Hand peeler, hand rasper
2. Washing	Local Calabash bowl	Aluminium tank
3. Grating	Rough stone, prickly trunk of palms sheet/tin iron pierced with nail on one side	Mechanized grater, Motorized grater, Hammer mill, disk grater
4. Fermentation	Heavy stone on heavy weighed cloth or nylon bag	Batch fermentation in aluminium tank, locally made hydraulic or mechanical
5. Dewatering or Pressing	Heavy stone on heavy weighed cloth nylon bag (for several days)	Hydraulic jack press, screw press, parallel board press, upgraded traditional press for few minutes
6. Sieving	Woven basket, suspended cloth pieces holding mash	Improved pulverizer *e.g.* drum sieve, rotating sieve
7. Frying/ Revasting	Cast iron pan oven over wood fire	Upgraded roaster, solar dryer, kiln type dryer
8. Sifting	Woven basket	Improved pulverized and sifter

Contd...

Table 3.3–Contd...

Processing Stages	*Traditional Technology*	*Improved Technology*
B. LAFUN		
1. Peeling	Knife made of bamboo, flint or metal	Mechanized peeler, motorized peeler, hand rasper
2. Soaking	Local Calabash	Aluminum tank
3. Pulverizing	Woven basket	Improved pulverizer
4. Dewatering	Heavy stone on heavy weighed cloth or nylon bag	Hydraulic press, Mechanical press
5. Drying	Cast iron pan over wood fire	Drum dryer, solar dryer
C. STARCH		
1. Peeling	Knife made of bamboo	Mechanized peeler,Cassava filter, Motorised peeler
2. Washing	Calabash bowl	Aluminium tank
3. Grating	Sheet or iron pierced with nail on one side	Power grater, motorized grater, disc grater
4. Dewatering	Heavy stone on heavy weighed cloth or nylon bag	Hydraulic press, screw press
5. Drying	Cast iron pan over wood fire	Engraved fryer, solar dryer
6. Packaging	Local jute bag	Scaled polythene bags
D. FUFU		
1. Peeling	Local knife	Hand peeler – (mechanized)
2. Washing	Local Calabash bowl	Aluminium tank
3. Grating	Rough stone	Motorized grater, Rotatary grater
4. Dewatering	Heavy stone on heavy weighed cloth	Mechanized press,Hydraulic press
5. Packaging	Local jute bag	Hydraulic polythene bag

Source: Odebode, S. (2008), Journal of International Women's Studies vol.9 No.3.

Shea Butter Processing

Nigeria is one of the leading world producers of shea butter, which is a product from the shea tree (*Vitellaria paradoxa*). It grows wild mainly in the North Central states of the country, namely in FCT, Nassarawa, Kwara and Niger States. Traditionally, women and children process shea nuts into shea butter through a process that is time consuming, strenuous and inefficient in terms of oil yield.

In order to improve on the process technology of the shea butter production, the RMRDC intervened in two ways namely 1) to eliminate the drudgery (but mimicking traditional way) in the process and 2) to increase the production capacity of the women groups. While the Council aimed at improving on the indigenous technology, it was also careful not to alter the quality of the shea butter (only attainable using traditional method of processing), which has attained premium price at the international market.

Two Shea Butter women co-operative societies in Abaji, FCT and Abakweji Ararumi, Moro LGA, Kwara State were accessed. The method by the Abakweji women co-operative society was adopted for improvement. To appreciate the extent of the Council's intervention; the traditional shea butter production process will be described.

1. *Nut collection and cleaning*: While some of the women groups pick the ripe shea fruits, the pulp of which is eaten and the seeds dried and shelled to obtain the nuts, majority prefer to purchase the dried nuts from the market. Women processing shea nuts normally target the next market day (in four days time) when they will sell the butter. The nuts are washed to clean before processing.
2. *Roasting*: The cleaned nuts are roasted (optional)

Figure 9.7: Moulds of Shea Butter Ready for the Market

Figure 9.8: Shea Butter Decanting Basins

Figure 9.9: Shea Nut Hammer Mill

3. *Crushing*: The nuts are then reduced in size by crushing in a mortar.
4. *Grinding*: The crushed nuts are then ground in a mill to finer particle sizes.
5. *Kneading*: The slurry from the mill is allowed to cool and poured into a basin where kneading with the hand (alternatively, some women use their feet to knead) takes place. Kneading helps to extract oil from the milled nut.Kneading is continous until the slurry changes colour to cream as shown in the picture above.
6. *Evaporation*: The forming shea butter is heated on fire to remove excess water.
7. *Sieving*: The liquid obtained after heating is sieved to remove particles of dirt.
8. *Boil*: Water is added to the liquid to clarify the butter
9. *Second evaporation*: The boiled shea butter is evaporated again.
10. *Cool and solidify*: The evaporated butter is allowed to cool and solidify into a crude shea butter.

RMRDC Intervention in Shea Butter Processing

1. *Crushing*: RMRDC developed a crusher, precisely a hammer mill fabricated by the African Regional Centre for Engineering Design and Equipment Manufacturing (ARCEDEM), Ibadan, which was used to replace the traditional mortar and pestle used for crushing roasted shea nut. It could crush in 15 minutes as otherwise the women will take one day to crush. A disc mill was adapted for milling crushed shea nut by enlarging the apertures to the milling chamber. This process replaced the traditional method of using grinding stone.
2. *Kneading*: The Council developed a kneading machine which increased the yield of butter. The kneading machine replaced the use of hands or legs for kneading in the traditional method.

The improved technology introduced by the Council enabled the women to produce a batch of shea butter in a day as against four days it took earlier when using traditional method.

Cluster Development in Nigeria: RMRDC's Approach

Clusters are geographic concentrations of interconnected enterprises, specialized suppliers, service providers, and associated institutions in a particular field that are present in a region. Clusters arise because they increase the productivity, which the enterprises can compete with. The development and upgrading of clusters is an important agenda for governments, companies, and other institutions. Cluster development initiatives are an important new direction in economic policy, building on earlier efforts in macro-economic stabilization, privatization, and market opening, and reducing the costs of doing business.

The cluster initiative was born out of the desire to make Africa nations relevant in the contemporary global economy where innovation and technological applications play a significant role.

Cluster concept is an industrial development strategy, which will give rise to the creation of a 'community of business' located together, where members enjoy collective benefits. It is a strategy to achieve an enhanced environmental, economic and social condition for the survival of industrial/productive enterprises in Nigeria.

In fact, apart from the large industrial estates in the big commercial cities in Nigeria, the small clusters scattered in different parts of the country also develope spontaneously. In most cases these clusters came into existence over a period of time as entrepreneurs got increasingly drawn together due to availability of space, or skill or market in a given location. Examples include the foot-wears, leatherworks and garment clusters in Aba, the computer village in Otigba, Lagos, the auto and industrial spare parts fabricators of Nnewi, the leather tannery in Kano, etc.

Establishing Raw Materials Industrial Cluster centers in the 774 local government areas of the Federation is one of the public private partnership projects embarked upon by the RMRDC. Presently, the Council is spearheading cluster formations on the following commodities among others across the country:

- ☆ *Cashew*: A cashew processing project has been established in collaboration with the Kogi State University, Anyingba Local Government (LG), and North central Nigeria. Within the same State, the Council will be assisting the local government on a similar project. Similar cashew clusters are to be established in Okigwe, Imo State and Ogbomoso, and Abeokuta in Osun and Ogun States respectively.
- ☆ *Ginger*: The Council is to provide the Mbaitoli Ginger Farmers Co-operative Society, Imo State with ginger splitting and drying machines to be fabricated by the National Roots Crop Research Institute, Umudike, Abia State.
- ☆ *Leather*: This cluster in Kano State is being executed in collaboration with UNIDO, Kano State government and the Nigerian Export Promotion Council (NEPC).
- ☆ *Cassava*: The cassava cluster initiative of the Council is still at the conception stage.

In a nutshell, the roles of RMRDC in cluster formations[14] are:

- Research, consultancy and brokerage services
- Raw materials sourcing and development
- Machinery sourcing
- Capacity building
- Information services and networking
- Co-ordination of the small scale industrial cluster services
- Information generation and dissemination
- Monitoring and evaluation of the programmes
- Linking knowledge centers and end users to the clusters.

Besides these government efforts, small uncoordinated clusters of businesses, especially in handicrafts, have existed in the communities for long. All they require is the necessary assistance to make them better focused.

Entrepreneurship Development

Global trends have indicated that stimulating entrepreneurship is one of the most viable ways by which people can be empowered. Having realized this, many industrialized and newly industrialized countries (NICs) have committed resources and time to the promotion of entrepreneurship through various means including specific emphasis on the educational sector, especially at the tertiary level. On the basis that entrepreneurship is a culture that can be taught/learnt; a majority of these countries have introduced the study of entrepreneurship in their learning institutions. Until recently, the entrepreneurial development was not taken as a national strategy for development in Nigeria. Individuals had to struggle on their own with little assistance to make a success of their ideas. It has been more difficult in the villages especially with the lack of basic social and physical infrastructures. This has led to many frustrating business failures with its accompanying job losses and negative impact on the society. However, in the last two years entrepreneurship development programmes have been entrenched in the national development strategy of Nigeria. The educational sector has included it in the school curriculum, while government agencies have created units and departments solely for entrepreneurship development.

The Raw Materials Research and Development Council (RMRDC), once again is not left behind in this move. Besides organizing entrepreneurial workshops on small scale projects for village groups, it is also collaborating with institutions on promoting entrepreneurship in Nigeria.

Conclusion

The impact of science and technology in Nigeria's rural based economic enterprises is progressively improving. That notwithstanding, there is need for proper linkage of all structures currently on ground for better results to be obtained. The concept of occupational village development when fully embraced will facilitate the economic development of the rural areas, which will transform to better living standards for generality of Nigerians. Increased International collaboration will boost

Nigeria's quest to develop her occupational villages. To facilitate the emergence of strong occupational villages development concept in developing countries and specifically in Nigeria it is recommended that:

1. African governments should make efforts at meeting the AU (expand) proposed 1 per cent of national GDP on S&T development of which Nigeria is a signatory, on its capital expenditure on agriculture and natural resources development. This will enable the S&T for executing its statutory projects;
2. Raw materials development should be addressed in an integrated manner that tie together all policies of resource rich emerging economies and further promote a sustainable partnership with resource dependent economies such as EU, USA and Japan;
3. Technology Incubation Centers are no doubt veritable instruments for transmitting innovative ideas into projects. But for it to impact fully on a nation's economy, it is recommended that they be replicated in villages or rural areas. Government authorities at all level as well as the private sector have major roles to play in realizing this goal;
4. In order to encourage commercialization of indigenous technology, development banks should be established to assume an active role in the allocation of funds to projects aiming at the exploitation of novel approaches to identify technological problems. This active role could take many forms ranging from preferential interest rates to extended grace periods and longer repayment schedules;
5. A project sponsored by the Department of International Development (DFID) - United Kingdom and now the European Union (EU) with NRI and the University of Agriculture Abeokuta (UNAAB) on the commercialization of traditional processed products from cassava, such as *wet fufu and dried fufu* has proven to be a successful initiative with the potential to offer new opportunities to rural households - either through the sale of fresh roots or through processing and marketing[2]. Other economies can adapt this for other resources in their own environment to meet their needs;
6. Community development projects along with the co-operative movement should be further encouraged and given all support to enable them come up with products that can compete favorably in both the local and international markets;
7. Effects of globalization are felt even in the villages, consequently there is an urgent need to harmonize all development programmes in developing economies, starting from bottom –up, and position themselves properly to reap maximum from it;
8. Proper linkage between village level processing and marketing activities is inevitable in a nation's development programmes. Consequently, the role of extension workers should not be down played. They should be given all necessary tools to perform their roles in a country;
9. Partnerships with other developing countries help Nigeria to promote the occupational villages' development concept. Thus it should be encouraged.

Acknowledgement

I wish to use this opportunity to appreciate Prof. A. P. Onwualu, Director - General of the Raw Materials Research and Development Council (RMRDC) for giving me this singular opportunity to attend this workshop. My sincere thanks go to Dr. M. O. Jolaoso, Director, Agricultural and Agro-Allied Department of the Council for his contributions to this work. I also appreciate the diligence with which Dr. Seb Obasi, Director, Industrial Plants and Equipment Department vetted this work and made useful suggestions. I am grateful to Mr. B.A. Aluko, Director, Planning and Information Services Department for his assistance in getting me nominated for this workshop. I thank Er. Emmanuel Kwaya for making materials available to me, and to Mr. Inyang for the encouragement extended to me.

Last but not the least I wish to appreciate my husband Mr. Chinenye M. Nwanya for his patience in the course of working on this paper and his moral support.

References

1. Adewale, G. J. (2005), Socio-Economic Factors Associated with Urban-Rural Migration in Nigeria: A Case Study of Oyo State, Nigeria, Journal of Human Ecology, 17(1): 13-15.
2. Apantaku, S, (1998), Indigenous Knowledge and Development Monitor, Agricultural Extension and Rural Development, UNAAB, Nigeria.
3. Akewushola, R.O., Poverty, Unemployment and Growth in Nigeria: The Role of Entrepreneurship
4. Akor, Sylvester (2010), NBTI Trains 81 Entrepreneurs on Economy Dependence, CHAMPION PAPER, www.champion.com.ng/index.php?news=29452
5. Deji, O. F. (2005), Membership of Co-operartive Societies and Adoption Behaviour of Women Farmers: Implication for Rural Development, Journal of Social science, 10(2):145-147
6. El-Namaki, M.S.S., Developing and Promoting Technological and Technical Skills in Small-scale Rural Manufacturing Enterprises www.RURALIND\rural tech-fao.htm.
7. Levy, P. (2004), Nigeria: Handicrafts in Nigeria (2nd ed.) published by Marshal Cavendish, p.42.
8. Iganiga, B. O. and Asemota, A. (2008), The Nigerian Unorganized Rural Financial Institutions and Operations: A Framework for Improved Rural Credit Schemes in a Fragile Environment, Journal of Social science, 17(1): 63-71.
9. Nwosu, A. C., Rural Development Problems in Nigeria, www.cabastractsplus.org/abstracts/Abstract.aspx?AcNo=19926714607
10. Obi, B. O. (2007), Fiscal Policies and Poverty Alleviation: Some Policy Options for Nigeria, for African Economic research Consortium, Nairobi.
11. Odebode, S. (2008), Appropriate technology for Cassava Processing in Nigeria: User's Point of View, Journal of International Women's Studies, vol. 9 (3): 269-284

12. Ogidefa, I. (2010), Rural Development in Nigeria: Concepts, Approaches, Challenges and Prospects, in Issues, Mar 22.

13. Olayiwola and Adeloye, O.A. (2005), Rural Infrastructural Development in Nigeria: between 1960 and 1990- Problems and Challenges, Journal of Social Science, 11(2): 91-96.

14. Onwualu, A. P. (2010), RMRDC: Promoting Raw Materials Processing Clusters in Nigeria, being a presentation at the Leadership Conference at Abuja, Nigeria.

15. Okafor, L., 2003, Nigerian Alternative Trade Network in www.new-academy.ac.uk

16. Okonkwo, E. M. and Akpan, A. U. "Science, Technology and Innovation: Current Status in Nigeria" in http://webcache.googleusercontent.com/search?q=cache:Qv8CE31gROMJ:www.educationdev.net/educationdev/Docs/n4.ppt±science±and±technology±policies+.

17. Owojori, A.A. and Oladejo, M.O. "Re=engineering Co-operative societies for Sustainable Small and Medium Business Growth and Development in Nigeria" in www.thefreelibrary.com/RE-ENGINEERING+COOPERATIVE+SOCIETIES+FOR+SUSTAINABLE+SMALL+AND+MEDIUM+...a01073959640

18. A report on Women Programme and The Better life Programme for The Rural Women from 1987-1991 by Lagos State Commission for Women, Lagos State.

19. State of Plant Genetic Resources for Food and Agriculture in Nigeria: 1996-2008, a country report by the National Centre for Genetic Resources and Biotechnology, Ibadan/Federal Department of Agriculture, Abuja, Oct. 2008, pp.1-60. http://www.pgrfa.org/gpa/nga/Nigeria2.pdf

20. The Economy of Nigeria, http://www.essay-911.com/samples/economy-nigeria.htm

21. www.budgetoffice.gov.ng/2010 per cent 20budget per cent 20message/SCIENCVE per cent 20AND per cent 20TECHNOLOGY.pdf

22. www.fmst.gov.ng

23. www.unesco.org/bpi/pdf/memobpi54-sciencepolicies_en.pdf

24. www.afdevinfo.com/htmlreports/org/org_47075.html

25. www.fao.org/docrep/007/y5548e/y554e/08e/htm

26. http://nacetem.org/ik.htm

27. www.ficci.com/media-room/speeches-presentations/2007/march/sessional/MallamandAdamuBello=Nig.pdf

Chapter 10

Vidatha Resource Centres: A Sri Lankan Attempt towards Rural Development by Taking Technology to the Village

Jayasamara Gunaratne

Assistant Director,
Ministry of Technology and Research, Colombo 03, Sri Lanka
E-mail: jayasamarag@gmail.com

ABSTRACT

Sri Lanka is an agricultural country having a majority of rural population. With a literacy rate of more than 90 per cent, she is in a better position for application of science and technology towards rural development. Identifying this opportunity, a national development programme of science and technology was launched in the country with the combined efforts of the institutions under the Ministry of Technology and Research, universities and the national research institutions. Accordingly, a computer linked network called the "Vidatha Resource Centres for taking technology to the village" programme was established in 2004. The main objective of this national programme was to create small and medium scale enterprises, upgrade technology for existing enterprises and finally to improve the living standards of the rural community as a whole. There are four main missions under the said programme *viz.*, a) directing scientists and technologists to engage in research and evaluations appropriate to the country and thereby strengthening the development process in the country, b) building up the economy status of the rural folks and empowering them through transfer of technology, c) identifying traditional technologies of the rural communities and make modifications for to re-utilizing such indigenous knowledge in the development process, and d) ensuring that the rural community receive a sustainable income by supplying high quality goods and services to the market. This paper discusses in detail as to how the said programme was initiated, how it was progressed, the problems encountered, the solutions found, and lessons learnt in achieving the targets. It further presents four case studies conducted in various districts of the country utilizing S&T to uplift the economy of rural community through the Vidatha National Programme.

Keywords: *Rural community, Science and technology, Sri Lanka, Socio-economic development, Technology transfer, Vidatha resource centres.*

Introduction

Sri Lanka is an island located in the Indian Ocean close to the southern part of India with tropical beaches, ancient monuments, rain forests, gorgeous waterfalls and many other sight seeing places to delight any person. The country has a tropical climate with an average temperature of 27°C that fluctuates between 15° C in the highlands to 35°C in certain areas of the lowlands. Being an agricultural country, rice is the staple diet. Tea, Rubber and Coconut are the main plantation crops that contribute to its main economy.

Sri Lanka has about 70 per cent of its population living in the rural areas and their main income source is agriculture. With a Human Development Index (HDI) of 93 out of 177 countries and a literacy rate of 90 per cent, Sri Lanka is in a good position to start up micro and small enterprises (MSE) towards socio-economic development. The agricultural sector contributes to 12 per cent of GDP of the country (Central Bank, 2008) while 33.1 per cent R&D expenditure is spent on agriculture R&D (NSF, 2008). Although the contribution of Agriculture to GDP of the country has significantly decreased over the years, it is an important determination of GDP directly accounting for about 1/5th of the national output, directly employing more than 1/3rd of the labour force (36 per cent), providing inputs from the agricultural processing industry and supplying a large share of domestic food requirements. However, due to low income through the existing agricultural practices, the rural sector farmer falls below the poverty level with a less than monthly income of around 20 US$ (Central Bank, 2008). In this context, the concept of taking technology to the village for improving livelihood of the rural farming community becomes important.

The Vidatha Movement: Concept and Background

Sri Lanka has achieved a significant progress in the fields of health, education etc., since regaining the Independence. However, she has not been as successful as the regional countries in alleviation of poverty. The main reason behind this could be the fact that technologies developed through research have not reached the villages. Mostly, the traditional technologies have been used in the production process throughout and therefore; the quality of rural products has not been as competitive as the products in the international market. Although it had proved sufficient for the local market, a good profit was not achieved due to the use of the same pattern, traditional features and the use of traditional methods with a high production cost. As the productions had to be done on a minimum profit, the producers did not succeed in earning an income beyond the level of subsistence. As a result, the producers were not left with funds to reinvest and many abandoned their self employments.

Also, there had not been a continuous flow of goods thus produced in an organized manner, the trading communities who did not want to let the customers down, tended to rely more on products that are imported from the foreign countries. The demand that existed for locally manufactured products decreased as large stocks were imported and such imports were done in exact intervals. This has resulted in further decrease of locally manufactured products to the level that the local economy was limited only to the provision of services. The consumers were subjected to price fluctuations occurring in the international market due to the country's economy being dependent on the supply of foreign goods. The economy was not made stable owing to the fact that the country's production industries had not been developed and the consequent rise in the unemployment levels had been unavoidable. Due to decrease in income levels; the local people were deprived of the economic strength to purchase these imported goods when the prices have gone up. In the past, this paved way for public unrests as well. The only solution for this crisis is the strengthening and the development of local manufacturing industries. In order to accomplish this, the following tasks had to be carried out.

1. Production of goods with a high standard and quality.
2. Production of goods at a minimum production cost.
3. Presenting the goods to the market at a price that could compete with the foreign goods.

It is not easy to overcome these challenges. For this purpose, the use of Science and Technology to the maximum assumes importance, though it is a task that may incur large expenses. Fortunately, a good education level and a better health condition prevail in the country's population. Due to free education, Sri Lanka enjoys high education standards and the general education level has also been of a good standard. The consequent governments too relied on many welfare programmes over the years contributing to this effort.

The Vidatha Technology transfer programme was first started in 2004 to take technology to the village. But unfortunately, it did not encourage technology transfer as expected, despite whatever micro and small enterprises were introduced to the rural areas, the country was not successful in industrial development. Most importantly, the technological knowledge required for any micro, small and medium scale industries were however, available at the scientific institutions located in the suburbs of Colombo, the country's capital. Highlighting the gaps in technology transfer and commercialization, His Excellency, the President being the Hon. Minister of Finance as well, mentioned the following in his Budget Speech of the year 2005.

"Economic development cannot take place without technological advance. Appropriate technology can be developed only by encouraging local science and technology institutions to provide necessary research and development. At present Sri Lanka has one of the lowest levels of funding of about 0.18 per cent of GDP for R&D in the world. I propose to raise this over the next five years to at least 1 per cent of GDP. The reversal of brain drain, international collaboration in research, the popularization of science, the exploration of the full extent of our natural resources, the promotion of innovation and ICT will therefore need funding. For technology to be transferred effectively to the micro, small and medium level entrepreneurs ..."

With the above vision, the preliminary investigations were done and initially, it has been expected to create a production culture among the people who are not engaged in any production before. It is not an easy task to compel the impoverished people to embark on the production process. There are many challenges as follows to overcome in achieving the above objective.

1. Being accustomed to relying on relief and/or other aids.
2. Lack of an initial capital.
3. Lack of funds for raw materials.
4. Lack of technical and technological knowledge.
5. Lack of knowledge of quality control.
6. Lack of knowledge of labour laws.
7. Lack of knowledge of handling the market.
8. Burdened with daily problems in life.

Further, the banking system of the country has been designed to offer loan facilities to the rich only who could provide securities. Schemes that provide the poor with capital aids are hard to come by. Therefore, the poor encountered many hardships in obtaining capital aids. Their inability to find funds for the purchase of raw material is a repercussion of the said hardships.

Another reason was seen as an acute shortage existed in the flow of technology to the village, since the country's education system focused mainly on other subjects. Although technology was provided by schools to a certain extent, it was not professionally adequate to earn money by putting it into practice. Even though technological subjects were taught in Technical colleges, the opportunities to earn from it directly had been scarce in the country. Although goods were manufactured, many did not have an idea of quality control. This resulted in the instant rejection of these products from the market. The producers had to face many inconveniences due to the lack of knowledge of labor laws.

It was, however, observed that the simple technology could be obtained from the various institutions free of charge, the medium technology, for a minimum price while advanced technologies for a reasonable price and various personnel as well.

Such institutions and personnel were listed as follows:

1. From research institutes under the purview of the then S&T ministry.
2. From other government owned research institutions.
3. From non-government organizations.
4. Through science and technology officers.
5. Through entrepreneurs already created.
6. Through persons with traditional knowledge.

After the preliminary investigations, the then Ministry of Science and Technology planned to establish a computer linked network called "Vidatha" Resource Center, one in each of the 320 Divisional Secretary Areas. Each of these centers was to have a science and technology graduate, a computer operator and a link agent.

At the village level; science and technology societies were formed to transfer technology to the village and promote SMEs with a view to provide employment and to reduce poverty. These resource centers were linked to the SME banks and Regional Development Banks to assist in development of banking and transaction, and formulating information to the private sector. The assistance from the Industrial Development Board (IDB) and the Export Development Board (EDB) was sought for technical support.

This programme was planned to be implemented during a three year period 2005-2007 with a proposal to provide an additional Rs. 300 million in 2005 to the then Ministry of Science and Technology in support of its 'Gamata Thakshanaya' – the Vidatha Programme."

As per the above proposal approved by Parliament, 365 graduates recruited under graduate trainee scheme, were sent to the said Ministry by the Ministry of Finance. Since they had to be effectively employed, they were recruited as Science and Technology Officers (STOs) to the Divisional Secretary areas to function as the officers in charge of Vidatha centres.

This programme was not only included in 2005 Budget Speech, but it was well recognized by the Mahinda Chintanaya, the policy document of the present government, which states the Vidatha programme will be carried forward with vigour in order to take science and technology to the village. This programme aims at filling the technology-gap prevailing between developed countries and developing countries while strengthening the lives of village community through application of Science and Technology in Sri Lanka.

Establishment of Vidatha Resource Centres

Within the mandate of the Ministry of Science and Technology, the Vidatha Resource Centres play a vital role in transferring technology to the village, *viz.*, utilizing science and technology for the development of the country. These Centres endeavour to solve the vital problems prevailing, especially in rural areas of the country, such as high unemployment and poverty because development of the village can be achieved if science and technology is applied to village activities sufficiently. Unfortunately, this is not happening. The research results of the Science, Research and Development Institutes were not generally available to villages since there was no suitable transfer mechanism for the purpose. Therefore, Hon. Prof. Tissa Vitarana, the then Minister of Science and Technology attempted to address this problem and introduced the idea of a special technology transfer mechanism namely 'Vidatha' which means Vidya dana thakshana

This is how this word was coined:

Vi/da/tha

- Vi : Vidya - Science
- da : danaya - Transfer
- tha : thakshanaya - Technology

With his efforts through many discussions with the Treasury officials, it was included in 2005 Budget Speech and became a locally funded project by the name of "Gamata Thakshanaya" (Technology to Village), to be executed by the Ministry of Science and Technology. This project is expected to pave the way to a long standing programme of technology transfer to the village.

Main Components

Main components of the programme are:

- Vidatha Resource Centres;
- Vidatha Societies;
- Co-operating Bodies;
- Steering Committees, and
- District Unit.

Vidatha Resource Centre (VRC)

A Vidatha Resource Centre (VRC) is a Centre opened one for each Divisional Secretary area and it comprises of a small office, a technology equipment store and a knowledge and training centre. This is called a 'resource centre' rather than a 'service centre' which is the normal practice of naming Govt. centres, established for various purposes. This name itself indicates that it is a resource centre, which should be utilized by the people of the area. What are the resources available in this centre?

It has a team of three competent officers. A science graduate as Science and Technology Officer (STO), a Computer Operator and a Field Coordinator. As an office it has information and communication facilities among which the computers and the Internet connectivity are prominent. There are five computers. They can be used for computer training, internet surfing, information sharing and data storing. It has other common office facilities like telephone and mail facilities.

These Centres also have various technology demonstration equipments depending on the requirements of the area, which can be used for training purposes as well as trial productions by new entrepreneurs. Some equipments available are as follows:

- Fruits and vegetable dehydrator;
- Yoghurt incubator;
- Virgin Coconut oil extractor;
- Coconut de-husker;

- ☆ Efficient fire wood stove
- ☆ Bakery oven and
- ☆ Set of electronic testing equipment.

This set of equipment is to be revised and up graded from time to time as and when new technological equipments are introduced by Research and Development institutes. The availability of these equipments for trial purposes is a very valuable resource for the prospective entrepreneurs.

As a knowledge centre, it has a small library of books and CDs on available technologies and visuals like wall hangings, banners and leaflets depicting various new technologies and services offered by VRCs and Research and Development Institutes. An important feature in this knowledge centre is the Data Bank with information on locally available Small and Medium Enterprises (SMEs), self-employments, other organizations providing services to entrepreneurs as well as the human and physical resources. This Data Bank was set up through a survey carried out by VRC utilising volunteers and will be updated from time to time. Data will be shared by all VRCs and the useful information to prospective entrepreneurs will be published in the Internet.

The Vidatha Resource Centres have facilities to function as training centres as well. Apart from Computer training that will be provided free of charge, it will implement an annual technology training and awareness building programme prepared by the Divisional Technology Steering Committee. Resource persons for these programmes will be drawn from R&D Institutes in addition to the STOs functioning as resource persons. These training centre facilities are available not only for the Institutions under the Ministry, but also to the Institutions under other Ministries, Departments and Provincial Councils etc. From the Ministry's budget funds will be allocated annually for these Centres and that will be utilized for general office expenditure as well as for technology transfer activities to be done. With all these resources and facilities; the VRC is prepared to serve the people in the villages.

Vidatha Societies

As mentioned above, the Vidatha Resource Centre has to be utilised by the people of the area and for this purpose they should be organized, motivated and lined up behind the VRC. The mechanism for this is the forming of Vidatha Societies around the VRC. Through these Societies, a people's movement will be built up and a momentum will be gathered for technology transfer activities as well as for the development of the village.

Vidatha Societies are self help organizations formed on voluntary basis. They are formed at least at three levels *viz.* primary, secondary and tertiary levels. Vidatha Societies at "Grama Niladhari Wasam" (GNW), an administrative unit at grass-roots level in Sri Lanka, are the primary societies and the membership of these societies will be given to self employed and other entrepreneurs as well as to those who get services from and/or connected with the work of the VRC. The role of these societies is to find ways and means to solve their member's entrepreneurial problems at village level and to inform the secondary societies on problems that need higher level attention.

The secondary society is formed at Divisional Secretary area level and is called Vidatha Divisional Union. Each of the Vidatha Societies at GNW level sends a representative to form this Union. In addition to finding solutions to attending to problems brought up from primary Vidatha Societies, the main responsibility of this Union is to take part in Vidatha Divisional Steering Committee through some representatives.

The tertiary society is the Vidatha District Federation formed at District level. This will have individual Vidatha entrepreneurs, Divisional Unions and Vidatha Societies as members. In addition to the traditional society activities like discussing matters referred to it by its members, it will undertake business activities in support of its members, such as procuring inputs, marketing products and providing credit facilities for them. Hence this will have a share capital, unlike other lower level Vidatha Societies. In fact, this Federation will run business activities too, to support its members. Another type of Vidatha Society has to be formed at national level when all other level societies become functional; that is Vidatha National Council, the Apex body of all Vidatha Societies. Its structure or functions should cover representing Vidatha movement at national level and finding solutions to the problems of its affiliated members.

Main Objectives

There are two main objectives in this programme and they are:

- Reducing Unemployment and
- Alleviating Poverty.

Reducing Unemployment

One of the acute problems prevailing in the village population is the unemployment or under employment and due to this situation they tend to go to urban areas or towns in seeking employment aggravating the problems of high population in the towns. If employment can be provided within the village, this problem can be solved. Many people prefer to seek for the government employments and they all cannot be employed by the Government. Hence, the solution is the Self-Employment. In the village, agricultural products go waste in glut periods and these can be used to generate self employment by producing industrial products. But the self employments in the village are not productive since science and technology is not available to them sufficiently. Because of this situation many efforts taken in the past to promote self employment were failed. Therefore, an efficient technology transfer programme is a must to generate self employment in the village.

Alleviating Poverty

Poverty can be reduced by increasing the income of the families and utilising the earned income efficiently in the family.

Increasing income can be achieved by:

- ☆ Motivating people to start self employment and
- ☆ Introducing technology to existing production ventures of the villagers.

Utilizing the earned income efficiently in the family can be achieved by:

- ☆ Providing employment in or around the homes and
- ☆ Appling the technology to domestic activities.

Motivating People to Start Self Employment

For this purpose, Vidatha Programme offers training in awareness building as well as in skill building in different technologies for the villagers. This will increase the income generating ventures in villages. The programme also promotes village enterprises by coordinating and supporting the efforts of the entrepreneurs in obtaining the other vital inputs such as credit, and marketing facilities for the products.

Introducing Technology to Existing Production Ventures

When appropriate technology is applied to production of goods, it enhances the quantity as well as the quality of the products. More products bring more income and high quality attracts high prices and more competitiveness in the market. All these phenomena contribute to the increased family income.

Providing Employment in or Around the Homes

Earned income will not be spent on travelling up to the work place and/or food and lodging when employed in his/her own house or in the vicinity of it; also no extra expenditure on special clothing and cosmetics is necessary. This means greater portion of the earned income will be available for family affairs and it will reduce poverty level. Therefore, the Vidatha programme encourages people being employed in the village itself.

Application of Technology To Domestic Activities

Technologies introduced to village will not only increase the productivity of enterprises and trades, but also increases the productivity of the household activities, saving time and money and increasing quality of the food prepared at home. This will not only reduce the poverty but also uplift the health of the families. For example, a Virgin Coconut Oil Extractor provides healthy oil for the family use and Energy Efficient Firewood Stove will increase the health of kitchen users while saving the fire wood.

In addition to the two main objectives mentioned above, there are other objectives, *viz.*, enabling Science Research and Development Institutes to focus their research on village requirements as well as upgrading the available traditional and indigenous technologies. For both these activities, the Vidatha programme makes the required information available to research scientists.

Main Activities and their Progress

Vidatha has two types of activities *viz.*, (*i*) Vidatha Project activities and (*ii*) Vidatha Programme activities.

Vidatha Project Activities

- ☆ The opening of the computer linked VRCs, one per each DS Division;
- ☆ Appointment of required staff and human resource development;
- ☆ Providing office facilities and equipment;
- ☆ Establishment of VRC Data Banks and interlinking them with each other;
- ☆ Forming Vidatha Societies at GNW, DS and District levels, and
- ☆ Setting up of Vidatha Divisional Steering Committees in each DS Division.

The Opening of Vidatha Resource Centres

Vidatha Resource Centers are opened in the Divisional Secretariat (DS) areas one for each Division. At the commencement of the programme in year 2005, there were 320 DS Divisions and today the number of Divisions has increased up to 331.

As the Divisional Secretary is the immediate Staff Officer who supervises the VRC staff and all the funds of the Ministry are channelled through the DS office to the VRC. Since the DS office is the most attended Govt Office by the general public at divisional level, it was decided that the ideal place to locate the VRC is in the premises of the DS Office. In consultation with the DS, every possible endeavour is taken to get sufficient space in the DS office. An extent of 1200 sq feet is required for all the activities in the VRC. If the VRC is in the DS Office, and also if the DS agrees to conduct all training programmes of VRC in the Auditorium, an extent of 600 sq. feet is considered to be sufficient for other activities of the Centre.

Appointment of Required Staff

As mentioned earlier the following officers are appointed to a VRC:

- ☆ One Science and Technology Officer (STO),
- ☆ One Computer Data Entry Operator,
- ☆ One Field Coordinator,
- ☆ One Office Aide (Karyala Karya Sahayaka) (KKS), and
- ☆ One Watcher/Caretaker.

Science and Technology Officer (STO) is the Officer in-charge of the VRC who is responsible for all VRC activities as well as good office keeping and administration. He/she is supervised by the Divisional Secretary of the area. Other staff members are working under the STO. Computer Data Entry Operator carries out all the work connected with ICT activities of the Centre while keeping office correspondence active. He also conducts the basic computer training offered by the Centre free of charge. Field Coordinator is responsible for all field work under the supervision of the STO and his main responsibility is the establishment and efficient functioning of the

Vidatha Societies. KKS, a minor staff is appointed to each VRC, while a Watcher or a Caretaker will be appointed to VRCs that are situated away from DS Offices.

Vidatha Technical and Human Resource Promotional Assistants (THRPA) are appointed to Districts to function under the close supervision of the District Secretaries. One, two or three of them are appointed to a District depending on the size of the District and they are allocated different zones in the District when there is more than one in the District. As mentioned earlier, their main responsibility is to carry out Vidatha promotional activities while attending to the District and Provincial level Vidatha matters.

Vidatha Programme Activities

Transferring Science and Technology to the Village

Under this programme, the Ministry of Technology and Research (then Ministry of Science and Technology) has recognized the importance of entrepreneurship development in the economy and has introduced a whole range of financial incentives and programmes directed at enhancing their competitiveness particularly in the areas of science and technological development; information technology development; and the quality and productivity improvement. Accordingly, many training programmes were implemented by the Technology Transfer Unit of this Ministry for entrepreneurship development in the rural areas.

Year	*No. of Programmes*
2006	316
2007	698
2008	1253
2009 (up to June)	1140

The Research and Development Institutions coming under purview of the Ministry, other external institutions and the trained Science and Technology Officers have been assigned to handle this task. These programmes are based on agriculture (agro– based), food based, industrial based, technology based and the handicraft sector.

The participants have the opportunity to obtain the know-how on new technologies including dehydration of fruits and vegetables, rice based products, rice based bakery products, Kitul (*Caryota urens*) tapping technology, soil cement block production, pre-stressed concrete door and window frame production, cost effective housing construction, mushroom cultivation, Jam and cordial production, papadam (food item) and candle production, ornamental fish farming etc. (Table 10.1).

Technology Transfer Programmes 2006-2009

Conducting the above programmes/workshops (Figures 1 and 2) has been able to develop a mechanism to link villagers, especially rural entrepreneurs with research and development institutions. The quality assurance of the products and programmes

Table 10.1: Programmes Conducted by Institutions

Institute	*No. of Programmes*				*Total*
	2006	*2007*	*2008*	*2009*	
Industrial Technology Institute (ITI)	237	62	38	14	351
National Engineering Research and Development Centre (NERD)	62	44	7	6	119
Arthur C. Clerk Institute for Modern Technology (ACCIMT)	17	20	4	6	47
No. of trained Science and Technology Officers	-	-	499	490	989
External Institutions		572	705	624	1901
Total	316	698	1253	1140	3407

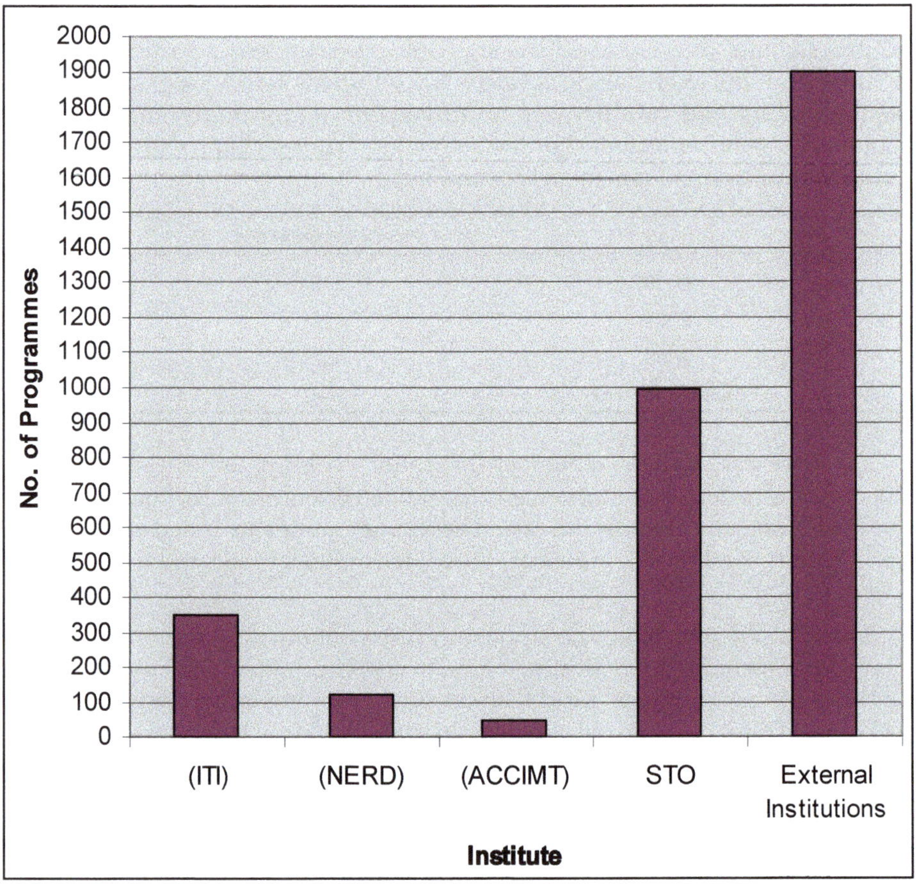

Figure 10.1: Number of Programmes Conducted by Institutions

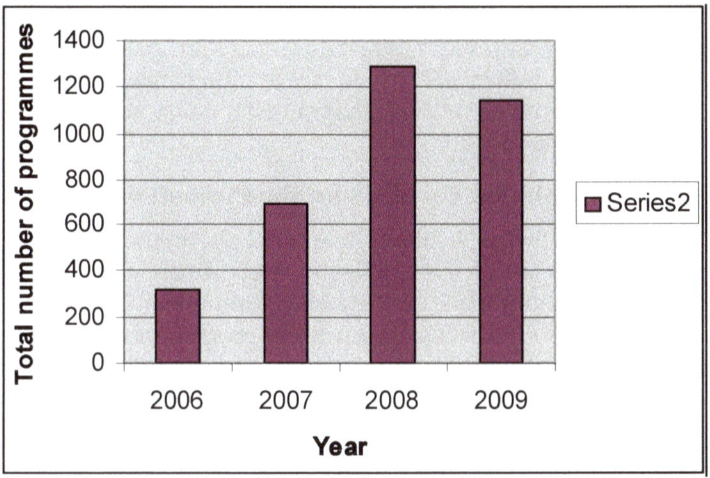

Figure 10.2: Number of Programmes Conducted during 2006-2009

for achieving the GMP status will continuously help enhance the quality, competitiveness of products at market and support the growth of the export markets. These programmes also trains the entrepreneurs who often lack the skills, knowledge and networks to get their ideas off the ground.

From mid 2005 to mid 2009 there are 5455 entrepreneurs who were established under the technology transfer programmes. This number includes both male and female entrepreneurs and the programme helped promoting gender equality and empowering women in the enterprise sector.

Assisting in Obtaining Credit and Market Facilities

Technology is transferred to villages with the aim of establishing successful enterprises, but technology alone is not sufficient for establishing a successful enterprise. In addition to technology, these micro and small enterprises should get other inputs such as credit and market etc. Therefore, the Vidatha programme, while providing technology, assists entrepreneurs in obtaining the said aspects from relevant organizations.

Assistance in Finding Credit Facilities

Following the discussions held at the ministerial level with the Peoples' Bank, Bank of Ceylon and Lanka Puthra Development Bank, steps were taken to provide loan facilities on relief interests to those who expect to embark on businesses after obtaining technology from Vidatha entrepreneur training programmes and also to the entrepreneurs who wish to further improve their business but encountering financial constraints. Amongst the above mentioned state banks, the Peoples' Bank responded positively by actively contributing to this cause. By the end of September 2009, more than Rs.100 million was granted as loans.

The Ministry is holding discussions with other entrepreneurial banks to launch similar loan schemes through VRCs. When these loans are granted to entrepreneurs, Science and Technology Officer (STO) helps the granting organization for follow up, evaluation and recovery of loans. The loans granted this way have a very satisfactory recovery rate of 100 per cent.

Assisting in Finding Market Facilities for the Promotion of Enterprise

In order to assist in finding a suitable market for the promotion of businesses, the Ministry held discussions with the cooperative Department under the sponsorship of the peoples' Bank and initiated a Vidatha Entrepreneur Cooperative Loan for Marketing and Production Societies. Presently this programme is being implemented in Kegalle, Kurunegala, Ampara, Kandy, Matale, Badulla and Rathnapura districts. Societies are being formed around every Vidatha Resource Centre. If the number of members is low, steps will be taken to establish district societies. The loans are provided through these societies utilizing the deposits of the members. Loans amounting to Rs. 200,000.00 are granted and two members should sign as guarantors for these loans. Bulk loan/working capital will be provided by the people's Bank. The membership of these societies includes Vidatha entrepreneurs, while the STOs are appointed as the secretaries.

Apart from this, action has been taken through the above cooperative societies, to establish Vidatha shops at the rate of one for each Div. Secretary's Division. Accordingly, these shops have already been established in Rambukkana and Warakapola Divisional Secretaries Divisions. Steps have been taken to keep these shops open from 7.00 am – 8.00 pm.The shops are constructed by the Vidatha Societies. The Vidatha society members could obtain the membership of the cooperative society. The accounts of these shops will be supervised by the Peoples' Bank.

After holding discussions with the Ministry of Export Development and Investment Promotion, the steps have been taken to introduce Vidatha products with G.M.P (Author – please give full name) certification, to foreign markets. Following discussions held with the Ministry of Tourism, the steps have also been taken to market local food products at tourist hotels.

Upgrading and Collecting Information on Traditional and Indigenous Technologies

One of the vital activities of a VRC is to collect information on traditional and indigenous technologies. This attempt will be done during the field surveys carried out by volunteers, but STOs take action to update such information whenever possible, which in turn should be sent to Research and Development Institutions for upgrading purposes. This will be a continuing process.

Assisting Science Research and Development Institutes

One of the vital aims of the Vidatha Programme is to assist Science Research and Development Institutes in obtaining the village requirements on Science and Technology. This is done mainly through the field surveys carried out to establish the Database and the information is made available via internet for the scientists and their organizations. In addition to this, R&D Institutes can utilise VRCs either to collect the required information for a particular research or to carry out research field work. During the past, VRCs are seldom used for this purpose.

Table 10.2: Entrepreneur Training Programmes 2006 – 2009

Resource Contribution	*Programme/Project*	*Number*
	ENTREPRENEUR TRAINING PROGRAMS 2006	
ITI	Mushroom cultivation, Kitul Tapping, Yoghurt production, Dehydration of fruits and vegetables, Rice based products, Bakery products, Cultivation and Processing of medicinal plants, Dustless chalk, Goraka Phaste, Bites production, Fruit drink preparation, Sauce and chutney production, Confectionary items, Rice based bakery products, Cashew based Products, White coir production, Awareness Programme on, Cinnamon Processing, Awareness Program on fruit based products, Herbal Tea, Ice cream Production.	237
NERD	Virgin coconut oil extraction and NERD technologies, Bio gas digester construction, Cost effective Housing construction, Soil Cement Block production.	62
ACCIMT	Workshop on Practical Electronics, Workshop on antenna designing, Workshop on Mobile Phone Repairing, Workshop on TV and Radio repairing, Workshop on Computer Hardware Engineering, Safety from lightening.	17
	ENTREPRENEUR TRAINING PROGRAM 2007	
ITI	Kitul Tapping, Cultivation and processing of medicinal plants, Preparation of bites, Processing of Herbal plants, Rice based Products, Rice based bakery products.	62
NERD	Awareness program on Biogas Digester construction, Soil cement block production, Cost effective housing Construction, Concrete Door and Window Frames Manufacturing.	44
ACCIMT	Mobile phone repairing, Computer Hardware Engineering, Antenna Designing, Practical electronics, TV and Radio Repairing, Protection from lightening	20
Other institutions	Entrepreneur development program, Food technology programme, Coconut based products, Papadam production, Dairy products, Candle production, Banana cultivation and Banana Products.	572
	ENTREPRENEUR TRAINING PROGRAMS 2008	
ITI	Kitul Tapping, Rice based Products, Rice based bakery products.	38
NERD	Soil cement block production, Cost effective housing Construction, Concrete Door and Window Frames Manufacturing.	7
ACCIMT	Mobile phone repairs, Computer Hardware Engineering, Antenna Designing.	4
TRAINED STOs	Yoghurt production, Candle production, Dehydration of fruits and vegetables, Hand made papers, Soap production, Computer training, Ornamental fish farming, Rice based bakery products, Jam and Cordial production, Bites production, Confectionary items production, Kitul tapping, and Mushroom cultivation.	499

Contd...

Table 10.2–Contd...

Resource Contribution	*Programme/Project*	*Number*
Other institutions	Entrepreneur development program, Food technology programs, Coconut based products, Papadam production, Dairy products, Candle production, Banana cultivation and Banana Products.	705
	ENTREPRENEUR TRAINING PROGRAMS 2009	
ITI	Video conferencing pilot project, Workshops on production of Kitul Honey and Juggery, Noodles production program, Awarded quality assurance reports.	19
NERD	Pre stressed Concrete Door and Window frames production, Cost Effective housing Construction, Manufacturing of Dehydration machine, Manufacturing of Bakery oven.	6
ACCIMT	Mobile phone repairing, Computer Hardware, Antenna Designing.	6
TRAINED STOs	Yoghurt production, Candle production, Dehydration of fruits and Vegetables, Hand made papers, Soap production, Computer training programs, Ornamental fish farming, Rice based bakery products, Jam and cordial production, Bites production, Confectionary items production, Kitul Tapping, Mushroom cultivation, Screen printing etc.	490
Other institutions	Entrepreneur development programmes, Handicraft, Food technology programs, Packaging, Coconut based products, Papadam production, Dairy products, Screen printing	624

Case Study 1 – Project on Fruit Drinks

Mr. Trilantha Silva, who visited the Vidatha Resource Centre from Bibila, obtained the fruit drink production technology from the VRC with the aim of embarking on self employment. With the dream of establishing a fruit drinks' production enterprise, he obtained a quality inspection report from the Sri Lanka Standards Institution after submitting his products for inspections.

At the initial stages, his monthly profit had been within the range of Rs. 2500-3000. Further, he was provided with packaging technology needed for the enhancement of his products and was guided by the VRC in preparing project proposals for the purpose of obtaining funds for his product project from the United Nations Development Programme, and Sri Lanka Chamber of Commerce respectively. These project proposals were approved by those institutions and he was granted funds to embark on his business.

Subsequent to his obtaining technology from the VRC and receiving quality inspection report, he was given the opportunity to compete with the products in the market manufactured by other major companies. Within a period of 04 years, his income increased by 60 per cent and he was able to provide with employment opportunities to 10 unemployed youths. It is his visit to the VRC that paved way for him to embark upon an income generating project, using the fruits that were being wasted in the Bibila area which is a remote and a dry zonal area in Sri Lanka. He has now been able to produce it to the market.

The VRC has played a vital role in providing accurate technology for his project from the initial stages and in assisting in upgrading his products and popularizing his products by organizing various exhibitions and thereby the VRC has been instrumental in accomplishing his goal.

Manufacture of fruit drinks from fruits collected rurally and according to a technological process is depicted in the Figure 10.3.

Case Study 02 - Dehydrated String Hoppers

Mrs Malani is resident of the Heliyagoda Division belonging to the Warakapola Divisional Secretariat. She was a beneficiary of the Samurdhi programme, a programme conducted by the government to support the poor but despite that, with her three children and her husband who was employed in the private sector for a small salary, she was fighting hard against economic problems in the family. As a means of overcoming her economic problems, she started making string hoppers and selling them to the shops around her residence but that too did not bring in much profit as often string hoppers were left unsold in large numbers.

The Samurdhi officer of her area directed her to Science and Technology Officer of the Warakapola Vidataha Resources Centre (Kegalle District). Malani was introduced to the technology of dehydrating and the experiment of dehydrating string hoppers was a success. Thereafter, Malani continued to do her work using the dehydrator at the Warakapola VRC and after some time with her savings and a loan of SL Rs. 50,000/- from the Vidatha – Peoples' Bank loan scheme, she bought a dehydrator to herself. At the beginning, Malani sold her products at the village fair at Alawwa, and to the villagers only. The Vidatha Officers introduced her to the bigger markets and got her the opportunities to present her products at the Vidatha exhibitions as well.

Thereafter, the said VRC worked to get the GMP certification from the Sri Lanka Standards Institution (SLSI) to Malani's products. First, she was instructed on the areas she was to improve and the VRC continued guiding her to improve up to the required standards. The Sri Lanka Standards Institution granted Malani, the GMP certification through the Ministry of Technology and Research. This certification got her products a good recognition and reputation and the demand increased mush higher. Malani was directed to the exhibitions held at the Bandaranaike memorial International Conference Hall (BMICH) and there she got large scale orders. Vidataha Co-operative Society gave Malani Rs. SL100,000/- as transport and other expenses during these exhibitions. With the support of the VRC officer, Malani had her packaging designed and produced, and the Vidataha Centre helped her throughout the whole process. Because of all the above reasons, the demand for Malani's products rose even higher causing problems in meeting the demand. Vidatha Centre collaborated with the Industrial Technological Institute (ITI) and produced a machine for making string hoppers, thereby solving Malani's problem of production capacity. In 2008 with the help of Vidatha Centre, Malani directed her second son to vocational training. In 2009, Malani returned her Samurdhi card to the Samurdhi Authority. At present Malani employs four women of her village who were previously unemployed but now earn a considerable income.

Figure 10.3: Fruit Drink Manufacturing Processes

The process of producing rice flour based dehydrated string hopper are depicted in the Figure 10.4.

Figure 10.4: Production of Dried String Hoppers (Food item)

Case Study 03 - Coconut Shell Creations

Before Mr Nishantha Dilruk was guided to the Vidataha Resources Centre at Bandaragama (Kalutara District) by the Kamburugoda East Grama Sewaka Officer, he made his living by making spoons with coconut shells. When the officer at the Vidataha Center saw Nishantha's beautifully crafted spoons of various sizes and designs, he was guided to explore the new areas in the field where he would make the best use of his creative talent. Nishantha was shown the photos of crafts made with coconut shells on the internet and the Vidatha officers introduced him with new equipment too.

Nishantha, after having convinced with the new creations he could do with his talent, continued to experiment and his products were shown to the Vidataha officers who guided them pointing out the areas to improve. Vidatha helped him to get a loan from the Peoples' Bank and he took part in the entrepreneurship development and positive thinking workshops conducted by the Vidataha Centre. Nishantha was given a chance to sell his products at the stalls of Eco-friendly Products exhibition organized by the Ministry of Technology and Research and he was also directed to other entrepreneurship development exhibitions at the BMICH. Through these exhibitions, not only he could sell a large amount of his products but also he got the opportunity to make contacts with a few companies who wanted to buy his products on wholesale basis. One of these companies was a reputed exporter.

Nishantha faced with problems in flattening the surface of the coconut shells and Vidatha Centre directed Nishantha to ITI where his problem was solved. With his problems solved, Nishantha's production capacity increased and gradually he paid attention to making new creations with natural material. The creations made with Bamboo are one such example and he was introduced to the bamboo boron treatment by the Vidatha Centre.

Vidatha Centre's support to Nishantha was not limited to providing guidance and introducing new technology. Nishantha was given a basic knowledge about marketing his products and a few companies were contacted to buy his products. At present with the help of Vidatha Centre, he is trying to increase the speed of his production process to meet the demand. Nishantha informed that before he came to Vidatha he earned around SL Rs. 7000/- per month by selling coconut shell spoons and now he earns around SL Rs. 25000/- per month.

The Figure 10.5 show various handicraft created from coconut shells, after subjecting them to technological process.

Case Study 04 – Jam and Fruit Cordial Producer

Mrs. Siriwardana, with her knowledge of the technology for making Jam, started her business at a place with two rooms in the year of 1998 with a capital of SL Rs. 5000/- only. At the beginning, a domestic grinding machine was used for making Jam and the entrepreneur went from house to house in the neighborhood to find markets. Later a motor bike was used for this as the business expanded including the demand from the shops and bakeries.

This entrepreneur joined the Biyagama (Gampaha District) Vidatha Centre in 2006 in order to improve her technological know how about the problems she had with making jam but it was not limited to that. At the Vidatha Centre, she also learned about making chutney, fruit cordial, pickles etc., and expanded her business significantly in the next few years. She started business with a grinding machine with 8Kg capacity but soon she bought another one, to meet the increasing demand. She expanded her market and started distributing her products with a hired van next and soon bought a van. At present she does her distribution using a lorry she has bought. The business place which was just two rooms at the beginning has now been moved to a new two storied building.

Figure 10.5: Production of Handcrafts Using Coconut Shells

Mrs. Siriwardana attended the workshops of the Vidatha Centre and equipped herself with the technological know how as well as knowledge about the business. She got the opportunity to market her products at the exhibitions organized by the Vidatha Center and exhibitions namely INCO, Techno etc., that were held in Colombo which gave her a good exposure and she got an opportunity to increase her sales. She is now supplying her products to both normal and super markets and her participation at the Eco-friendly Product Exhibition organized by the Ministry of Technology and Research that was held at the Hilton Hotel, Colombo let her to try her samples with the hotel. Despite her business growing each day, she still keeps contact with the Vidatha Center and her next aim is to get the GMP certification for her products. The Biyagama Vidatha Resource Centre hopes to give her the fullest support to achieve her targets.

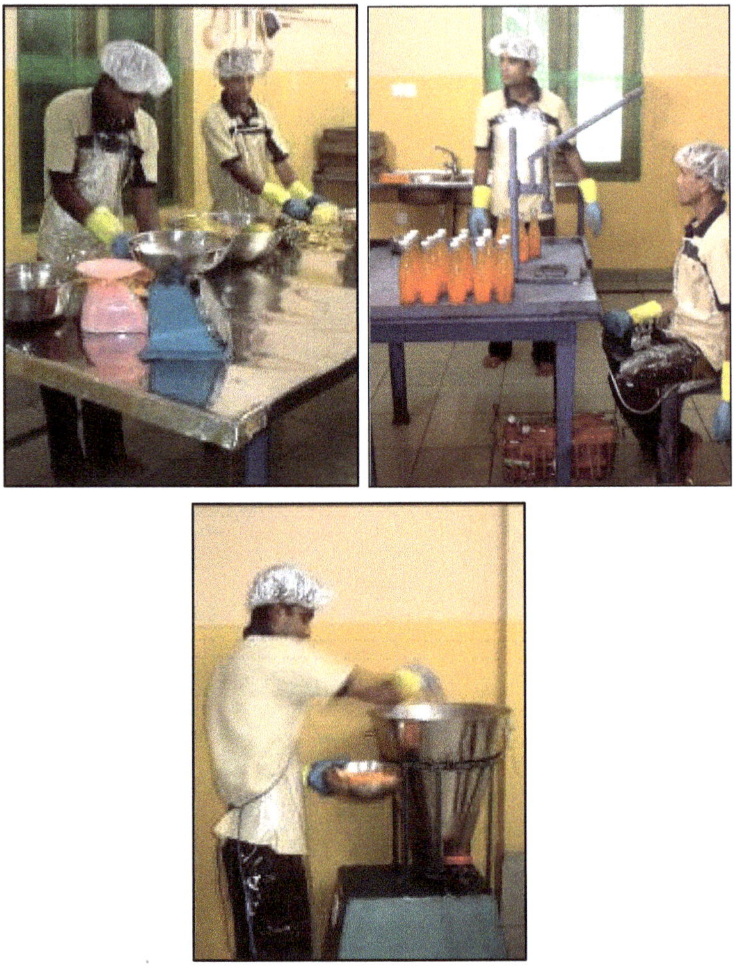

Figure 10.6: Project on Dehydrated Fruit Products and Manufacturing Jam and Cordial

Experience Gained Whilst Working In the Vidatha Programme

The author had the chance to gain much valuable experience while activating the Vidatha programme at a rural level. In transferring the technology to village level, it was learned that the simple and medium technology that showed best results. The major reason for this was that the rural community could get used to these technologies easily and the investment at the beginning of the business was at a low level. It is emphasized that the science, technological and research institutes must assist in adding value to the resources and material that are abundant and wasted in rural areas that can be conducted through the Vidatha Resouce Centres established under this programme. It was also observed that the high technology is not productive when implemented for rural self employment and small scale businesses.

Before transferring technology, the general public must be made aware of the attempt of the Vidatha Concept, and their attitudes must be changed and their confidence must be built up. Therefore, it is important to provide them with the knowledge, training, and other relevant information for building up the entrepreneurship ideas before technology is transferred to them. Further, when transferring the technology to the public, simple language must be used without any complications, technological terms and words from languages other than their mother tongue. If the communication medium is clear enough for the public that will encourage and motivate the rural community to actively take part in the programme. On the other hand, the officers with a science background who bear the responsibility of transferring the technology to rural community must be trained to understand the public needs and social issues interlinked with such needs. They also must interact well with the public.

Another interesting feature observed by the author was that among the rural entrepreneurs, women play a major role and they were more committed than men when participating at the Vidatha programmes as well as in handling their business. This tallies with the observation made by Wickremasinghe (2010) that women, as micro and small entrepreneurs are found to be more reliable and more responsible than men, and are more conscious of paying back money loaned to them by the banks, though women start business at a disadvantage because of prevailing cultural and social norms.

The rural women's enthusiasm towards taking up entrepreneurship is clearly visible, throughout the implementation of Vidatha programme. Currently, projects are being initiated to add value to many resources in rural areas, which are found in abundance by blending them with technology. During a period of 5 years, various technologies have been transferred to rural level and about 7000 Vidatha entrepreneurs have been created. It has been confirmed that more than half of these entrepreneurs are women. This makes obvious the fact that the women represent the majority of those who embark on entrepreneurship, when transferring technology to the village. This has resulted in the subsequent increase of the family income and systematic and gradual enhancement of their economic and social levels. On the other hand, by adding technology to a resource, which is going to waste in large quantities in rural areas, a value added good can be produced. This paves way to the increase of income of rural folks and also strengthens the rural food security. The nutritional levels are also upgraded as the people tend to retain food for their consumption. Also, youths who failed to enter universities tend to start business in their own, by undergoing training on electronics, computer software and hardware, mobile phone repairing, UHF,VHF antennae manufacturing etc.

Challenges Ahead of Vidatha

The biggest challenge faced by the Vidatha programme is the finding of market opportunities for the products of the micro, small and medium Vidatha entrepreneurs. It is important to establish Vidatha shops at a district level and the Vidatha Sales Corners must also be established in the existing super markets.

The shortage of Science and Technology Officers to start new Vidatha centres and the present officers resigning from the job is another problem faced by the programme. This has mainly affected establishing the Vidatha Centres in the provinces of East and North, which needs fast development programmes after the end of 30 years of civil conflicts.

Lack of funds is also another problem faced by the programme where the government attention is needed. People's attitudes are hard to change as this problem comes up when making the rural population get used to the new technology. It's rather difficult to change their traditional views and habits.

Lack of sufficient advertising and publicity about the programme is another gap observed because the people get little information about the programme and this in return has led to a communication gap. Also there is no proper mechanism to flow the information to the people.

Conclusion

Vidatha "Gamata Thakshanaya" (Technology to Village) Programme, strongly endorsed by the Maninda Chintanaya, the government policy document as a tool to uplift the Village, has been a new thinking of the former Minister of Technology and Research, Hon. Prof. Tissa Vitarana, which has enormous opportunities to uplift the lives of the rural people. Now it has become a major development project at rural level, providing job opportunities for unemployed and an extra income for poor families, not only in increasing their monthly income, but also in foreign currency, when products get into international markets. Through introduction of new technologies, it can help micro, small and medium entrepreneurs to produce quality goods that are suitable for foreign markets as well as competitive local markets. It is a programme that needs to go ahead by learning through mistakes and adjusting to the new requirements as required for the socio-economic development of the country along with the global trends.

Acknowledgement

The efforts of Hon. Prof. Tissa Vitarana, former Minister of Technology and Research (then Minister of Science and Technology) in establishing the Vidatha Resource Centre Programme in Sri Lanka are gratefully acknowledged. The author appreciates and thanks Dr. Seetha I. Wickremasinghe, Head/Science and Technology Policy Research Division, National Science Foundation, Colombo, for having useful discussions with her expertise in S&T interventions in rural development programmes, and for close guidance during the preparation of this paper.

References

1. NSF. (2010). Sri Lanka Science, Technology and Innovation Statistical Handbook 2008. National Science Foundation, Maitland Place, Colombo- 07, Sri Lanka, 11p.

2. The Central Bank of Sri Lanka. (2008). Annual Report. Central Bank of Sri Lanka, P.O. Box 590, 30, Janadhipathi Mawatha, Colombo 01, Sri Lanka.
3. Wickremasinghe Seetha I. and Dilrukshi P.R.M.P (2010). Women entrepreneurship and innovation system: a case study on mushroom enterprise in Sri Lanka, Third World Organization for Women in Science (TWOWS) Abstracts, International Conference on Women in a changing world held from 26-30 June, 2010, Chinese Academy of Sciences, Beijing, China, 391p.

Chapter 11

Environment Protection for Sustainable Development of Occupational Villages

Nguyên Van Mien,

Deputy of Chairman,
Vietnamese Occupational Village Association, Hanoi, Vietnam
E-mail: ngvanmien@yahoo.com

ABSTRACT

Violation of environmental standards is a serious issue among occupational villages, causing pollution and spoiling public health, as well as affecting the existence and development of occupational villages themselves. The occupational village is the life earning way of farmers and it is an important policy in economic development of the country. The most concerned issue of the occupational villages at present is the environmental pollution. Productions in occupational villages has polluted the environment at different levels and has also made changes to micro-climate conditions that has given birth to many diseases. This has further led to the reduction in average life span of people living in the occupational village and is causing bad impact on the agricultural cultivation of the region. Environmental pollution in occupational villages also affects the tourism activities and causes conflicts among the community. Environment protection and pollution treatment in occupational villages can be done by recognizing the responsibility of the occupational villages and strengthening the State role in environment protection in occupational villages. To contribute to the environmental protection and sustainable development of occupational villages; Vietnam Association of Occupational Village has been conducting many beneficial activities.

Keywords: *Environmental pollution, Vietnam, Occupational villages, Economic Development, Public Health, Environment protection.*

Introduction

The environment protection must be taken into the account for sustainable development of occupational villages. At present, the violation of environmental standards is a serious issue among occupational villages, causing pollution and spoiling public health, as well as affecting the existence and development of occupational villages themselves. Therefore, the current environment protection activities for development of occupational villages must abide by the environmental standards while addressing the serious environmental pollution caused due to the activities of the occupational villages. These are obstacles and challenges of the occupational villages and must be carefully considered during the development of occupational villages.

Overview of Current Environment Protection in Occupational Villages

About the Occupational Villages in Vietnam

Currently, there are about 2.790 occupational villages in Vietnam, providing jobs for 30 per cent of the rural labour force. With more than 1.4 million rural households involved, the occupational villages are mainly located in the northern plain.

The present development of occupational villages not only provides the job opportunities for farmers during the harvest intervals, but also it is a life earning solution for farmers. During the oncoming economic restructure progress, this is an important solution for farmers, helping to re-distribute the labor force, provide job opportunities and improve the living standards of the farmers, when cultivated land is narrowed down due to industrial and urban development. Development of rural occupations is a key element in forming a new rural structure in Vietnam.

There were ups and downs in the history of occupational villages, but they always continue to develop, especially at the time of renovation and integration. The products of occupational villages are becoming more attractive and have been sold nation-wide and abroad. Export turnover of occupational villages has increased sharply.

Besides the positive effects of becoming more and more important in socio-economic development, the occupational villages face difficulties and obstacles. In general, the development of occupational villages is still spontaneous, in particular, the use of minor portative production methods. There is a little linkage and co-operation among the occupational villages and households. The results of this weakness in linkage are: low competitiveness of products, high production cost and difficulty to introduce the products to the market due to low attractively products, inadequate production sites and the capital, out-of-date technology and low quality of the labour force. The recent global financial crisis and economic depression have made these obstacles harder and more difficult to overcome. But the most concerned issue of the occupational villages at present is the environment pollution, which is becoming more and more serious, spoiling the rural environment and the public health, as well as threatening the existence and development of occupational villages.

Outlines of Current Environment Pollution in Occupational Villages

Researches and studies have been already carried out to identify the status and impact to environment in occupational villages, and the common result is the productions in occupational villages which has polluted the environment at different levels, even very often exceeding the allowed standards. The main reason for this pollution is poor hygiene activities: all kinds waste including solid, gas and liquid (solvent and wastewater) containing toxic elements are discharged into the natural environment without prior treatment. This causes terrible air pollution, water pollution (including surface and undergroundwater) and soil pollution where toxic inorganic chemicals remain in the soil forever causing long term pollution. Another problem is the production in the occupational villages which makes changes to microclimate conditions *viz.*, air, dust, temperature, light, noise and vibration etc., by often exceeding the allowed standards.

The immediate and long term impact of environmental pollution in the occupational villages is clear, and must be considered seriously.

Impact of environment pollution in occupational villages to public health is bad because the production sites of occupational villages locate among the residential areas, and therefore, not only the direct producers/workers but all residents in the area and the surrounding areas are affected by the environmental pollution.

At the occupational villages, the most common disease and symptom are mental stressed disorders (headache, insomnia, vomiting); ear-nose-throat problems; eye problems; skin diseases; digest related ailments; metal intoxicated ailments and cancer. The percentage of affected people in occupational villages is in the rising tend. The average life span of people living in the occupational villages is reducing, 10 years shorter compared to the average life span of people in the country and 5-10 years shorter than that of people living in villages without occupational production.

Environmental pollution in the occupational villages also causes bad impact on the agriculture cultivation of the region because the polluted wastes from these occupational villages in turn spoil water, air and soil, threatening the existence of animals and plants living in that area, and also reduce the productivity of crops and the aquaculture products.

Environmental pollution in occupational villages also affects the tourism activities badly. Many villages receive tourists but the polluted environment; narrow pathways, stinking smell and dusty air and smoke coming out of the production workshops send them away. Due to this reason, the number of tourists coming to the occupational areas gets reduced sharply.

Environmental pollution in occupational villages causes conflicts among the community (*i.e.* among the occupational and un-occupational households, and villages). However, the environment protection activities receive less concern than the economic benefits gained by the people and the country as a whole. The followings are the main reasons:

1. The people's awareness about environment protection is very poor at both management authority level and the producing household level. This is the result of a traditional, long lived habit of the individual manufacturers who only care for short term but not long term effects and public benefits.
2. The organization and management ability of the occupational villages is poor due to: small scale and closed manufacturing process, mainly carried out in households and small enterprises; the linkage and co-operative manner in producing and selling products is limited; the existing obstacles faced by the occupational villages; and the backward infrastructure in rural areas etc., jointly cause the high production cost with low benefits. This makes it difficult for occupational villages to expand their production or apply new manufacturing technology, especially investing on a proper pollution treatment system.
3. Organizing and managing role of the state to occupational villages and environment protection is inadequate and unsuitable. For example:
   - ✰ The development of occupational villages is spontaneous with no planning strategy and policy for a sustainable development, resulting in the weakness and obstacles to be faced by the occupational villages at the moment.
   - ✰ Due to this spontaneousness, the environment protection in the occupational villages is not controlled from the beginning, resulting in large scale and serious pollution, about which is now well aware of and must be put under control. The establishment of the Law on Environment has just met the demand, but it also put the occupational villages under more difficulties: environment pollution which is spread and serious, is illegal but can not be solved immediately and occupational villages can not be kept functional in a continuing polluted condition. There must be a solution and a process to solve the current pollution in occupational villages.
   - ✰ The state management in production and environment protection in the occupational villages is inadequate. The legal documents on environment protection do not identify the standards for each kind of occupations. There is no instruction for supervising the environment protection of occupational villages from the construction stage and during their operation. The toxic and polluted occupations must be considered as the ones that are under conditional operation. Besides, the propagation activities on environment protection are not appreciated and/or conducted regularly. Responsibilities of the local authorities at all levels are not properly recognized, and there is no liaison and cooperation between manufacturers, researchers and social career organizations in regard to environment protection.

Some Solution for Environment Protection for Sustainable Development of Occupational Villages

Due to the serious and continuous pollution, despite the continuing efforts of local authorities and the occupational villages themselves, despite the actions that have been taken, the environmental pollution situation has not improved, but yet got worst. In the coming years, the occupational villages must be kept developed, because it is the life earning way of farmers and it is an important policy in economic development of the country. The environmental pollution in occupational villages must be soon solved because it threatens the health and lives of people living in the villages as well as in the whole community. What should be done by the villages and the society? The solution for this environment pollution in occupational villages must be defined and must take action.

Environment protection and pollution treatment in occupational villages need comprehensive solution: propaganda programmes must improve its management and the public awareness; improve the legal awareness on environment protection; solution on production management, appropriate policy and economic mechanism; and application of advanced science and technology. Therefore, responsibilities of management authorities and the manufacturers themselves must be properly defined. It is necessary to call for closer community cooperation between the manufacturer and S&T agent with the participation of career associations

Recognizing Responsibility of the Occupational Villages

Environment protection is the first responsibility of each occupational village, of each producing establishment and of each resident in the villages. The regulations of environment protection must be strictly followed from the set up of any producing establishment and during their operation progress. These organizations must be aware of the danger of pollution and must fulfill their duties relating to environment protection policies, such as paying the relevant taxes and fees on the environment and contributing to the environment protection funds.

With current pollution status in the occupational villages, each of the manufacturing establishment and each of the village (through the local career associations and with instructions from local authorities and professional agencies) should define and categorize the types of violations and pollution to find the suitable solution for each.

Each manufacturing establishment should define their direct responsibilities to solve the problem, from improving their awareness to changing their production technology and equipment used. They should join the clean water program, and recycling wastes to produce gas and other sub-products. The manufacturing units, unions and community in the occupational villages who are benefited by the environment should co-operate to protect the environment through regulations and conventions, such as the convention of keeping the producing and living environment clean along with the places they work and the public pathways. The garbage should be disposed at the collection sites only, but should not be disposed at the public places. Regular cleaning times for public places, drainage and sewage system must be planned while changing the traditional bad habit of "let it be, there will be someone else does the cleaning" by a more positive attitude "always make things clean".

Construction of the environment protection infrastructure such as drainage and sewage systems; local solid waste bury sites; encouraging the individuals and organizations to invest on infrastructure of the occupational villages and on waste treatment system; forming the waste management co-operatives; and making use of the waste to produce biological fertilizer are some of the solutions suggested.

For the works that can not be done at the producing establishments, a consultancy service is needed and support should be given by the higher authorities. This is because of the fact that, the producing establishments do not conduct means for environment pollution treatment, partly because of their poor understanding and awareness, and partly because of too high financial investment even when they are aware of the situation. Only some occupational villages are affordable for a proper treatment system for water, gas and solid waste to reduce environmental pollution, or changing their production technology, applying clean technology.

Strengthening the State Role in Environment Protection in Occupational Villages

1. Vietnam already has a system of functional authorities to manage and protect the environment in occupational villages, but their management functions are not defined clearly. There is overlap in activities in managing the occupational villages between the Ministry of Rural and Agriculture Development and the Ministry of Industry and Trade. The environment management functions of the central government and the local authorities are not clearly identified. There are no specialized personnel to manage the environment issues in occupational villages. Hence, it is time for functional authorities to work practically and more effectively to help improve environment conditions in the occupational villages. Policies and laws on environment protection must be put in place and strictly followed. There should be policies and guidelines for financial support to help occupational villages to function better in the problem of environment protection. A comprehensive and long standing solution including education, improvement of awareness, planning and changing policies on economic aspects, use of technology and act according to the laws is essential.

   Further, the environment protection works should not be done within one village but must be carried out in the whole region because in many cases, the pollution is spread. Now, it is necessary for occupational villages to evaluate and categorize the pollution types according to the nature and damages to the environment to find out a suitable solution for each category. Responsibilities of the occupational villages should be considered as the first priority. For the works that are beyond the capability of the villages, the state should provide support through projects and implement necessary supportive policies such as removing the pollution makers out of the residential area. The upgrading of production machinery and applying new technology need cooperative investment by the national state and the manufacturing establishments themselves.

2. It is necessary to reorganize and rearrange the villages, to protect the environment and ensure a sustainable development of occupational villages. There must be policies encouraging the establishment of enterprises in rural area and the occupational villages in large scale by planning the production in a specialized manner, by making each unit responsible for one producing state of the product so that there will be a close cooperation from aspects such as researching and designing, supportive services, producing as per consumer requirements and demand etc., and more importantly, there should be opportunities for the occupational villages to apply new science and technology, to gather the highly potential pollution makers for building pollution treatment system, and thus minimizing the environmental pollution in the particular area.

   To effectively implement the conservation and development of occupational villages, the planning works must be adhered to stop the traditional spontaneous development, which caused pollution and weakness of occupational villages. The planning works must also follow the business strategies of the villages and the market, based on the geographical location advantages, traditional occupations and the skills in each region. The planning of rural occupation must associate with the establishment of industrial and handicraft parks, making the concentrated manufacturing zones and helping to move the serious pollution makers out of the residential area. This is a comprehensive solution to develop the occupational villages sustainably and solve the current pollution situation by associating the development occupational villages while reorganizing a modern and civilized residence area and helping to build up new and modern countryside.

   The planning of the industrial and handicraft zones must be conducted synchronously in the production site, while improving the infrastructure such as roads, electricity and water supply, communication system, wastewater collection and treatment system, collection of solid waste for concentrated treatment etc.

   It is necessary to overcome the weaknesses in implementation management and illegitimate causes, when planning the occupational village zones only with the aim of taking land for both producing and living purposes while forming new clutter residential areas. In these areas, there are both producing and living activities with poor infrastructure conditions and pollution, which need strict regulations on management plus designation.

   The state should have supportive policies for: farmers to develop their occupations, forming new countryside by investing on the infrastructure of occupational village zones, encouraging the use of science and technology, training for improving quality of the labor force, reducing costs, implementing the environment standards, and increasing competitiveness of occupational villages' products.

3. Strengthening the legal education and better implementing regulations on environment protection in the occupational villages.

   The legal provisions on environmental protection were included in the Law on Environment in 2005 such as decree no.67/2003/ND-CP on environment protection charges for wastewater, Decree no. 174/2007/ND-CP on environment protection charge for solid waste and so on. However, the system of legal documents on environmental protection is not yet completed and specified. There have not been any legal documents specified for the environment protection in occupational villages based on the nature of each producing categories. Although there were general provisions, but so far almost all farmers in the villages have no environmental impact report or registration of environmental standards. Environmental protection function of the Ministries, authorities and especially the responsibility of the People's Committees at all levels are unclear and overlapped. The implementation of management tools is still weak, and there is a lack of human and financial resources for protection environment in occupational villages.

   As for the sanction of administrative violations in environmental protection, there should be serious and deterrent sanctions for the producing units and villages that intentionally violate the rules. However, specific characteristics of each village should be taken into account. That is, production establishments in the occupational villages are spontaneously formed for many years, mainly the households whose producing activities are conducted in their house interspersing within the residential areas. The environmental pollution has lasted for years, being aware by the people but due to the lack of capital, environment protection standards could not be applied. At present, if all the legal provisions are properly applied, most of these producing establishments are liable to be sanctioned.

Works implemented by the Vietnam Association of Occupational Village

To contribute to the environmental protection and sustainable development of occupational villages, as a social - occupational organization, the Vietnam Association of Occupational Village has been conducting the following activities:

Firstly, the Vietnam Association of Occupational Village has actively collaborated with other functional agencies to strengthen propaganda to raise awareness and law observance sense on environment protection in occupational villages. At the same time, the shared initiatives and experiences of the villages whose environment treatment works are done well, so that the other villages can study and actively apply same in their village. Each village must be aware that the implementing of the environment protection method is important to the sustainable development of each enterprise and of the whole village.

The Vietnam Association of Occupational Village has performed a bridging role by transferring the research results and environmental pollution treatment technology of researchers and experts to villages to apply in large scale, along with other functional authorities, research and implement projects on environment of occupational villages. The Association is planning to set up a support center for environment protection and climate change response, to execute this task.

Conclusions

It is important that the functional authorities enhance their support to the producing establishments, together with the unions and occupational associations, for actively carrying out the environment protection measures; conducting their production works associated with environmental protection for the sake of production requirements as well as community health. Through such approaches, the occupational villages, in future, will develop in a sustainable manner and promote their economic and cultural values, while contributing to the development of the rural area as well as the whole country.

Chapter 12

Rainwater Harvesting (RWH) to Generate Sustainable Solutions for Drinking Water and Create Local Employment in Karnataka, India

A.R. Shivakumar

Principal Investigator – RWH,
Karnataka State Council for Science and Technology (KSCST),
Indian Institute of Science (IISc), Bangalore – 560 012, India
E-mail : rwhkscst@vsnl.net

ABSTRACT

In most Indian cities the municipal water supply is very unreliable. Many villages in India do not have potable water supply. Hence Rainwater Harvesting (RWH) has been proposed as an ideal sustainable solution.

"Rainwater Harvesting in Rural Karnataka" a collaborative project between the Rural Development, Panchayat Raj Department (RDPR) and Karnataka State Council for Science and Technology is aimed at providing safe drinking water in rural schools and also train rural youths for establishing rainwater harvesting structures for the local needs.

All the schools of the Karnataka State were carefully studied for the water availability, quality (fluoride contamination) and proximity to the secondary water source. Under this programme 23,683 schools were identified in the first phase to provide drinking water.

The RWH system for schools is a collection of roof top rainwater and channeling it through the PVC pipes or gutters and stored in a fully enclosed surface tank. The so collected rainwater is filtered through a sand bed filter built over the surface tank before storing. The first flush separator is a part of the pipe line, which allows the first rain to be flushed out with the contaminants on the roof and subsequently, the relatively clean water is allowed to pass through the sand bed to filter and after that it is stored in a tank.

A systematic training and awareness programmes were conducted for 2,525 people in all 27 districts of Karnataka. These trained engineers, planners, contractors, plumbers and officers are involved in the installation of RWH systems in the schools and monitoring the performance.

This programme generated opportunities for local employment as well as raised enthusiasm among youth to take up the system of integration for rainwater harvesting

Keywords: *Fluoride contamination, Groundwater recharge, Local employment, Rural development, Rainwater harvesting.*

Introduction

In most Indian cities the municipal water supply is very unreliable. Many villages in India do not have potable water supply. Hence Rainwater Harvesting (RWH) has been proposed as an ideal sustainable solution. The Karnataka State Council for Science and Technology (KSCST) has undertaken steps to promote, implement and create awareness on sustainable water harvesting and groundwater recharge concepts among people. Rainwater harvesting principle entails the collection of rainwater where it falls in a scientific and controlled manner so that the water can be stored for future use (*1*). RWH consists of roof top water harvesting, water from open areas such as paved ways, parks, roads, fields and in lakes and ponds and so on.

Among the three projects that were initiated by the Council, the "Rainwater Harvesting in Rural Karnataka" was funded by the Rural Development, Panchayat Raj Department (RDPR) and the government of Karnataka in India, which is probably the largest Rainwater Harvesting Project in the world. All the schools of the Karnataka State were carefully studied for the water availability, quality (fluoride contamination) and proximity to the secondary water source. Under this programme 23,683 schools (*2*) were identified in the first phase to provide drinking water that was collected through roof top Rainwater Harvesting method, which is a very simple but rugged system. Given below are the baseline data relevant to the average annual rainfall and the schools considered under the project (Tables 12.1 and 12.2).

The Rainwater harvesting System (RWH)

The RWH system for schools is a collection of roof top rainwater and channeling it through the PVC pipes or gutters (PVC pipes slit in to two halves) and stored in a fully enclosed surface tank. The so collected rainwater is filtered through a sand bed filter built over the surface tank before storing. The first flush separator is a part of the pipe line, which allows the first rain to be flushed out with the contaminants on the roof and subsequently, the relatively clean water is allowed to pass through the sand bed to filter and after that it is stored in a tank.

Table 12.1: Average Annual Rainfall and Number of Schools Receiving Water through RWH by Districts in the Karnataka State

Sl.No.	*District*	*Average Annual Rainfall (mm)*	*No. of Schools for RWH in Each District*
1.	Bagalkote	573	831
2.	Bangalore Rural	879	1602
3.	Bangalore Urban	936	729
4.	Belgaum	738	847
5.	Bellary	588	529
6.	Bidar	867	550
7.	Bijapur	606	829
8.	Chamarajanagar	771	417
9.	Chikmagalur	1785	1337
10.	Chitradurga	570	701
11.	Davanagare	607	754
12.	Dharwad	682	455
13.	Gadag	597	353
14.	Gulbarga	717	1380
15.	Hassan	956	2115
16.	Haveri	721	812
17.	Karwar	2722	1226
18.	Kodagu	2497	131
19.	Kolar	754	1116
20.	Koppal	602	429
21.	Mandya	741	462
22.	Mangalore	3797	518
23.	Mysore	828	524
24.	Raichur	640	790
25.	Shimoga	1717	1188
26.	Tumkur	715	2768
27.	Udupi	4075	290
	Average/Total	**1173**	**23683**

Source: Rural Development and Panchayathraj Department, Government of Karnataka, India.

In the design of the tank and the sand bed filter, care has been taken to maintain good quality of harvested water. The surface tank is deliberately built without a direct manhole for entry inside and one or more taps have been attached at the bottom. This prevents the sunlight coming from entering the tank (presence of sunlight and air together supports growth of bacteria, algae etc. inside the water). The absence of a regular manhole prevents withdrawal of water by dipping vessels and also

Table 12.2: Other Baseline Data Relevant to the Project

Number of schools for RWH programme in Karnataka	23683
Number of school days per year	210
Average number of student strength in each school	118
Average roof area of each school	147 m^2
Average rainfall of 27 districts of Karnataka (12yr.avg)	1173 mm
Average RWH storage tank capacity in each school	6300 L
Drinking water requirement per student per day	1.50 L
Rainwater potential from 23633 schools per year	4084 mL
Drinking water requirement per year for 23633 schools	880 mL
Total RWH storage tanks capacity	149 mL
Minimum number of times the RWH tanks required to be filled to meet the requirement	6

Source: Rajeev Gandhi Rural Housing Corporation, Progress report, Government of Karnataka, India.

avoids undesirable elements (lizards, cockroaches, dust etc.) entering into the storage tank. The bottom of the sand bed filter itself is fabricated as a manhole with perforations and can be removed for cleaning the tank when the sand bed is removed for periodic cleaning (Figure 12.1).

The size of the tank and the pipes for collection of water from the roof are designed by using the parameters like local annual rainfall, seasonal variation, number of students in the school and available roof areas.

Implementation of RWH System in Schools

A systematic training and awareness programmes were conducted for 2,525 people in all 27 districts of Karnataka. Currently these trained engineers, planners, contractors, plumbers and officers are involved in the installation of RWH systems in the schools and monitoring the performance. The School Development and Management Committees (SDMCs) in the local villages and the teachers of the respective schools took part for the planning, implementation and maintenance of the RWH system. A RWH manual called "Amruthavarshini" (A guide for Rainwater Harvesting) has been published by the Karnataka State Council for Science and Technology (KSCST) which gives all the details for effective implementation of the programme.

This programme generated opportunities for local employment as well as raised enthusiasm among youth to take up the system of integration for rainwater harvesting. One advantage of this RWH system is that one could use locally available materials to build the structures for water storage. Innovative, but this cost effective techniques were taught by the research group to the local youths in building structures with sustainable concepts. The programme had multi-level approach for skill development as well as imbibing management skills to carryout multiple tasks. Hands on training on plumbing to integrate and assemble pipes and pipe fittings of various shapes and

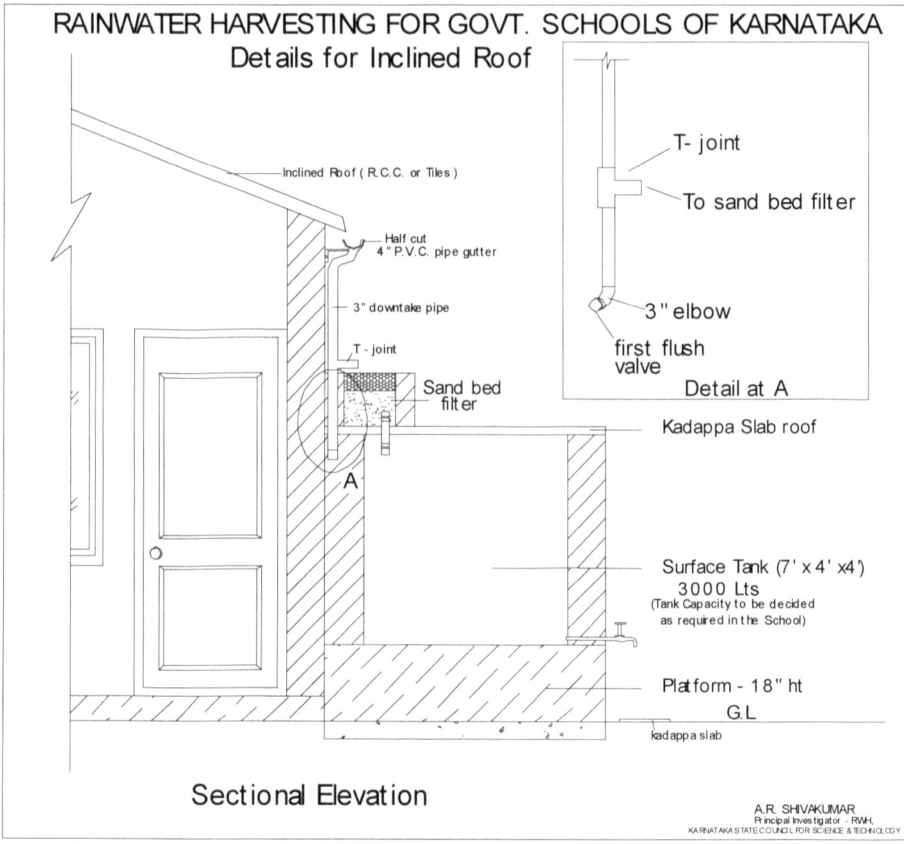

Figure 12.1: RWH System

sizes along with the techniques of filtering roof top rainwater was also provided to the youth. Multi-task approach to carry out plumbing and masonry work was provided to the youth who showed interest in learning and taking up local contract works for implementation (3).

At the same time, administration and monitoring techniques for the RWH Programme were devised to carry out large scale implementation of rainwater structures in the rural areas. Quality control and progress monitoring were undertaken by a combined group of teachers in the schools with the village development committees at the local level. School children were educated to monitor the water quality and also to keep an account of rainfall data along with the quantity of water collected from the roof in the rainwater tanks. The school children were directed in turn to educate their parents on the rainwater harvesting system through the practical experience, which they learnt in their schools helped in adopting rainwater harvesting in each individual house too (Figures 12.2–12.5).

Figure 12.2: RWH System Adopted in a Village School with Tile Roof

Figure 12.3: RWH System Adopted in a Village School-Tile Roof on Either Side

Figure 12.4: Sand Bed Filter Over a Rainwater Storage Tank

Figure 12.5: Rainwater Harvesting System Adopted in a Village School with Sloping RCC Roof

Conclusion

The rainwater harvesting system adopted in the schools of the Karnataka State helped in finding a solution for availability of drinking water. The chain of practically educating elders through young children created social change in the villages considered while maintaining a sustainable method of roof top rainwater harvesting in each individual house too. RWH programme for rural villages is a win - win situation for all the players in the society and also an effective tool in conservation measures for both water and environment.

References

1. "Amruthavarshini" Guide for Rainwater Harvesting, Published by; Karnataka State Council for Science and Technology, Bangalore, India.
2. Rainwater Harvesting Project Training Programmes - RDPR and KSCST Collaborative initiative for safe drinking water in rural schools.
3. "Suvarna Jala" Programme of RDPR, for Rural Water Supply, Government of Karnataka.

Chapter 13

Spreading the Green Aura of Medicinal and Aromatic Plants for Improving Rural Economy

Anil Kumar Singh

Scientist-G and Head,
Technology and Business Development Division (TBD),
Central Institute of Medicinal and Aromatic Plants (CIMAP),
Lucknow – 226 015, India
Email: ak.singh@cimap.res.in

ABSTRACT

India is the homeland of several economically important medicinal and aromatic plants (MAPs) which are being grown, processed and marketed by a large number of farmers and entrepreneurs. The country is in an advantageous position for cultivation of such medicinal and aromatic plants because of the existence of many types of climates and soil conditions. These plants gain popularity day by day owing to people's inclination towards use of naturals/ botanicals in various formulations useful as flavor and fragrances and also for general health care. Several government owned research organizations and institutions in India, especially Council of Scientific and Industrial Research (CSIR), Indian Council of Agricultural Research (ICAR), National Medicinal Plants Board (NMPB), Department of Biotechnology (DBT), Department of Science and Technology (DST), Indian Council of Medical Research (ICMR), state and central agricultural universities and even private organizations are now focusing their efforts towards development of know-how and technologies for profitable cultivation and value addition to the MAPs raw materials followed by systematic dissemination for socio-economic gains especially in the rural areas. Such activities have significantly contributed in generating employment opportunities, thereby improving the quality of life of the poor farmers. The work carried out by the CSIR's premier laboratory, Central Institute of Medicinal and Aromatic Plants (CIMAP), over the last about 50 years has clearly established that medicinal and aromatic plants based know-how and technologies could empower the poor farmers and unemployed youth including women towards self reliance with an additional income and improved livelihood. These plants can improve the agri-economic growth of small growers in India by strategically introducing some economically viable high value medicinal crops (as cited as model example set by CSIR-CIMAP in 'Bio-village' mode). The major work done in the area of MAPs, based on Science and Technology (S&T) applications for empowering rural people has been presented in this paper.

Keywords: *Medicinal and aromatic plants, Rural development, Bio-villages, Employment generation.*

Introduction

In India, medicinal and aromatic plants (MAPs) collected from wild source have been used since ancient times either for curing various ailments or providing fragrances and flavors. There has been tremendous amount of genetic erosion of such plants owing to over exploitation of MAPs, thus necessitating their systematic conservation and utilization. India is blessed with various types of agro-ecological zones making it possible to grow and nurture different MAPs in one or the other parts of the country.

It is estimated about 100 millions of farming families in India have the lands less than a hectare and farming as such has witnessed acute problems of dwindling natural resources and diminishing net return from conventional crops. Therefore, it was pertinent to introduce some economically viable specialty crops such as MAPs.

Cultivation of medicinal and aromatic plants (MAPs) besides providing sustainable means of natural source of high value industrial raw material for pharmaceutical, agro-chemical, food and cosmetic industries, also opens up new possibilities for higher level of gains for farmers with a significant scope for progress in rural economy (Khanuja *et al.*, 2005). Though these plants have been known and used since ancient times to heal and cure diseases, recently with technological advancements, validation of traditional knowledge and usage is leading to consumer inclination towards naturals and high market value of these crops. Such crops in India covering an area of merely about 0.4 million hectares find much higher place in the international agri-business with an estimated annual growth rate of 10-15 per cent. Currently, some of the important MAPs cultivated in larger scale in the country are Mints (*Mentha* sp.), Damask rose (*Rosa damascena*), *Ocimum* sp. (Tulsi) in the Indo-Gangetic belt; Lemongrass (*Cymbopogon flexuosus*), Geranium (*Pelargonium graveolens*) and Patchouli (*Pogostemon patchouli*) in the southern parts; Lavender (*Lavandula* spp.), Pyrethrum (*Chrysanthemum cineraefolium*), Clarysage (*Salvia sclarea*) in the temperate region of the Northern parts; Citronella (*Cymbopogon winterianus*) in the north eastern regions; and Ashwagandha (*Withania somniferua*), Senna (*Cassia angustifolia*) and Palmarosa (*Cymbopogon martinii*), Isabgol (*Plantago ovata*) in western and central parts of India. Some of the other medicinal plants namely Satavar (*Asparagus recemosus*), Kalmegh (*Andrographis paniculata*), *Rauvolfia serpetina*, etc., also cover small to larger areas in different states of India.

With the increase in demand of quality material for pharma, perfumery and flavour industries; the research and development (R&D) activities in MAPs have found place in the mandate of several research institutes, government organizations and promotional agencies (CSIR, ICAR, NMPB, DBT, DST, ICMR, etc.), the state and central agriculture universities and even in the private sector. Central Institute of Medicinal and Aromatic Plants (CIMAP), which is a pioneer institute of the Council of Scientific and Industrial Research (CSIR) of India, has an exclusive mandate for R&D in MAPs contributing significantly to the overall development of these plants in the country for about last five decades. The mission of CSIR-CIMAP 'To be ultimate in green technologies for better health' is the reflection of its vision and commitment towards social empowerment through plant wealth and taking up for the cause providing India's global leadership in this sector for world trade while simultaneously enabling rural India economically with *life par excellence* and healthy humans, life and environment. CSIR-CIMAP's significant contributions in the area of rural development have been recognized nationally with two prestigious awards, Federation of Indian Chamber of Commerce and Industry (FICCI) Award for Rural Development in 2005 and thereafter, CSIR Award for Innovation in Rural Development (CAIRD) in 2008.

Few other laboratories of CSIR namely Institute of Himalayan Bioresource Technology (IHBT), Palampur; National Botanical Research Institute (NBRI), Lucknow; Indian Institute of Integrative Medicine (IIIM) (formerly RRL, Jammu) and North East Institute of Science and Technology (NEIST) (formerly RRL, Jorhat) also have some projects/activities relating to medicinal and aromatic plants suitable for growing in the respective areas where these institutions are located and made significant contributions in providing know-how, processing technology, quality testing, literature and other related services to the growers and entrepreneurs.

MAPs Offering High Economic Return

Medicinal and aromatic plants (MAPs) are currently cultivated by the farmers in India provide high economic return ranging from about Indian Rs 35,000 to about Indian Rs 80,000/- per hectare. These crops can also be integrated into the existing cropping system enabling farmers to earn more income per unit area. Even different types of marginal and wastelands are also utilized through cultivation of some of the MAPs in India. Besides cultivation of MAPs, several ancillary activities can also be started *viz.*, establishment of nursery/seed production unit, distillation of aromatic crops, making of herbal extracts and powders, product formulations of common use and vermi-composting utilizing agro-waste, thus opening new avenues for job creation and additional income.

The current status of production and economic returns through cultivation of some of the important medicinal and aromatic plants and major R&D initiatives are undertaken at CSIR-CIMAP and in other institutions during the last few years are summarized below:

Ashwagandha (*Withania somnifera*)

Asgandh or Aswagandha (*Withania somnifera*) is an important medicinal plant whose roots have been employed in Indian traditional system of medicines, Ayurveda

and Unani. This shrubby bush plant grows well in the dry and subtropical regions. Therefore, ashwagandha is a potential cash crop for greening the dry land zones, making productive and beautifying the wasteland. It is cultivated in an area of about 5000 hectares of north western region of the Madhya Pradesh *i.e.* marginal lands in and around the villages of Manasa, Neemuch, Jawad and Bhanpura in the Mandsaur district.

Ashwagandha is also grown in adjoining villages of Kota, Jaipur, and Jodhpur districts of Rajasthan and Jammu forest (Yadav *et al.*, 2008). One hectare plantation of Ashwagandha yields an average six quintals of dried roots which are sold at about Indian Rs. 6000/- per quintal giving a net return of Rs 27,000 from a 5-months old crop. CSIR-CIMAP's initiatives in popularizing this important crop in Ananthpur district of Andhra Pradesh have shown a new hope of agri-economic development in otherwise a dry land area. The variety 'Poshita' developed by CSIR-CIMAP is presently grown in about 300 hectares of land by about 500 farmers due to the continued efforts made by CSIR-CIMAP Research Centre, Hyderabad under the institute's ongoing rural development programme. The State Medicinal Plants Board of Andhra Pradesh has also supported the project which enabled the farmers under the technical guidance of CSIR-CIMAP to produce and distribute approximately 3000 kilogram seeds, which will further facilitate the expansion of the crop and subsequent rising of income of the growers. Another variety named 'NMITLI- 118' developed by CSIR-CIMAP and NBRI was released by CSIR in 2009 for cultivation by the farmers.

Sarpagandha (*Rauvolfia serpentina*)

Rauvolfia serpentina roots, commonly known as serpentine roots or Sarpgandha, is one of the most important drug used in the traditional as well as in the modern system of medicine. It is a drug for various types of ailments, ranging from disorders of central nervous system, such as maniacal behaviour, insanity, epilepsy and insomnia to intestinal disorders. In the modern system, it is used as an antihypertensive and sedative drug. It is an erect, evergreen 0.60 to 1.0 meter high, shrub and grows in the Indian sub-continent, Sri Lanka, Thailand and Indonesia. It is generally collected from forests. Its cultivation is on the limited scale and is currently being promoted by different organizations and institutes as the plant has become rare in most of the accessible areas of its natural occurrence due to over exploitation. To facilitate the quality production of this plant, CSIR-CIMAP has developed an improved variety 'CIM-Sheel' along with agro-technology, which is being popularized presently through regular training programmes conducted by the institute.

The plants raised from seeds; give better yield of roots ranging from 100g to 400g per plant. It is estimated that with a spacing of 60 cm x 30 cm and survival of 80 per cent, the yield of roots per hectare from seed raised plants could be about 1,175 kg (dry basis). Average root yield/ha under irrigated condition from 2- year old plantation is about 1,000-1,200 kg. The roots could be sold at Indian Rs. 100 per kg, and a grower may get a net return of Indian Rs. 50,000 from one hectare.

Kalmegh (*Andrographis paniculata*)

Kalmegh is an important medicinal plant, which is employed in the Indian traditional system of medicine, mostly to cure liver disorders. It is widely distributed throughout the plains of India from Uttar Pradesh to Assam, Madhya Pradesh, Jammu, Tamil Nadu, Kerala, Orrisa etc. The whole herb is used as medicine. Its major bitter principle is andrographolide used in the bitter tonic. The whole plant has astringent, anodyne, antipyretic, anti-inflammatory, immunosuppressive and anthelmintic properties. Recently, the cultivation of Kalmegh has been started in India due to efforts made by CSIR-CIMAP and other organizations. Considering its pharmaceutical potential, there is a need to increase its large scale systematic cultivation in India. CSIR-CIMAP has developed a high yielding variety 'CIM-Megha' and released to the farmers for cultivation. Well maintained crops are grown in one hectare area during monsoon season yield 3.5 to 4.0 tonnes of dried herb giving a net income of about Indian Rs. 23,500 from a 3-month crop.

Satavar (*Asparagus racemosus*)

Asparagus racemosus, popularly known as Sataver, is an indigenous medicinal plant used in Siddha and Homeopathic medicines. Sataver roots are used mainly as

lactogogue, which promotes secretion of breast milk. It is useful in improving the body weight and is also considered as an aphrodisiac. In general, it helps to maintain the health by providing immunity to diseases. The demand for Sataver roots are on the increase and therefore, the destructive nature of harvesting the whole plant has resulted in shrinking of natural population. It is estimated that more than 500 tonnes of Sataver roots are required every year for medicinal preparations in India.

CSIR-CIMAP promotes cultivation of Sataver by providing seeds and saplings of the improved variety 'CIM-Shakti' developed recently, which takes about 18-20 months for the crop to mature. However, better yield can be obtained after two years of planting. The crop should be harvested during October and December (dormancy period) of the year. On an average, 6-10 tonnes of dried root could be obtained from one hectare and a net profit of about Indian Rs 300,000 from a two year old crop.

Quinghao *(Artemisia annua)*

Artemisinin, derived from the plant *Artemisia annua* (commonly known as Quinghao), is the only cure from chloroquine resistant *Plasmodium falciparum*. The demand of this drug is on the rise due to the approval of artemisinin combination therapy (ACT) by the World Health Organization (WHO) for the treatment of cerebral malaria. CSIR-CIMAP has developed agro-technology of this plant and genetically tagged superior varieties like 'Jeevan Raksha' and 'CIMAP-Arogya' produce significantly higher yields of artemisinin (about 1 per cent).

The manufactures of artemisinin derivative of the drug have been importing artemisinin from international sources but the non-availability of sufficient raw material has forced the Indian industry to depend on cultivation of this crop indigenously for self reliance. It is estimated that to sustain >100 million doses requiring to meet global need of malaria affected people, the crop may need to cover an area of about 25,000 hectares and with the current production levels, farmers may earn a profit of about Indian Rs. 40,000 per hectare in a span of 6 months for single harvest. The pharma companies are coming forward to go for contract farming for the

***Artemisia annua* var. 'CIM-Arogya' in a Farmer's Field**

planned cultivation and sustained production of herbage for isolation of active compound artimisinin to meet the global demand and also the national need to combat malaria. The agro-technology for cultivation of improved variety of *Artemisia annua* was licensed by CSIR-CIMAP to seven leading companies in India that promoted the cultivation of this crop involving the farming community. With the high artemisinin content in the Indian variety of *Artemisia annua* developed by CSIR-CIMAP, it is expected that the crop would be promoted by the pharma companies in the near future. The IPCA Laboratory, Ratlam, Madhya Pradesh has organized contractual cultivation of CSIR-CIMAP variety by involving farmers of Uttar Pradesh and Uttarakhand this year.

Menthol Mint (*Mentha arvensis*)

The Menthol mint oil, obtained on distillation of the herb, is a rich source of natural menthol used in pharmaceutical, flavor and cosmetic industries. About 45 years ago, all requirements of mint oil and menthol in India were met through imports. CSIR-CIMAP developed its agro-technology and as a result of extension of the crop to the farmers' fields with the involvement of menthol manufacturing industries and growers, now around 16,000 tonnes of menthol mint oil valued at about Indian Rs. 8 billions are being produced in the country, which is not only consumed within the country, but also exported to meet world's demand.

The plant is cultivated covering more than 150,000 hectares of land in the Indo-Gangetic plains of North India and has emerged at present as a major industrial crop. It is believed that over 200,000 farming families grow this crop regularly contributing to 75-80 per cent global menthol mint oil produce bringing them a profit of about Indian Rs 30-40,000 per hectare within a period of 2-3 months. Per capita productivities of two superior late planted genotypes 'Kosi' and 'Kushal' developed by CSIR-CIMAP have enabled farmers to take this crop as a bonus between *Rabi* (after

Harvesting of Menthol Mint Crop in Progress

wheat crop) and *Kharif* (before rice crop). Introduction of the new rotation based upon these newly developed short duration genotypes have enabled growers to achieve net profits of approximately Indian Rs 61,000 per hectare per year (Kumar *et al.*, 2001). CSIR-CIMAP has introduced another variety named 'CIMAP-Sarayu' in January, 2010, which is a high oil and menthol yielder and thus could further augment the net returns. Menthol mint have really transformed the socio-economic status of poor farmers with all round improvement in quality of their life with visible impact in the form of housing, nutrition with balanced meals to the family, mechanization of their farms and machinery including post harvest processing capabilities and their visible move towards literacy with their children going to school. In the past, *Mentha arvensis* was cultivated largely in various districts of Uttar Pradesh. Now, some new areas in the states of Bihar, Madhya Pradesh, Punjab and Uttarakhand have come up as production centres of this crop.

Besides *Mentha arvensis,* CSIR-CIMAP is also paying attention towards popularization of *Mentha piperita* (Peppermint), which requires similar agro-climatic conditions as that of menthol mint. At present, CSIR-CIMAP is developing improved varieties suitable for cultivation and to meet the demand of industry especially for menthofuran-rich oil with less pulegone content.

Geranium (*Pelargonium graveolens*)

Geranium oil is produced by steam/hydro-distillation of fresh herb of *Pelargonium graveolens* belonging to the family Geraniacae. Geranium oil is rich source of rose alcohols and is processed to isolate 1-citronellol (rhodinol), which is used very extensively in fine fragrances for its sweetness and delightful freshness. The current international demand of above 600 tonnes is largely met by China, Egypt, Morocco, Reunion Island and South Africa. Presently, India's production is not more than five tones geranium oil per year against its annual requirement of around 150 tonnes. Considering the heavy demand of the cosmetic and perfumery industry; export

promotion/import substitution; yield potential and economics of its cultivation; it is high time to produce geranium oil in India.

With good soil and better management, a yield of 20-25 tonnes of fresh herb can be obtained per hectare. The herb is distilled fresh and about 0.1 per cent oil on commercial scale is obtained from the herb. A second harvest of 5-10 tonnes of herb is also possible. Thus, a total of 30-35 kg oil per hectare would be obtained giving a net profit of Indian Rs. 60,000.

During the year 2001-02, CSIR-CIMAP conceptualized and developed a model for technology dissemination and implemented it on the farmers' fields in Uttarakhand resulting into a path breaking achievement and visual impact on the rural economy. It all happened with the launch of mission geranium with the support of Technology Information and Forecasting Council (TIFAC) of the Department of Science and Technology (DST) using CSIR-CIMAP Bourbon cultivar and thereafter 'CIM-Pawan', another improved genotype to sustain the quality production of geranium oil. CSIR-CIMAP had shown the way by not only introducing the improved cultivars of geranium capable of producing higher yields of oil with better quality, but also providing the processing facilities in the vicinity of the production areas and establishing linkages with the industry for buy back facilitation of the essential oil produced by the farmers. This was possible through concerted team efforts of CSIR-CIMAP scientists who worked in a mission mode adopting bio-village technology dissemination approach. In this mission programme of establishing a value chain through CSIR-CIMAP's geranium technology, the farmers in Uttarakhand could practically achieve an earning of approx. Rs. 1200/- per 200 m^2 area through geranium cultivation as compared to approximately 800-900 Indian rupees obtained through potato cultivation prevalent in the project area, which was taken as economic bench mark by the local farmers of the area. With the efforts of CSIR-CIMAP, the state of Uttarakhand has emerged as the best location of geranium oil production in India.

Also, from Uttarakhand hill, the crop spread to plains of Uttar Pradesh and CSIR-CIMAP has demonstrated that good quality geranium oil could also be produced in Uttar Pradesh. Realizing the economic returns that could be gained from cultivation of this crop, the State Horticulture Mission in Uttar Pradesh is at present planning to promote geranium cultivation among the farmers during this year.

Rose (*Rosa damascena*)

The rose oil obtained from the flowers of *Rosa damascena* is the costliest oil used in high-grade perfumes. The institute has also developed distillation technology for rose oil extraction. The recovery of oil was obtained from 0.025 to 0.03 per cent as against 0.01 per cent obtained using country-made stills. The area under rose cultivation in Uttar Pradesh alone is estimated as ~ 1500 hectare from where an estimated amount of about 100 kg oil that worth around Indian Rs. 20 million is produced annually. A Bulgarian strain of rose was introduced in Kashmir valley, which gives higher yields than those obtained from Bulgaria. CSIR-CIMAP has also developed a superior strain of Damask rose named 'Noorjahan' that is suitable for cultivation in the subtropical areas of Uttar Pradesh and the adjoining states.

Another variety 'Rani Sahiba' was also developed and released by CSIR-CIMAP. Two varieties named 'Jwala' suitable for plains and 'Himroz' suitable for hilly areas were also developed by Institute of Himalayan Bioresource Technology (IHBT), Palampur. About three to four tonnes of flowers are produced from one hectare of rose plantation in a year, which upon distillation yields about one kilogram of oil worth of about Indian Rs. 200,000 to 250,000. The State Horticulture Mission of Uttar Pradesh is planning to initiate a massive programme of rose planting this year for increased rose oil production. Once established, the rose plantations remain in commercial production for about 15 years.

Lemongrass (*Cymbopogon flexuosus*)

The oil of lemongrass, obtained from the leaves of *Cymbopogon flexuosus*, is used in the soap perfumery and is a source of citral, a chemical used in the perfumery and flavour industries. It is also used as a raw material for the manufacture of vitamin-A. About 250 tonnes of the oil valued at about Rs. 100 millions are being produced annually for internal consumption as well as for export. CSIR-CIMAP developed and improved agro-technology for cultivation of this plant. A superior variety, 'Pragati ', a citral-rich strain 'Praman' and a *geraniol*– rich strain, GRL-1 have been developed. Two other high yielding varieties, 'Cauvery' and 'Krishna' have been evolved at the CSIR-CIMAP's field station in Bangalore. 'Krishna' variety has a potential to yield up to three times more oil than the existing varieties, which is now being popularized among the farmers.

The lemongrass oil has a high demand in the world market. Guatemala, China, Sri Lanka, and Brazil are the other major producing countries. The production of lemongrass oil in India has declined in recent years but there are immense possibilities for production of good quality lemongrass for the domestic market, as well as for export. One hectare plantation of lemongrass yields about 175 kg oil giving a net profit around Indian Rs. 50,000 per annum.

The CSIR-CIMAP has made available the know-how for cultivation of this crop in the form of complete agronomic package, especially for those areas where adequate irrigation facilities are not available and have degraded soils or wastelands.

Palmarosa (*Cymbopogon martinii*)

The palmarosa oil is obtained from the flowering tops, leaves and over ground parts of the plant *Cybmopogon martinii* variety 'motia'. The oil is mostly obtained from wild sources and is used in the perfumery, cosmetic and flavor industries. Brazil, Indonesia and Guatemala are the major competitors to India in the world market.

The CSIR-CIMAP developed agro-technology and provided quality palmarosa seeds to the farmers in various parts of the country. It is estimated that 50-60 tonnes of palmarosa oil worth about Indian Rs. 35 million is being produced annually in the country. Two high yielding varieties namely 'Trishna' and 'PRC-1' have also been developed that give 40-50 per cent higher yield than the existing variety. Palmarosa oil is exported to France, Germany, the Netherlands, USA, etc., and ample opportunities are available for extending cultivation of the crop to areas where salt affected, cultivable wastelands are available. Brazil is the competitor of India in the world market. One hectare crop of palmrosa gives about 100 kg oil that results in a net return of about Rs. 45,000/- per annum.

Java Grass (*Cymbopogan winterianus*)

The citronella oil of Java quality is obtained by steam distillation of leaves. It is an important source of perfumery chemicals, such as citronellal, geraniol and hydroxycitronellal. Until 1973-74, the oil was imported to meet the domestic requirement of the country.

The institute introduced the plant to the country in 1962-63 and developed the indigenous agro-technolgy for the crop. As a result of transfer of technology by CSIR-CIMAP and CSIR-NEIST (Formerly RRL, Jorhat) more than 300 tonnes of citronella

Java oil worth about Indian Rs. 80 million is produced annually. The plant is grown in the north eastern states and in south India. Newer areas are coming under citronella crop in the north Indian plains owing to popularization of the improved technologies and cultivars by CSIR-CIMAP.

It has a great potential for cultivation in Uttar Pradesh, Madhya Pradesh and Maharashtra where good irrigation facilities are available. The CIMAP has also developed five high yielding varieties in citronella Java namely 'Manjusha', 'Mandakini', 'Manjari', 'Jalpallavi' and 'Bio-13'. These varieties are becoming popular among the citronella growers of the country. The oil is also produced in Indonesia, China and Sri Lanka, which are the major suppliers to the world market. One hectare crop of citronella under normal conditions yields about 250 kg oil per year worth of about Indian Rs. 50,000 as net returns to a grower.

Vetiver (*Vetiveria zizanioides*)

The vetiver oil is obtained by steam distillation of roots of *Vetiveria zizanioides.* The oil is used in high- grade perfumes and is extracted from the roots of wild varieties. The CSIR-CIMAP has developed agro-technology plus improved processing technology for the vetiver oil production and also developed two high yielding varieties, 'KS-1' and 'KS-2', yielding 0.7-0.8 per cent oil. An artificial tetraploid having upto 1.4 per cent oil in fresh roots was also developed.

Three new cultivars of vetiver, 'Gulabi', 'Kesari' and 'Dharini' were developed and released by CSIR-CIMAP. Another variety named 'CIM-Vridhi', which matures in about twelve month time, was developed and released recently to the farmers. The CSIR-CIMAP, over the last about three years, established a vetiver bio-village involving about 40 farmers and an area of about 10 hectares was brought under the cultivation of 'CIM-Vridhi'. The cultivation is likely to increase further owing to high yield with lesser inputs as compared to other crops. Currently, about 20 tonnes of oil is produced mostly from the roots collected from the natural growth. The plant offers immense possibilities for expansion, especially along the river beds. About 10 kg oil valued at about Indian Rs. 100,000 is produced from one hectare from a12-15 month crop, thus giving a net profit of about Indian Rs. 65,000 to the farmer.

Patchouli (*Pogostemon patchouli*)

The patchouli oil is obtained from leaves of the plant *Pogostemon patchouli* and is one of the most important essential oils used practically in most of the perfumes as the base because of its fixative property. Indonesia and China are the major exporters of the oil to the world market. Indigenous agro-technology for cultivation has been developed at the CSIR-CIMAP Research Centre of Bangalore and it has been found that the crop could be grown in coastal areas as an intercrop in the coconut plantations. India imports substantial quantities of patchouli oil to meet her requirement and there is a good scope to produce patchouli oil in order to save valuable foreign exchange to the country.

The patchouli plantation raised in one hectare gives about 50 kg oil valued at about Indian Rs. 80,000 and a net profit of about Rs. 60,000 can be obtained per year. The CSIR-CIMAP Research Centre of Bangalore has also launched a bio-village programme in a few villages in Siddapur Taluk, Uttara Kannada district in Karnataka in 2003 with the aim to provide an alternate source of income to the small and marginal farmers. These efforts have led to the cultivation of patchouli crop in nearly 300 acres of area involving about 350 farmers. The CSIR-CIMAP has also developed an improved variety of patchouli named 'CIM-Samarth' which is being popularized among the growers at present. This crop has also been included under the activities of the State Horticulture Mission in the Uttar Pradesh.

Tulsi (*Ocimum basilicum*)

The Indian basil (*Ocimum basilicum*) is a short duration aromatic crop, which is cultivated in commercial scale in India for the production of essential oil. The oil of basil owes its importance due to its extensive use in the perfumery and cosmetic industries. It is a short duration crop (45-60 days) in the mint growing areas as a rainy season crop. The average herb yield of basil was found to be about 20-25 tonnes giving oil yield of about 80-100 kg per hectare. The crop thrives well on moderate fertile and well-drained sandy loam oil. It can be grown in subtropical and tropical climate conditions, but temperate climate is not suitable for this crop. The CSIR-CIMAP has developed two improved varieties 'Vikarsudha' and 'CIM-Saumya'. One hectare crop of tulsi gives about Indian Rs. 12,000 net return and is very much popular among the growers because of its short duration, which permits the farmers to grow other crops from the same piece of land.

CIMAP's Bio-Village: A Mission towards Rural Entrepreneurship

In rural areas, economic empowerment demands that the farming community should make progress towards entrepreneurship. In this direction, a novel programme named as Bio-village Mission was started by CSIR-CIMAP about 9 years ago. This programme was given special impetus to popularize medicinal and aromatic plant (MAPs) cultivation so that farmers could be linked with industry with the aim of increasing production and ensure market opportunities and mitigating the unemployment problem in rural areas (Khanuja and Singh, 2007).

The Bio-village mission has expanded the industrial cultivation of the selected medicinal and aromatic plants from North to South India and in Uttar Pradesh from Lucknow to all directions. As a special feature of this mission, newly developed varieties, agro-technologies and the processing were made available to these bio-villages so that the medicinal plants are propagated, conserved and utilized in the long term basis.

The major bio-village mission programmes that were undertaken by CSIR-CIMAP include Geranium Bio-village in Uttarakhand, Artemisia Bio-village in Uttar Pradesh and Uttarakhand, Aroma Bio-village in Uttar Pradesh, Patchouli Bio-village in Karnataka, Withania Bio-village in Andhra Pradesh and Vetiver Bio-village in Uttar Pradesh.

The Bio-village constituted of one or more than one villages in which the team of CSIR-CIMAP scientists provided the improved cultivation tips to the farmers and thus made them capable to produce quality material. The agreements were signed with interested industries or buyers to ensure market at the farmers' doorstep. The farmers were trained for quality production and motivated for establishing the processing and value addition units in these bio-villages.

Through Bio-village mission of CSIR-CIMAP, on one hand, the people belonging to lower income group were empowered in cultivation and processing of the medicinal and aromatic plants and on the other hand it facilitated the unemployed youth including women becoming entrepreneurs. These efforts had visible impact on the socio-economic status of the rural people. These Bio-villages established by CSIR-CIMAP over the years of sustained efforts have now proved to be replicable models of innovative technology dissemination and have opened new avenues for establishing MAPs based micro-enterprises in villages.

Enabling Farmers Through Training and Skill Upgradation

The CSIR-CIMAP through support of various funding agencies and the government bodies such as Department of Biotechnology (DBT), Department of Science and Technology (DST), National Medicinal Plants Board (NMPB), Small Industries Development Bank of India (SIDBI), National Bank for Agriculture and Rural Development (NABARD) and several other state government departments have conducted more than 60 advanced training -cum -demonstration programmes in various parts of the country during last seven years or so through which more than 3000 farmers and entrepreneurs were provided training on cultivation and processing of MAPs. Of them many have adopted the technologies and have been benefited substantially. The women farmers in Uttar Pradesh, Karnataka and Andhra Pradesh were provided the know-how for nursery technology of MAPs and utilization of agro-waste generated through vermi-composting for production of superior quality organic fertilizer. It is envisaged to develop scientifically trained human resources to the tune of over 10,000 in the rural sector in the next five years for quality production of MAPs.

The MAPs have also made available the socially acceptable employment avenues to the women in villages. These include nursery making and seed production, drying of herbs and making of products of daily use. A novel initiative of CSIR-CIMAP with

Village Women Trained by CSIR-CIMAP Making Incense Sticks

the support of Department of Biotechnology (DBT) has enabled about 100 women residing near the two prominent places of worships situated in the districts of Lucknow and Barabanki in Uttar Pradesh in making incense sticks (agarbattis) and rose water utilizing the flowers offered by the pilgrims and devotees who visit these places daily in very large numbers.

These incense sticks are sold in the local shops. Such efforts have provided additional job opportunities to the poor women residing in the vicinity of the religious places. This initiative is serving as a replicable model of unique entrepreneurship at other places of worship in the country. It is envisaged that re-use of offered flowers or bio-resource will serve as a mean of economic gains rather than leaving the same to be spoiled and create environmental pollution.

Conclusion

Looking at the global trend towards preference for natural materials derived from medicinal and aromatic plants for health and beauty care, and the R&D capabilities and improved technologies available in various organizations and institutions in India, the time is now ripe for making significant strides in the MAPs sector by undertaking area specific mission programmes for cultivation and processing plus value addition and marketing. Such initiatives could be the launched pad for establishment of the medicinal and aromatic plants based micro-enterprises in the villages to cater to the domestic as well as global market and thereby improving the rural economy. Research and development initiatives in collaboration with the industry will be desirable in this endeavour. Such efforts will pave a long way in generating ample job opportunities in the rural sector of the country.

Acknowledgements

Author is thankful to Director, CSIR-CIMAP and Director General, CSIR for giving an opportunity to participate at the International Workshop in Hanoi, Vietnam as the official nominee of CSIR, to the organizers of the Workshop, NAM S&T Centre of New Delhi, Department of International Cooperation (ICD), Ministry of Science and Technology,Hanoi and the Centre for Regional Research and Development (CRD), Hanoi for providing logistical support for participation and to the colleagues at CSIR-CIMAP for their help in preparation of the paper.

References

1. Khanuja, S.P.S., Kalra, A, and Singh, A.K. 2005. Medicinal and Aromatic Plants: Gains from entrepreneurship. The Hindu Survey of Indian Agriculture, 2005, pp.191-194.
2. Khanuja, S.P.S. and Singh, A.K., 2007. Learning from bio-villages: A case study of medicinal and aromatic plants. Consulting Ahead, vol. 1 (2), pp. 76-79.
3. Kumar, Sushil, Srivastava, R.K., Singh A.K., Kalra A., Tomar V.K.S. and Bansal R.P., 2001. Higher yields and profits from new crop rotations permitting integration of mediculture with agriculture in the Indo-Gangetic plains. Current Science, 80 (4), pp. 563-66.
4. Yadav R.P., Kumar S., Kalra A. and Singh A.K., 2008. Production and trade of medicinal and aromatic plants in Uttar Pradesh. I.I.A. News, Vol.9, September, 2008.

Chapter 14

Digital Villages Project: A Recent Strategy towards Socio-Economic Empowerment of Rural Communities in Kenya

S.M. Guchu and G. Gichini*

Ministry of Higher Education, Science and Technology,
Department of Research Management and Development,
P.O. Box 30568, Nairobi, Kenya
**E-mail: salomeguchu@scienceandtechnology.go.ke*

ABSTRACT

Science, Technology and Innovation (STI) play a critical role in socio-economic development of a country. The development and sustainability of the necessary STI infrastructure, as well as technical and entrepreneurial skills are essential pre-requisite for transformation of a nation into a knowledge-based economy. The Kenya development strategy, Vision 2030 identifies STI as the foundation of the three pillars of development namely; economic, social and political pillars. Therefore, various projects have been identified to link STI to the country's development plan. Digital villages initiative is one of the projects recently rolled out by the Government, which is applying STI to empower the wider rural population in Kenya. The project is utilizing the global fibre optic network to harness the vast untapped potential of the rural areas by making Information Communication Technology (ICT) more accessible and affordable to the wider population. This is aimed at bridging the gap between urban and rural communities through countrywide access to ICT facilities. Implementation and financial support of the project is done by an innovative public-private partnership. This partnership involves community based organizations, public, private sector, development partners and other relevant stakeholders. Creation of the rural digital villages is a key strategy that will enhance equity and wealth creation opportunities through improved access to information, provision of employment both directly and indirectly and enhanced provision of Government services. The success of this project is a showcase of how application of STI in an enabled environment can enhance the process of eradicating extreme poverty in developing countries. This paper highlights the noble initiative by the Government on utilization of STI to empower the rural communities for socio-economic development.

Keywords: *Information communication technology, Science, technology and innovation, Socio-economic empowerment, Digital villages.*

Introduction

Among the main priorities of the Kenyan Government towards the realization of national development goals and objectives is the achievement of a knowledge-based society for the benefits of economic growth, employment creation and poverty–reduction. In this respect Science, Technology and Innovation (STI) is identified as the foundation for sustainable development in the Kenya's development strategy, Vision 2030. It is however recognized that there is wide disparities in STI capacity and capability between the rural and urban communities in Kenya. Such glaring disparities among other factors lead to various national development challenges of economic growth and equity concerns. To address such challenges, the vision identifies seven key strategic thrust areas under the STI sector plan (2008-2012). Among the key thrust areas; is strengthening the technical capacities and capabilities through implementation of programmes and projects, which will upgrade skills and technical competencies and consequently increase productivity and level of technology utilization. The Government recognizes that Information Communication Technology (ICT) is one of the fastest growing areas of science and technology thus harnessing it in turn, will help in realizing the national development goals and objectives for economic growth and poverty reduction. Digital villages' project is one of the STI projects strengthening the technical capacities and capabilities specifically in ICT in order to realize various key public policy objectives for socio-economic development. Details of this project and benefits to the communities are discussed in this paper.

Digital Villages Project

The 'digital villages' project comes under the Kenya Transparency Communications Infrastructure Project, which is designed and facilitated by the Kenya ICT Board in collaboration with key partners. The project is branded 'Pasha', a Swahili word with the meaning 'to inform'. The main objective of the project is to establish and maintain a countrywide network of ICT facilities, referred to as digital villages or 'Pasha Centres'. These are e-centres, which provide a collection of services to the public using computers connected to the internet, digital cameras, printers, fax machines and other communication infrastructure. The centres are run by the private entrepreneurs who, under an innovative public-private partnership, obtain training through a certified training programme run by the Kenya ICT Board. The training programme provides essential skills in business and information technology that are required to successfully run a digital village/Pasha centre. The provision of the training is supported by the project and the prospective entrepreneurs who obtain certification are eligible to apply for the Pasha Development Loan from a revolving fund, which is already put in place.

The project is addressing various issues, which include:

1. Adoption and diffusion of ICT
2. Attaining millennium development goals (MDG) through ICT
3. Emergence of knowledge based society
4. ICT infrastructure for public administration and reform
5. E-Governance
6. Innovations in capacity building for development

By distribution of ICT facilities across the country, the project is levelling out the disparities between the rural and urban communities. It is thus addressing the challenges of equity concerns and facilitating the realization of national development goals and objectives for economic growth and poverty reduction.

The development of the project and the increased desire to utilize the global fibre optic network has encountered some challenges and also has various opportunities. Some of these challenges include; (*i*) lack of electricity power supply in the rural and marginalized areas, (*ii*) limited local content, and (*iii*) literacy levels. The opportunities for the success of the project include: (*i*) utilization of solar energy for development of digital kiosks, (*ii*) making use of the rural electrification programme to ensure power supply for the digital villages, (*iii*) the rolled out initiative by the Government to invest and develop local Digital Content Industry.

Benefits to the Communities

The digital village's project promotes digital literacy through training and connectivity particularly in the rural and marginalized areas. This benefits the communities by:

1. Enhancing both their business skills and knowledge.
2. Exposing them to world news and trends that may positively enhance their lives including; e-banking such as money transfer services, e-markets such as agriculture commodity pricing and exchange, among others.
3. Creation of employment, directly through generation of economic activities by the digital villages/Pasha centres, and indirectly through conception of opportunities from the information provided.
4. Enhancing provision of government services, for example, online applications of government jobs, online access of National Social and Security Fund statements, driving licence application forms, tax return forms, etc.

In addition, the centres are envisioned to become one-stop-shops for business incubation within their neighbourhoods thus enhancing the socio-economic development of Kenyan communities especially in the rural and marginalized areas.

Lessons Learnt and Conclusions

The running of the digital villages/pasha centres by rural entrepreneurs under the public-private partnership gives a strong sense of ownership among the communities. The massive demand for the entrepreneurship training course for the prospective centre managers are indications of communities desire to be empowered with information in order to acquire decision making power for socio-economic development.

References

1. Digital inclusion: Pasha Components; http://www.ict.go.ke
2. Kenya Vision 2030: A Globally Competitive and Prosperous Kenya (2007).
3. Kenya Vision 2030: Sector Plan for Science, Technology and Innovation 2008-2012: A Globally Competitive and Prosperous Kenya (2008).
4. Ministry of Higher Education, Science and Technology- Science, Technology and Innovation Bill (2009).

— Section III —

Government Policies and Initiatives in Promoting Rural Industrialization

Chapter 15

Uganda Situation Analysis

Eng. Balu Tabaaro

Director,
Geo Engineering Technologies, Entebbe, Uganda
E-mail: eng_balutabaro@yahoo.com

ABSTRACT

Uganda's national plan stipulates Uganda's medium term strategic direction, development priorities and implementation strategies. In addition, it details Uganda's current development status, challenges and opportunities. In line with the national vision framework, six(6) five-year NDPs (National Development Plan) will be implemented. The first of these is Growth, Employment and Socio-Economic Transformation for Prosperity. The thrust is to accelerate socio-economic transformation from a peasant to a modern and prosperous country within 30 years. This will be supported by the environment necessary for sustainable development which will entail making continuous improvement to the political and social economic conditions.

The post Independent Uganda has gone through periods of political stability and turmoil. It has also witnessed a relatively satisfactory economic progress and an economic decline as well. Regardless of economic trends, the post independence period has been hampered by common binding constraints. This chapter provides a review of Uganda's economic performance since independence in 1962. It is structured into two subsections: microeconomic performance and national binding constraints. The subsection on macroeconomic performance examines performance in seven specific areas namely: the overall macro economy, primary growth sectors and complementary sectors, social sectors, enabling sectors, private sector and the civil society perspectives.

Keywords: *Uganda, National Development Plan, Socio-economic Transformation, GDP.*

Macro-economic Performance

Overall Macro-Economy

Uganda's economy performed remarkably well from the Independence to 1970 when the country experienced a relative political stability. The inflation was maintained at an average of 3 per cent per annum, real gross development product (GDP) grew by 5.2 per cent per annum and fiscal deficits rarely exceeded 2.5 per cent of GDP. The real Interest rates were positive for most of the period; the current account balance was in surplus and domestic savings with an average of 15 per cent of GDP. In the same period, the growth in exports averaged to 5 per cent per annum while that of imports was 6.2 per cent.

From 1971 to 1979, Uganda's economy was seriously damaged by economic mismanagement and civil conflicts that negatively impacted on the gains made during the previous periods. The average rate of inflation was 30 per cent per annum between 1970 and 1980 as the government financed public expenditure through bank borrowing. During the same period, GDP declined at 1.6 per cent per annum, exports by 8.5 per cent and imports by 9.8 per cent. This led to a decline in export revenue which subsequently impacted negatively on the balance of payment and external debt positions.

The growth in GDP was temporarily restored between 1981 and 1983, when an average annual growth rate of 5.6 per cent was registered. The inflation declined from 111 per cent to 25 per cent. The overall budget deficit reduced from 2.8 per cent of GDP in 1981 to 0.6 per cent in 1983 but rose to 11.9 per cent of GDP again in 1984. Thereafter, the civil strife and political instability that ensued especially in the central part of the country negated the achievements made. As a result, negative GDP growth rates were recorded during the period between 1984 and 1986.

In 1987, the government launched a minimum Economic Recovery Programme (ERP) followed by a series of other reforms aimed at restoring macroeconomic stability to provide a favourable environment for economic growth and private sector development. The key reforms included a currency reform, changes in tax and fiscal polity geared towards improving revenues and restraining expansion in the government expenditures, while maintaining a strong focus on economic recovery and growth. Between 1987 and 1996, the GDP grew at an average annual rate of 6.5 per cent, translating into 3.4 per cent growth in per capita terms. The decline in monetary growth led to substantial reduction in inflation.

Over the period 1997/1998 to 2000/01, the average GDP growth was 7.2 per cent per annum. Between 2000/01 and 2003/04 it averaged 6.8 per cent and between 2004/05 and 2007/08 it was 8 per cent. As a result of the global recession, which reduced the demand for Uganda's exports to Europe and America, the GDP growth declined slightly in 2008/09 to 6.2 per cent at basic prices but was projected to increase to 6.4 per cent at basic prices in 2009/10. The fiscal deficit stood at 10.2 per cent and 7.9 per cent of GDP on average during 2000/02 to 2003/04 and 2004/05 to 2007/08 respectively. Inflation rate was kept at single digit for most of these periods; foreign reserves covered at least five months of imports and the exchange rates were competitive.

There has been a significant increase in both foreign and local investment flows into the economy. A percentage of GDP, private investment rose from 12.2 per cent in 2000/01 to 20.6 per cent in 2006/07. Public investment averaged 5.1 per cent of GDP over the same period but was more or less stagnant over the last decade. Furthermore, the investment pattern of recent years shows rising private construction especially of residential buildings, modestly increasing machinery and equipment investment and low public construction. This indicates issues with low capital investment in industries, services and labor productivity.

The global downturn is expected to negatively influence investments through reduced private remittances, foreign direct investment and loans, although recent global economic trends coupled with the resilience of the country's economy have subdued the initially perceived risks. Foreign Direct Investment (FDI) dropped from 5.3 per cent of GDP in 2007/08 to 4.5 per cent GDP in 2008/09 and it is further projected to remain slow in 2009/10. However, the anticipated negative effects on aid have, at worst, been marginal as most development partners have continued to meet their obligations.

Uganda's trade deficit has been widening despite improvements in the composition and value of exports. The trade deficit as a percentage of GDP declined from an annual average of 12.9 per cent for the period 2000/01 to 2003/04 and from 13.5 per cent for the period 2004/05 to 2007/08. The balance of payments has also been unfavourable with a deteriorating trend in recent years. These results could partly be due to lower demand for Uganda's exports in the advanced economies, although this is partly being compensated by the increased regional exports. Table 15.1 indicates that regional trade is taking an increasing share of Uganda's exports while the Middle East is also emerging as a major export destination. The European Union, which in the past was the leading destination for Uganda's exports now accounts for about one quarter of total exports.

Table 15.1: Percentage Share in Total Exports by Region of Destination

Region	*2003*	*2004*	*2005*	*2006*	*2007*
COMESA	27.7	26.8	30.7	29.5	37.9
Other Africa	8.6	5.7	4.8	3.9	6.6
European Union	26.3	27.3	31.1	27.4	24.3
Other Europe	14.8	17.1	10.1	5.1	6.8
North America	2.7	2.9	2.3	1.7	1.8
Middle East	3.5	5.6	10.8	20.6	14.3
Asia	9.3	8.9	7.5	7.8	5.4
South America	0.1	0.1	0.1	0.1	0.2
Rest of the World	0.4	0.0	0.1	0.0	0.0
Unknown	6.6	5.7	2.5	3.8	2.8

Source: UBOS Statistical Abstract (Various).

Whereas imports are becoming more technologically advanced, exports are still dominated by primary commodities, indicating that the country's consumption patterns are changing faster than transformation of its technology of production. This contrasts with some East Asian economies where exports were dominated by manufactured products. Therefore, the growth of the service and manufacturing sectors in the country will have to undergo transformation in future with the inclusion of new technologies in order to be more competitive in the export market.

The growth in GDP has been accompanied by changes in its broad sectoral composition. These changes reflect structural transformation in the economy. Figure 15.1 indicates the trend in the changing shares of GDP for agriculture, industry and services. Between 2000/01 and 2008/09, the share of agriculture to GDP fell rapidly while that of industry registered notable growth between 2004/2005 and 2007/2008 before converging with the share of agriculture in 2008/2009 at about 2.3 per cent.

While there have been changes in the sectoral composition of GDP, there has not been a commensurate change in the distribution pattern of the labour force. The GDP share of the emerging modern sectors is increasing but their share of the labour force is falling. The share of the labour force employed in manufacturing and services sectors decreased from 6.8 and 26.8 per cent to 4.2 and 20.7 per cent respectively despite the rise in the GDP shares of these sectors. However, the share of labour force engaged in the agriculture sector increased from 66.4 per cent in 2002/03 to 75.1 per cent in 2005/06 while the share of agriculture GDP declined over the same period. This may be attributed to a variety of factors including: a mismatch between skills acquired and the requirements of employers; the development of low skilled services and industries; the high rate of growth in the labour force and the inability to absorb it in the growing sectors. These trends contribute to low productivity in agriculture, which undermines the growth potential of the economy and contributes to issues related to food insecurity.

As the GDP growth rate increased with country gaining peace in most parts in 1986, the per capita GDP also increased. Figure 15.2 shows real per capita GDP between 1970 and 2008. It depicts a declining trend in per capita income until 1988 after which the real capita GDP began rising, reaching its 1970 level in 2003.

In spite of Uganda's economic recovery from the down fall of the 1970s and 80s, its size and the country's per capita income is still very low compared to other economies in Africa and Asia. While Uganda's economic performance was on a par with that of countries such as Kenya, Ghana and Malaysia in the early 1970s, these economies have since improved significantly over Uganda's economy (Figure 15.3 and 15.4). For example, Malaysia is now only a role model to Uganda as opposed to being on a par with, as it was in the 1970s.

The relatively high economic growth, which has been sustained since early 1990, has contributed to a reduction in the levels of poverty. In 1992/93, 56 per cent of the population was below the national poverty line; this dropped to 44 per cent in 1997/98 and to 31 per cent in 2005/06. A greater proportion of the population living under the poverty line is located in the northern part of Uganda (Figure 15.5). This is largely attributed to the insecurity experienced in the region for over the past 20 years.

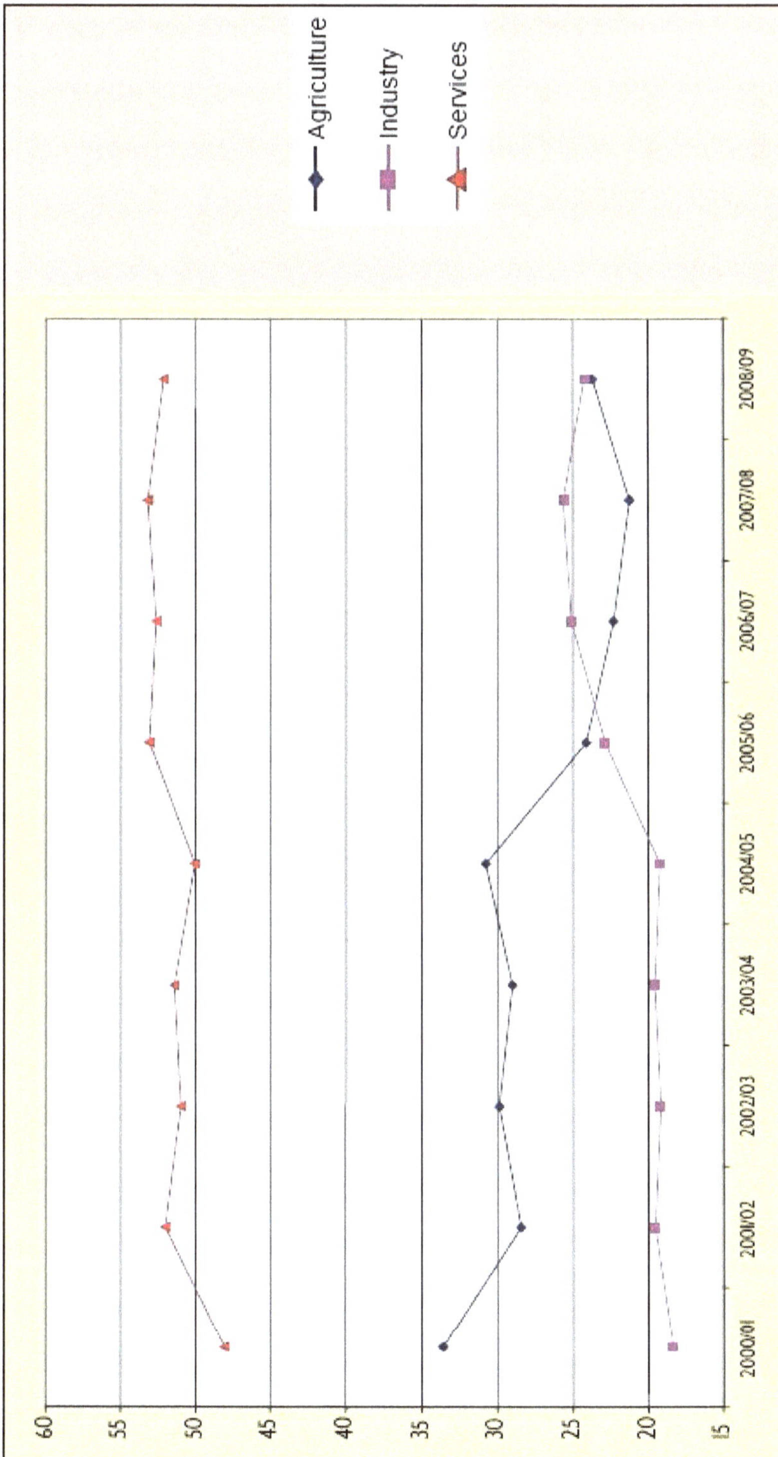

Source: Uganda Bureau of Statistics Database

Figure 15.1: Sectoral Contribution to GDP (Per cent) 2000/01–2008/09, at Market Prices

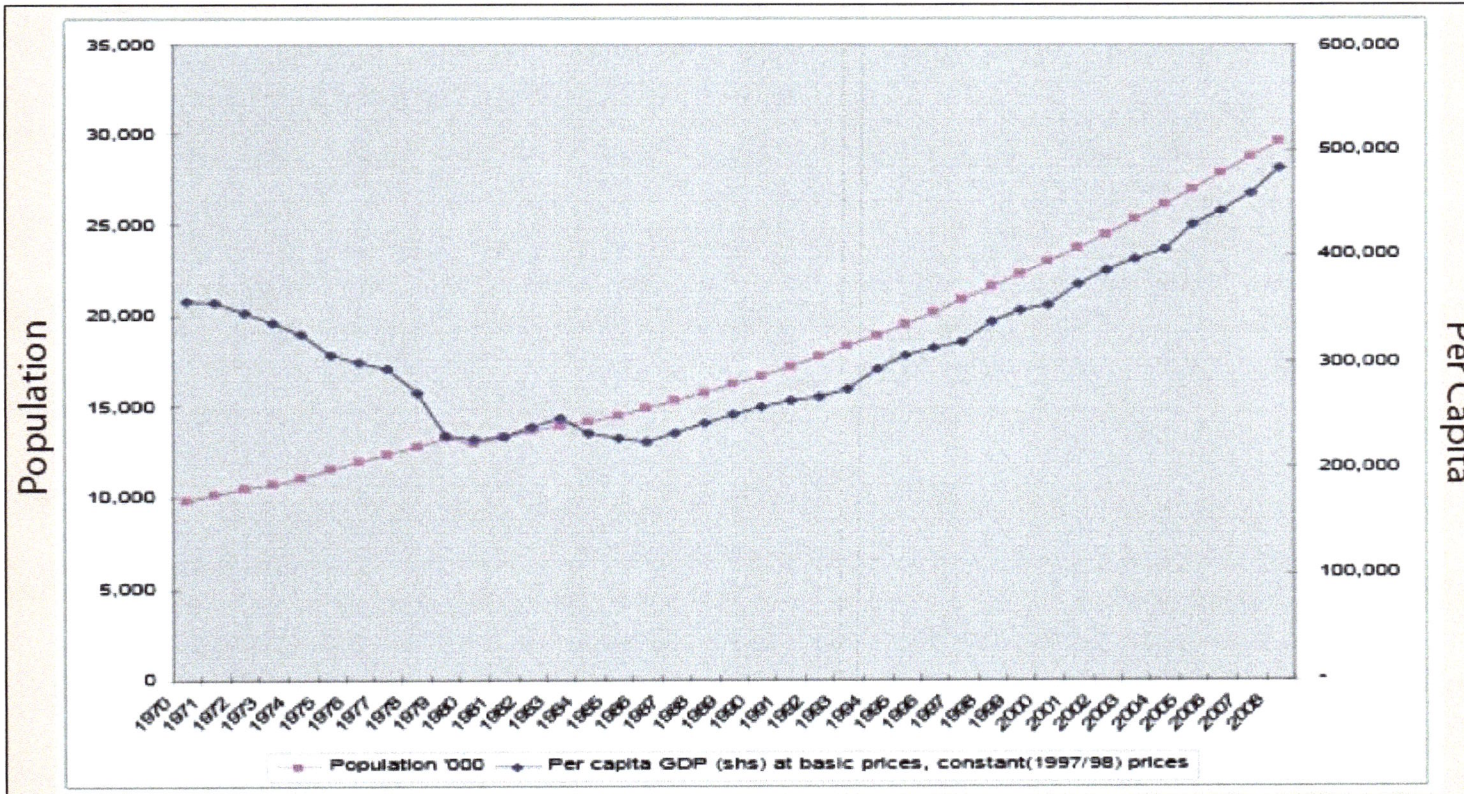

Population figures ('000s) are on the left axis while the real GDP per capita are on the right axis.
Source: Uganda Bureau of Statistics Database

Figure 15.2: GDP per Capital and Population

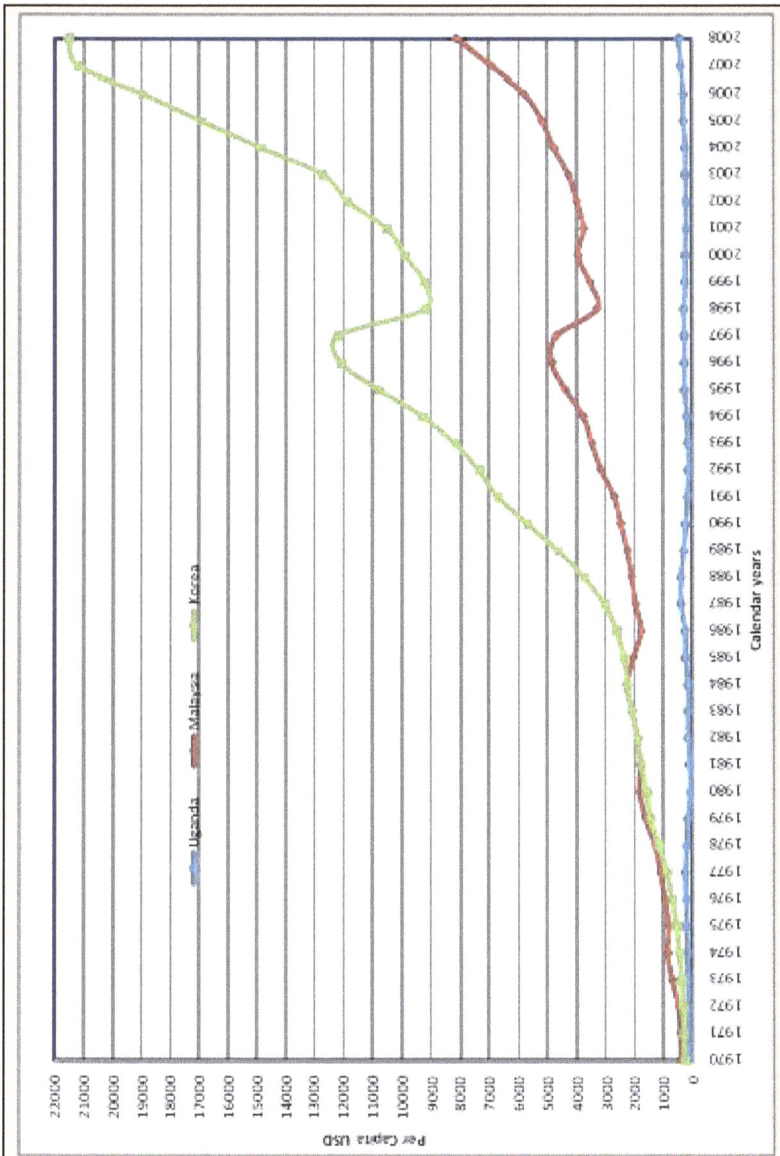

Source: World Development Indicators Database

Figure 15.3: Comparison of Per Capita Income (USD) of Uganda, Malaysia and Korea 1970–2008

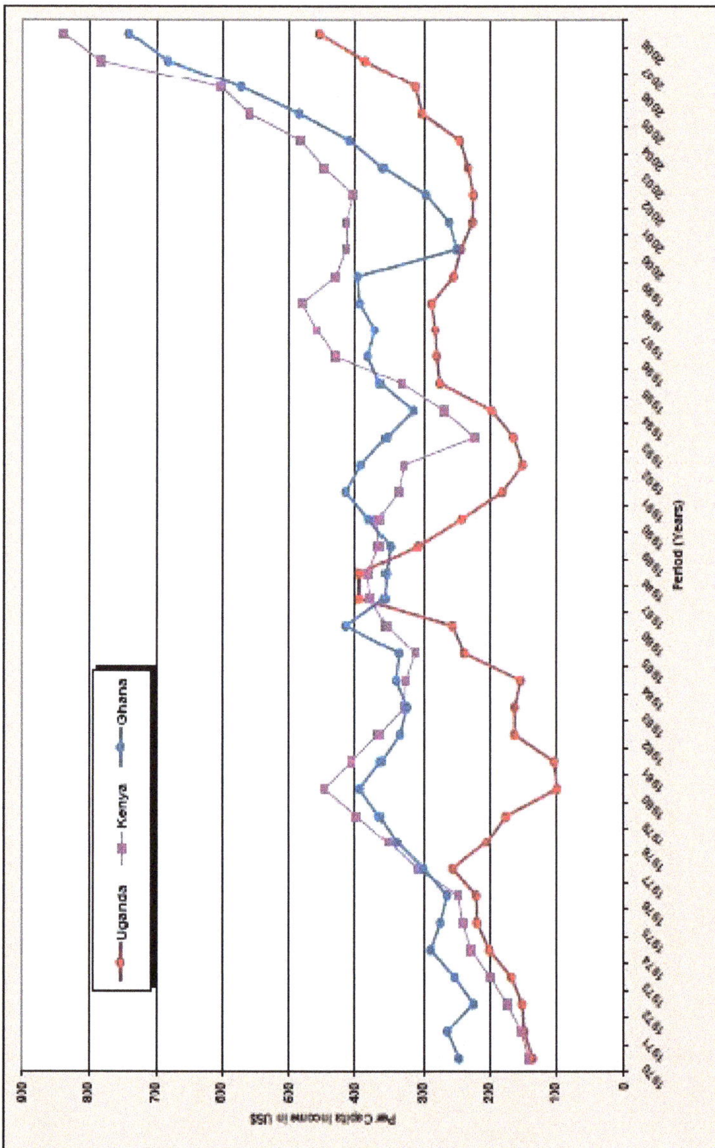

Source: World Development Indicators Database

Figure 15.4: Comparison of Uganda's Per Capita Income with Regional Peers

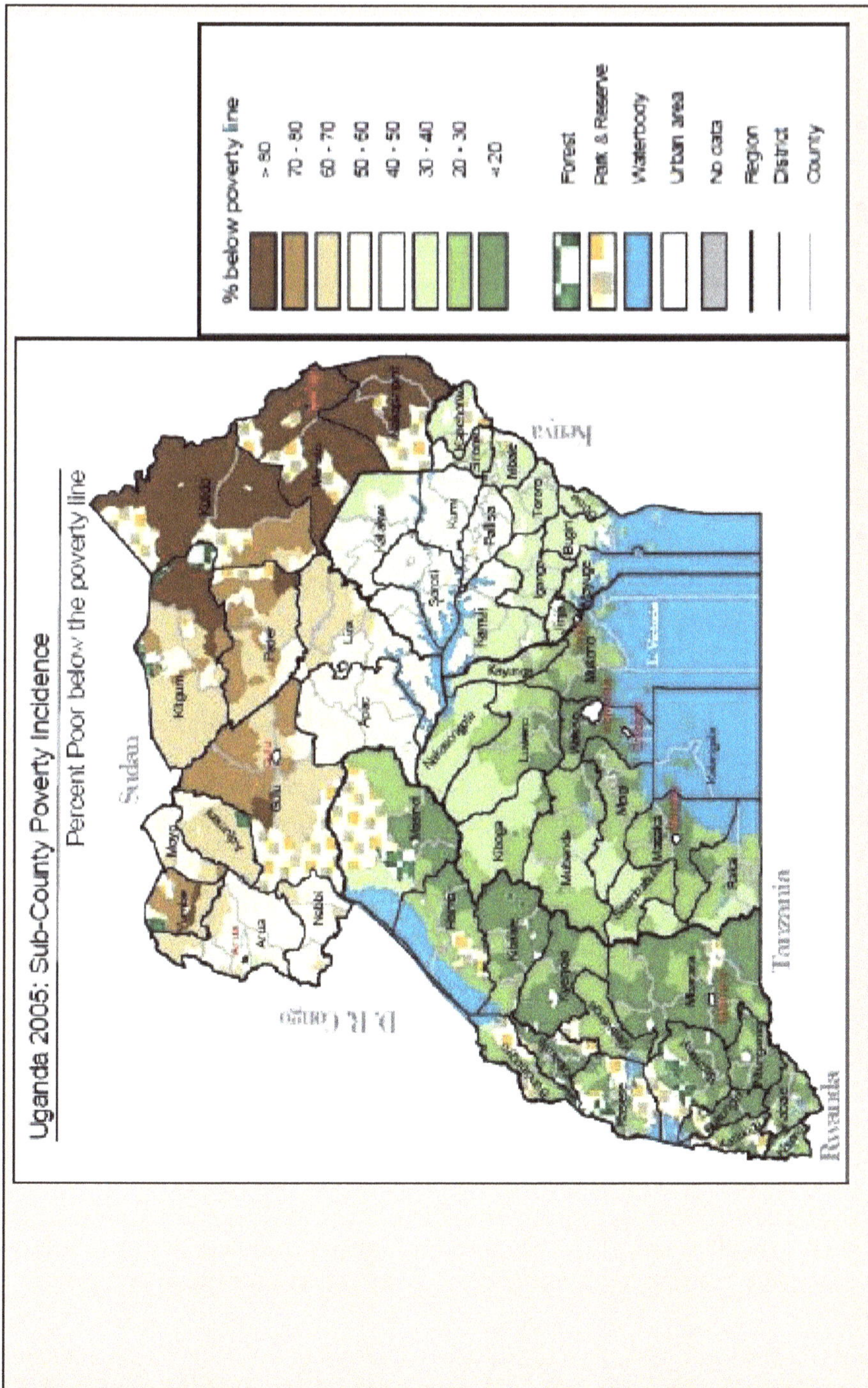

Source: Uganda Bureau of Statistics 2008

Figure 15.5: Poverty at Sub-National Level

Inequality in consumption is high in Uganda by the African standards. In addition to regional disparities (Table 15.2), there is also a disparity between rural and urban income levels with the mean consumption of the richest area (Kampala) being 2.5 times that of the poorest area (northern region). The data from 1992 to 2006 indicates that whereas rural inequality is declining after a period of an increase in inequality is yet to revert to its level of 1992/93.

Table 15.2: Inequality Gini Coefficient for Uganda

Region	*Year*			
	1992/93	*1999/00**	*2002/03*	*2005/06*
National	0.37	0.40	0.43	0.41
Urban	0.33	0.33	0.36	0.43
Rural	0.40	0.43	0.48	0.36
Central	0.40	0.42	0.46	0.42
Eastern	0.33	0.35	0.36	0.35
Northern	0.34	0.34	0.35	0.33
Western	0.32	0.32	0.36	0.34

Source: Comprehensive UNHS 2005/06 and World Bank calculations, 1999/00 data exclude four districts.

Primary Growth Sectors

The primary growth sectors consist of sectors and sub-sectors that directly produce goods and services. These include: agriculture, forestry, manufacturing, tourism, mining, oil and gas, ICT and housing development. The performance of these sectors has been varying (Table 15.3)[2]. Table 15.3 presents the share of primary growth sectors in GDP at current prices. Added values in agriculture and industry need to become engines of growth but the agriculture sector has been declining in its growth rate and share of GDP and the industry sector has not been expanding at a notable rate. While there has been growth in hotels, the restaurants (tourism) and communications, they are not yet significant sectors in terms of GDP.

The share of agriculture[3] in GDP was 51.1 per cent in 1988 and 33.1 per cent in 1997, declining further to 15.4 per cent in 2008 as shown in Table 15.3. The sharp decline in the share of agriculture in GDP represents significant structural transformation in the economy. Whereas this is a positive development, the challenge is to ensure that this shift is at the same time accompanied by productivity in the agricultural sector and value addition in the industry and service sectors in order to absorb excess labour from agriculture. Also there is a compelling need to ensure that productivity growth in agriculture supports the high population growth, in addition to reversing the continued dominance of production and export of primary commodities.

Between 1988 and 1997, agriculture grew at an average rate of 3.9 per cent and 5.4 per cent between 1998 and 2002. However, from 2004 to 2008, the growth of the

Table 15.3: Share of Selected Primary Growth Sectors in GDP and Growth Performance

Primary Growth Sectors	*Percentage Share in GDP (Current Prices)*					*Growth Performance (Per cent)*				
	1988	*1997*	*2004*	*2007*	*2008*	*1988/1997*	*1998/2008*	*2004/2008*	*2007*	*2008*
Agriculture	51.1	33.1	17.3	14.5	15.4	3.9	5.4	1.1	1.7	2.2
Forestry	2.2	1.7	3.3	3.5	3.4	4.7	7.0	3.9	2.2	4.2
Manufacturing	5.9	8.4	7.0	6.9	7.2	13,2	7.2	6.3	7.6	6.7
Hotels and Restaurants (Tourism)	1.1	1.9	4.0	4.1	4.1	13.1	3.8	9.6	9.2	12.5
Mining	0.1	0.6	0.3	0.3	0.4	34.6	8	13	5	10.4
Post and Telecommuni-cations (ICT)	0.2	0.6	2.0	3.0	3.4	10.1	22.8	26.2	16.1	39.6
Construction	4.1	6.5	11.9	12.2	6.5	6.3	6.3	6.3	4.8	5.8

Source: UBOS Statistical Abstract (various years).

sector slowed down markedly to an average of 1.1 per cent, with a growth of 1.8 per cent recorded in 2006. This performance raises issues of productivity in the agricultural sector and the need for radical strategies to improve the productivity levels and modernize the sector. The employment pattern has also been changing but not in a way that reflects transformation in the structure of the economy. The share of the labour force in agriculture increased from 66.4 per cent in 2002/03 to 75.1 per cent in 2005/06.

The share of the forestry sub sector in GDP was 2.2 per cent in 1988 and 1.7 per cent in 1997, rising to 3.4 per cent in 2008. The increasing share of forestry in GDP is a positive development. However, the improvement in the contribution of forestry has been characterized by recent declining performance. Between 1988 and 1997, forestry grew at an average rate of 4.7 per cent per year and between 1998 ad 2002 at an average rate of 7 per cent. From 2004 to 2008, the sector grew by 3.9 per cent per annum, a trend that needs to be accelerated. This trend is partly due to declining forest cover, which decreased from 4,933,746 hectares in 1990 to 3,604,176 hectares in 2005, representing a 27 per cent reduction.

The share of the manufacturing sector in GDP was 5.9 per cent in 1988 and 8.4 per cent in 1997, peaking in 2002 and declining to 7.2 per cent in 2008. Between 1988 and 1997, the manufacturing sector grew at 13.2 per cent per year, recording the highest growth of 18.3 per cent in 1995. The sector grew by 7.2 per cent on average between 1998 and 2002 and by 6.3 per cent between 2004 and 2008. The share of the labour force employed in manufacturing sector decreased from 0.29 per cent to 0.14 per cent between 2002/03 and 2005/06.

The share of tourism in GDP measured by the share of hotels and restaurants in GDP was 1.1 per cent in 1988 and 1.9 per cent in 1997, rising to about 4 per cent in 2008. This represents a shift in the economic structure of the economy in terms of its GDP. Between 1988 and 1997, tourism grew by 13.1 per cent per annum on average and 3.8 per cent between 1998 and 2002, increasing to 12.5 per cent between 2004 and 2008.

The share of mining in GDP was 0.1 per cent in 1988 ad 0.6 per cent in 1997, declining to 0.3 per cent in 2008. Between 1988 and 1997, mining grew at 34.6 per cent per annum on average and 8 per cent per annum. The share of the labour force employed in the mining sub-sector remained almost constant over the period 2002/03 to 2005/06, increasing from 0.93 to 1.0 per cent respectively.

The share of ICT in GDP measured by the share of posts and telecommunications was 0.2 per cent in 1988 and 0.6 per cent in 1997, increasing to 3.8 per cent in 2008. Between 1988 and 1997, ICT grew at 10.1 per cent and 22.8 per cent between 1998 and 2002. From 2004 to 2008, the performance of the sector was 26.2 per cent. There has been a rapid rate of telephone penetration with the number of fixed and mobile lines per 100 people rising from 0.24 and 0.02, in 1996 to 0.56 and 27.68 in 2008 respectively. Internet usage has been slow to pick up partly because of very limited internet support infrastructure available in the country and the low computer literacy rates amongst Ugandans.

The share of construction in GDP, which is used as a proxy to measure housing performance, was 4.1 per cent in 1988 and 6.5 per cent in 1997, rising to 11.9 per cent in 2008. Between 1988 and 1997 the housing sub-sector grew at 6.5 per cent and 6.3 per cent between 1998 and 2002. From 2004 to 2008, the performance of the sector was 6.3 per cent. The share of the labour force employed in construction remained nearly the same between 2002/03 (0.08 per cent) and 2005/06 (0.07 per cent).

The oil and gas sector is relatively new but with huge potential. The oil and gas reserves in the country are estimated at 2 billion Barrels of Oil Equivalent (BOE)[5]. To exploit these resources, large investments will be required for further exploration, development and extraction. In addition, government will invest in the necessary physical infrastructure.

Complementary Sectors

The complementary sectors consist of sectors and sub-sectors that provide institutional and infrastructural support to primary growth and other sectors, namely: transport, energy, land management and administration, physical planning, urban development, trade development, co-operatives, science and technology and water for production. Table 15.4 shows the share of some of the complementary sectors in GDP growth performance.

The share of the transport sector in GDP has been without a clear growth pattern. While the performance of the sector improved between 2004 and 2008 compared to 1998 and 2002, its share in GDP has been about 3 per cent (Table 2.4). The share of the labour force employed in transport sector decreased from 2.84 per cent to 1.61 per cent between 2002/03 and 2005/06. The moderate performance in the transport sector is evident from the low share of the paved road network (4 per cent); limited flights and limited functionality of the rail system (26 per cent).

Until 1997, the contribution of financial services in GDP was not reflected in the economy's development, its contribution has reached significant proportions accounting to 3.2 per cent of GDP in 2008. Despite the growth performance of the sector declining to 3.9 per cent in 2007, the overall growth performance has been impressive. However, the share of the labour force employed in the financial services sector decreased from 2.11 per cent to 1.98 per cent between 2002/03 and 2005/06.

Achieving socio-economic transformation requires continuous improvement in the way goods and services are produced within an economy. Measurement in terms of percentage exports with high technology content to the total manufacturing exports, the Science, Technology and Innovation sector is still underdeveloped. The expenditure share in R&D as a percentage of GDP is 0.1 per cent; Total Achievement Index (TAI)[6] was 0.24 in 2005 compared to the minimum required of 0.35 per cent. The ratio of arts to science graduates is 5:1 and the country has one researcher to 1,000 employees.

While Land Management and Administration, Physical Planning and Urban Development are not directly reflected in GDP computations, they play an important role in GDP generation and growth as well as employment and prosperity for all. In the recent years, the level of urbanization in Uganda has been rising rapidly. In 1991,

Table 15.4: Share of Selected Complementary Sectors in GDP and Growth Performance

Complementary Sectors	*Percentage Share in GDP (Current Prices)*					*Growth Performance (Per cent)*				
	1988	*1997*	*2004*	*2007*	*2008*	*1988/1997*	*1998/2008*	*2004/2008*	*2007*	*2008*
Transport	3.0	3.9	3.2	3.3	3.4	9.3	5.3	6.8	9.6	6.9
Energy and Water	0.6	1.3	3.5	4.5	4.1	7.6	6.2	2.0	5.3	4.0
Trade	14.7	10.0	12.7	14.1	14.3	7.6	6.5	10.3	13.0	13.6
Financial Services	–	2.3	2.8	2.9	3.2	-	5.5	11.2	-3.9	11.1

Source: UBOS Statistical Abstract (various years) (*Figures adjusted to 2002 prices).

the population living in urban areas was 11.3 per cent, increasing to 12 per cent in 2002 and it is estimated an increase of 15 per cent in 2009[7]. The urban population growth rate currently stands at 5.9 per cent per annum. The increasing levels of urbanization in Uganda and congestion in Kampala city creates considerable pressure on housing, transport, water, health, education, social welfare and employment, which need to be addressed through systematic physical and infrastructural planning.

Social Sectors

Social sectors consist of sectors and sub-sectors that provide services required for maintaining a healthy and quality population, and developing the required human resource for effective engagement in profitable economic activities. These sectors include: education, health, water and sanitation, social development and gender and population.

The national population size and its growth have a powerful impact on the achievement of socio-economic goals. Uganda has a fast growing population, which has expanded from 9.5 million in 1969 to 24.2 million in 2002 and estimated to be 30.7 million in 2009. At 3.2 per cent growth rate per annum (1991-2002), Uganda's population is projected to reach 38 million in 2015.

The high fertility rate coupled with the young population means that the population will continue rising even if fertility declined (Figure 15.6). Nearly half of the population is below the age of 15 years and the population structure is expected to remain youthful for the next fifteen years. The population trend described above represents several challenges to future growth and structural transformation unless serious measures are taken to convert it into a population dividend. Already Uganda has one of the highest dependency ratios in the world (above 1.5), which is expected to rise under the current growth trends. The causes of high fertility include low levels of education, poor access to family planning services with an unmet demand estimated at 41 per cent, a low contraceptive prevalence rate of 24 per cent and early child bearing with 25 per cent of adolescents being pregnant before the age of 19[9]. Other causes include the prevailing cultural and religious beliefs and preferences for large families as a source of social security at old age.

There is a high growth rate of 3.7 per cent per annum in the unskilled labour force but what the economy requires is skilled labour hence contributing to high levels of unemployment and under-employment. Public sector employment has not grown significantly and can only absorb a few people leaving agriculture to absorb the bulk of the labour force (which was estimated at 75.1 per cent of the labour force in 2005/06). (not clear!)In 2005/06 Agriculture employed 75.1 per cent of all labour -75.1 per cent of Uganads labor force is employed in agriculture.

Table 15.5 gives the share of education and health services in GDP between1988 and 2008 and their performance over the same period. The share of education services in GDP significantly increased between 1988 and 2004 where it peaked at 7 per cent but declined thereafter. Health services on the other hand have maintained an almost constant share for the most part of the period 1988 to 2008 with exception of 1997 where their share peaked at 1.9 per cent of GDP. The growth performance of both

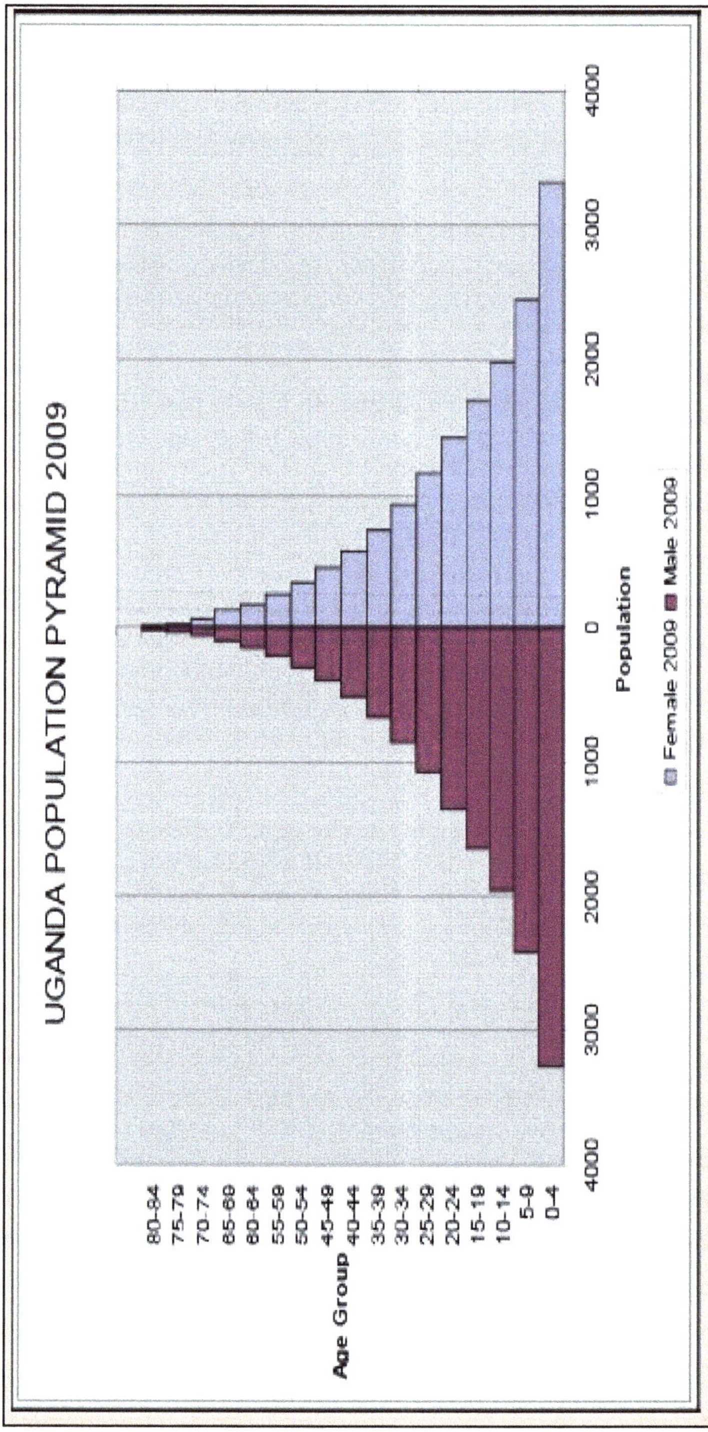

Source: UBOS Statistical Abstract 2009

Figure 15.6: Population Distribution by Age

Table 15.5: Share of Selected Social Sectors in GDP and Growth Performance

Social Sectors	*Percentage Share in GDP (Current Prices)*					*Growth Performance (Per cent)*				
	1988	*1997*	*2004*	*2007*	*2008*	*1988/1997*	*1998/2008*	*2004/2008*	*2007*	*2008*
Education	2.8	4.7	7.0	6.3	5.5	5.7	7.5	4.5	1.5	-2.4
Health	0.9	1.9	1.6	1.3	1.1	4.5	4.5	3.8	-1.3	0.2

Source: UBOS Statistical Abstract (various years).

sectors has significantly declined with contractions in 2007 and 2008. The share of the labor force employed in the health sector moderately decreased from 2.79 to 2.66 per cent between 2002/03 and 2005/06 respectively while that in education decreased from 0.8 to 0.68 per cent.

Uganda has made a notable progress in increasing literacy and access to education at all levels. Access to Universal Primary Education (UPE) increased from 2.5 million students in 1997 to 7.5 million in 2008, reaching a level of 82 per cent of eligible pupils enrolled. The introduction of Universal Secondary Education (USE) and Universal Post Primary Education Training (UPPET) in 1997 increased the secondary school enrolment by 25 per cent from 814,087 in 2006 to 1,088,744 in 2008, with girls consisting 46 per cent. In the same period, enrolment of (author-pl give full name of TVET-Business,Technical and vocational Educational and Training) TVET increased by 46 per cent from 25,682 to 47,298. The adult literacy rate has improved from 69 per cent in 2005/06 (UBOS, UDHS 2005/06) to 73.6 per cent in 2009 (HDR 2009).

In spite of the improved performance, Uganda's progress towards reaching the Millennium Development Goals (MDGs) by 2015 in respect of educational and health status is just as fast as desired. Table 15.6 sets out the results from the UBOS Statistical Abstract 2008. The MDG target for net enrolment ratio in primary education is 100 per cent by 2015 while the outturn in 2005/06 was 84 per cent.

Table 15.6: Progress on Health and Education Related MDGs by 2006/06

Goal/Indicator	*Progress (2005/06)*	*Target (2015)*
Goal 2: Achieve Universal Primary Education		
Net enrolment ratio in primary Education	84	100
Goal 4: Reduce Child Mortality Rate		
Infant Mortality Rate (per 1000 live births)	76	41
Under five Mortality (per 1000 live births)	137	60
Proportion of 1 year old children immunized against measles	68.1	90.0
Goal 5: Improve Maternal Health		
Maternal Mortality ratio (per 100,000 live births)	435	131
Proportion of births attended by skilled health personnel	41.1	90
Total Fertility rate	5.0	60
Contraceptive prevalence rate among women 15-49 years	23.9	
Goal 6: Combat HIV/AIDS, Malaria and other diseases		
Condom use among higher risk sex group 15-24 year olds.	52.9	

Source: UBOS Statistical Abstract, 2009.

Technical, vocational and university education is not yet adequately supporting the development of a workforce with appropriate skills. In 2006, student enrolment in science and technology at both private and public universities was less than 27 per cent out of the total registered students. This is below the minimum requirement of 40 per cent recommended for rapid accelerated economic growth and effective contribution in the global knowledge based economy.

The current status of Uganda's health outcomes is mixed as measured by four core indicators set out in Table 2.6. While the trend in all the four outcome indicators was positive, the country fell short of all its defined national targets as articulated in the PEAP (Poverty Eradication Action program)author-pl. give full name) and risks missing the international targets. Generally, Ugandan health standards remain poor, even by Sub-Saharan Africa Standards. Although, there has been a decline in the maternal mortality ratio from 505/100,000 live births in 2001 to 435/100,000 in 2006 (UDHS, 2006), it is unlikely that Uganda will meet the MDG target of 131/100,000 live births by 2015. Access to maternal and reproductive health services for girls and women is still low.

Uganda's Human Development Index (HDI) score, which among others, considers life expectancy, literacy and education enrolment rates, improved from 0.272 in 1995 to 0.581[10] in 2006 before reducing to 0.514 in 2009. This is lower than the figures of Kenya (0.54) and Tanzania (0.530) in 2009 but within the range of the medium human development category of countries. Uganda's 2009 HDI score translates into a rank of 157th out of 182 countries with data.

The development of Uganda, however, continues to be constrained by gender inequalities and social vulnerabilities. In the case of gender, Uganda's constitution guarantees equality between women and men and has a number of affirmative action measures. Uganda is a signatory to various international commitments and has a Gender Policy, which provides for a framework for gender responsive development. Affirmative action has led to an increase in the percentage of female Members of Parliament from 18 per cent in 1995 to 29 per cent in 2009. Despite the presence of these measures, gender disparities still persist.

Gender based violence is estimated at 68 per cent for females against 20 per cent for males (UDHS 2009). Retention in primary education on the whole is low and exhibits gender disparity with 53 per cent of boys and 42 per cent of girls completing primary school in 2006. Enrolment figures for secondary level education also show gender disparities with only one third of girls who enrolled in primary education continuing in school up to the age of 18 years, compared to half in the case of boys (UDHS 2006). HIV prevalence rate is also higher among (7.5 per cent) women than it is among men (5 per cent) (UHSBS 2004/05).

Furthermore, though it is estimated that women comprise 70 per cent of the work force in agriculture, they experience unequal access and control over important productive resources such as land. This limits their ability to move beyond subsistence agriculture. In wage (paid) employment, women are concentrated in the lowest paying sectors, which is linked to their lower education levels; 50 per cent of employed women work in the three lowest paying sectors, compared to 33 per cent of men (2002/02 and 2005/06 UNHS). At the household level, women's participation in decision making is also limited; only 51 per cent of women reported participating in making major household purchases and overall men believed that a husband should play the major role in making most household decisions (UDHS 2006). The social vulnerabilities are generally associated with demographic characteristics such as age, disability, unemployment, ethnic minorities and other characteristics such as poverty ad disasters. The UNHS 2005/06 report reveals that more than 2.1 million vulnerable children, representing 13.7 per cent of young people below 18 years were under the care of older persons, putting a lot of stress and further impoverishing them. About 1.9 million people (7 per cent of the population), live with disabilities and about 7.1 per cent of older persons have access to pension, of whom 60 per cent are male. Analysis of household survey data (2005/06) reveals that over 80 per cent of the poor or about 7 million people are still trapped in the chronic poverty.

Enabling Sectors

Enabling sectors encompass all sectors and sub-sectors that provide a conducive environment and framework for efficient performance of all sectors of the economy. They include defense and security, justice, law and order, legislature, environment management, climate change, water resource management, public sector management, public administration, disaster management, accountability, statistics, meteorology and EAC integration.

For a long time, after the Independence, the defense and security sector was characterized by weak strategic leadership, unethical code of conduct and ineffective organizations and structures to regulate it. This created problems for the country resulting in animosity between the forces and the civilians, political instabilities and coups. Since 1986, efforts have been directed at addressing these issues. The statutes, which guided the forces' ethical and operational code of conduct, were established and the restructuring has been carried out. The security forces are subject to parliamentary oversight, making them accountable and properly regulated.

Uganda's relations with her neighbours have greatly improved, which limits the possibility of the re-occurrence of inter-state conflicts, reduces the likelihood of further disagreements over border points as well as eliminating the changes of dissidents using the neighbouring areas to destabilize the country. Peace and stability has been restored throughout the country, including Northern Uganda where armed conflicts had persisted for over 20 years.

A number of democratic institutions have been put in place in line with the 1995 Constitution. The three arms of Government namely the Legislature, the Executive and the Judiciary are now in place and functioning, albeit with varying degrees of respect for each other's constitutional mandate. Parliament continues to exercise its constitutional mandate and has strengthened a number of institutions for promoting democracy and good governance including the Electoral Commission (EC), Inspectorate of Government (IGG), the Auditor General (with enhanced powers), and the Uganda Human Rights Commission (UHRC), among others.

However, the country's culture of constitutionalism is not improving fast enough. Many of the institutions established in the Legislative, Executive and Judicial branches of Government are still struggling to fully embrace the ideals, principles and practices of democracy. The media has been liberalized and is relatively free with both the print and electronic media providing citizens with ample opportunity to express their views on a wide range of issues that affect their lives. At the same time, Uganda has seen a dramatic emergence of non-governmental and civil society organizations able and willing to engage the arms of Government in policy debate and advocacy on democracy and good political governance issues, albeit with challenges in the policy and regulatory framework. However, the citizenry in general is not yet empowered to engage effectively in demanding their rights and insisting that institutions meet their obligations.

The country has established a number of institutions to enable effective public sector management. These institutions are responsible for policy formulation and implementation and the public service delivery. However, there are opportunities to improve the structure of the public sector. There are issues with the allocation of roles and responsibilities which sometimes affects service delivery. Most of the public institutions were formed in the early 1960s before the liberalization policies of the 1990s. The organization of some public institutions is not suitable and impedes the delivery of policy, regulation and public services. Other issues include: overlaps and duplication of mandates, weak oversight of institutions, poor corporate governance and weak regulatory frameworks. The civil servants are still poorly remunerated and this limits their productivity. Furthermore, the coordination of the public sector institutions is still a challenge.

The progress made both in public administration and public sector management is being hampered by corruption at various levels of Government. The government recognizes the devastating socio-economic effects of corruption and is committed to taking corrective and deterrent measures in order to fully realize the potential for improving social and economic conditions. Uganda ranks 130th out of 180 countries on the Transparency International's 2009 Corruption Perception Index, while many African countries rank better (Tanzania 126, Zambia 99, Swaziland 79, South Africa 56 and Botswana 37).

In order to enhance accountability in the government, an accountability sector was established in 2007 with the goal of promoting, supervising and implementing accountability systems. The sector comprises of the Ministry of Finance, Planning and Economic Development; the Inspectorate of Government; Office of the Auditor General; Directorate of Ethics and Integrity; Ministry of Public Service-Inspection; Ministry of Local Government; Public Procurement and Disposal of Public Assets Authority; Uganda Bureau of Statistics; Uganda Revenue Authority and Development Partners. There is also a Stakeholders' Forum in which civil society views are presented and addressed. Civil Society includes private business, media, anti-corruption NGOs and community groups, which play a vital role in representing views and experiences of public service beneficiaries and holding public officials accountable.

The legal, policy and regulatory framework that guide the accountability sector include the Budget Act, Public Finance and Accountability Act 2003, the National Audit Act 2008, the Leadership Code Act 2002, the National Records and Archives Act 2001, Public Procurement and Disposal of Public Assets Act 2003 and the Access to Information Act 2005. Other laws pending legislation include: the Anti-Money

Laundering Bill, the Anti-Corruption Bill 2008 and the Whistle Blowers' Bill. A special court has also been established to try corruption cases.

There is a need to ensure that there is sustainable management of the environment and natural resources. The environment is characterized by a geographical variance in rainfall levels. The amount of rainfall ranges from 400mm in Karamoja to 2200mm on Ssese Islands in Lake Victoria. The total renewable water resources are estimated at 66 km^3 per year responding to 2200m^3 per capita per year. While this meets the current needs of the country, it is estimated that by 2017, Uganda will be a water stressed country.

There are serious challenges in reversing the reduction in the habitant cover, which decreased from 95 per cent in 1980 to 85 per cent in 2000. Associated with this, the native (plant and animal(indigenous?) species have been declining with the population index decreasing from about 60 per cent in 1980 to 40 per cent in 2005. There has been reduction in total fish production from 434,000 tonnes in 2004 to 374,000 tonnes in 2007. The biodiversity has also been decreasing as reflected in the change in the biodiversity index from 0.8 in 1980 to 0.7 in 2005 and the reduction in the species richness index from 95 per cent in 1980 to about 85 per cent in 2005. The ecological footprint has decreased from 2 in 1980 to 1.6 in 2005 and the bio-capacity has decreased from 2 in 1980 to 1.2 in 2005[12].

Private Sector Performance

The Private sector in Uganda largely comprises Micro, Small and Medium Enterprises (MSME) and these cut across sectors of the economy, contributing to 20 per cent of the GDP[13]. The number of registered businesses in Uganda was 25,000 in 2007 with the majority (11,003) located in the central region of the country. Of the firms across sectors, those in the industry constituted 17.7 per cent and were mainly engaged in beverages, sugar, textiles, building materials, foot wear, packaging and food processing.

Although the private sector has grown rapidly between 2001 and 2007, growth in the number of firms has been concentrated in small firms with low value addition, whereas larger firms with high value added per employee did not increase rapidly. The service sector is the fastest growing sector in Uganda (13 per cent per annum), with 32 per cent of these investments owned by Ugandans. Predominant in the service sector are telecommunications, hospitality and trade sub-sectors.

MSMEs employ over 1.5 million people of the total non-farm workforce. These include: retail trade, education and restaurants accounting for the bulk of the total employment in the new firms. With the exception of education, it was found that the rapid growth in enterprise is focused on low-value services and is, therefore, unlikely to be a platform for significantly transforming the economy.

The economic conditions and the business climate have significant impediments as indicted by various studies. The World Economic Forum's Country Competitiveness Index 2009 ranks Uganda 108 out of 133 countries. Many other African countries rank better (Tanzania 100, Kenya 98, Namibia 74, Botswana 66, South Africa 45) indicating better investment and business environments in those countries. In this study, Uganda deviates negatively compared to a peer group of factor economies with regard to health, primary education, higher education and training, infrastructure, institutions and technological readiness[14]. These are key sectors that need to perform well to support the primary growth sectors. As shown in Figure 15.7, there is a positive deviation from this group with regard to Uganda's micro-economic stability, financial market sophistication and labour market efficiency[15].

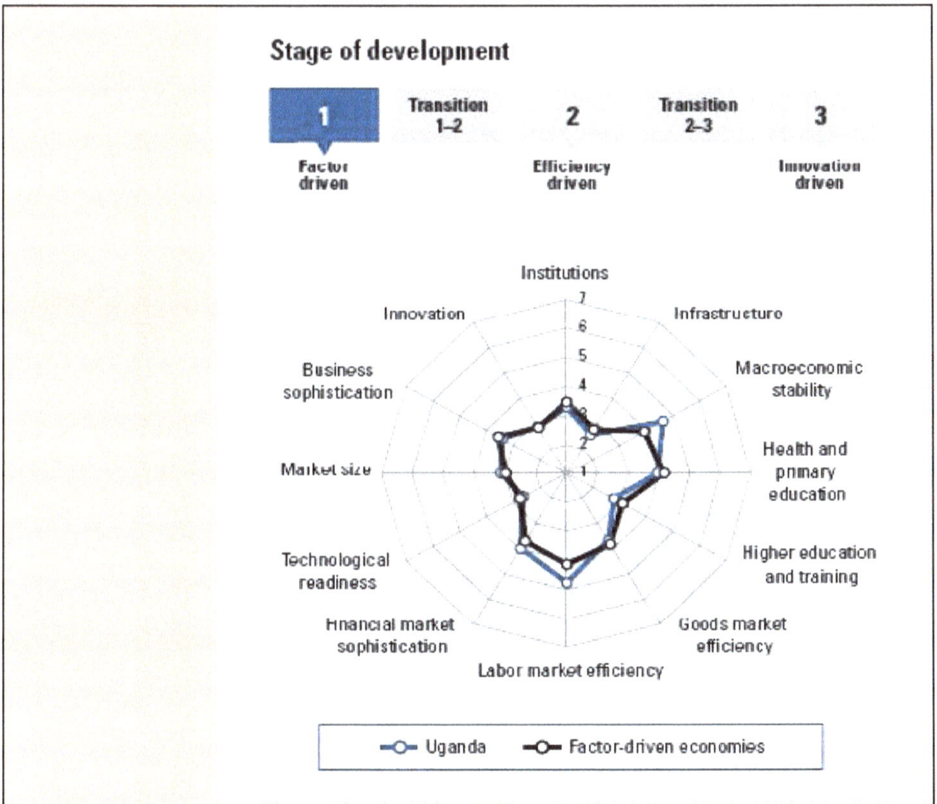

Source: World Economic Forum: World Competitiveness Index 2009

Figure 15.7: Comparison of Uganda's Competitive Position to Peer Group

The international "Doing Business Survey 2010" report ranks Uganda 112th out of 183 countries on a wide range of business indicators. The problems identified in particular, are (a) registering property, (b) trading across borders, (c) protecting investors, (d) starting a business, (e) enforcing contracts, and (f) getting credit. As Figure 15.8 shows, these impediments, among others, affect the competitiveness of the economy, with the main issues being access to finance, corruption, infrastructure, tax administration, work ethics and the government bureaucracy.

The Government has been implementing measures to address the barriers to investment through the competitiveness and the Investment Climate Strategy. However, these actions have been slow and often uncoordinated. The Third National

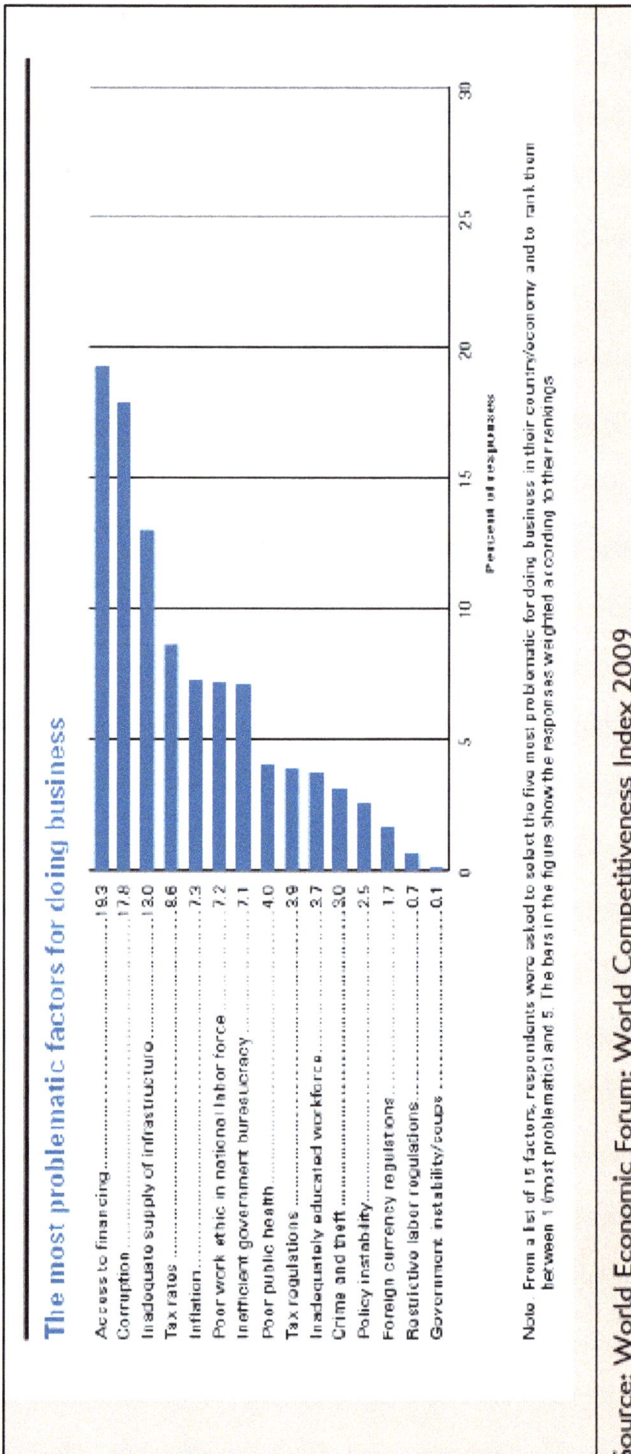

Figure 15.8: Significant Problems Impeding Competitiveness 2009

Competitiveness Forum noted similar issues to the ones identified, as well as a broader range of issues that included: high transport costs; lack of a policy framework for private-public partnerships; poor urban planning and the limited market size. Other issues that were noted include: the potential effects of the financial crisis in many countries and the world economic downturn reducing the Foreign Direct Investment (FDI) and remittances. Similar problems were identified in a survey of small and medium enterprises.

Civil Society

As acknowledged in past, the government policy framework is as follows: 'it is essential for the concept of civil society that their actions are not planned or dictated by the Government'. The civil society in Uganda is as diverse as the concept itself, making it quite challenging to work with the sector. However, the emergence and growing importance of the broadly inclusive non-governmental platforms presents an opportunity to engage with civil society in a more strategic and sustained manner. These platforms currently include the non-governmental organizations (NGOs) and other non-profit entities including the faith (religion) based organizations, trade unions, community based organizations, professional associations and interest groups.

When examining the structure of Uganda's civil society, a mixed picture emerges. The participation of the citizens in the (Civil Society Organizations?) CSOs appears to be very extensive. To a great extent, this is because rural life, in this largely agrarian country is often characterized by membership in various forms of community and other mutual help groups, which are socially inclusive and to which faith-linked organizations can be added. Most of these CSOs are donor-dependent. The sector is fragmented across many facets of development and tends to favour accountability towards donors over accountability to the local population.

Following the Paris Declaration, CSOs are increasingly seeing advocacy work as a legitimate undertaking. It is also an area of work that donors support. The government is opening avenues, especially at the district level, and some of the CSOs are being contracted for service delivery.

Despite the important role that civil society has and can pay in the development process, a number of factors directly and indirectly work against their full effectiveness. The National Development Plan (NDP) has identified a weakness in the lack of a comprehensive and consistent framework to institutionalize the interface between NGOs and various Ministries, Departments and Agencies (MDAs) at national and district level. The current NGO law also constraints the engagement between NGOs and Government, thus fuelling mutual suspicion and sometimes hostility, rather than cooperation and partnership.

Several agencies of the government have little understanding of what conceptually NGOs are and/or ought to be partly due to lack of comprehensive information on the nature, precise contribution and value of NGOs work to Uganda's development. Several agencies thus view NGOs as appendages of the government whose programmes and financing should be integrated in the government plans.

Furthermore, there is little empirical information available or known about what NGOs do, where they are operational, and the impact of their activities. Thus the outcome of their work is not meaningfully captured in the development statistics.

Most Binding Constraints

As observed in the Barcelona Agenda 2004, the government will adopt a range of policies and interventions in its development strategy in the NDP in order to address the most binding constraints to economic growth. This will entail implementing a mix of microeconomic and macroeconomic policies and interventions, with the government playing a more significant role. These actions will stimulate growth and productivity in the private sector and will improve performance in the public sector. In this context, a systematic methodology for undertaking country diagnosis and identifying the most binding constraints to growth was used in the course of the NDP preparation process. These constraints are presented below.

Constraint 1: Weak Public Sector Management and Administration

Analysis of the binding constraints to growth in Uganda reveals that public sector management and administration is characterized by weak policy, legal and regulatory frameworks; weak institutional structures and systems; weak civil society and civil participation; inadequate data and information; inadequate standards and weak quality infrastructure; limited social protection and support systems; and weak management of environment and climate change. Over 70 per cent of the government sectors have obsolete, absent or weak policy frameworks. The weak institutions, structures and systems take the form of inappropriate organizational structures, inadequate systems, understaffing, limited strategic oversight, overlapping and duplication of roles, protracted institutional infancy, weak client responsiveness and inefficient bureaucracy. This weak public sector management has led to low absorption of public funds and poor delivery of services.

Although considerable efforts have been made to reduce corruption, including putting in place an appropriate legal and institutional framework, it still significantly affects public service delivery in the country. Corruption is most rife in procurement, administration of public expenditure and management of revenue. The citizenry is not yet fully empowered to engage effectively in demanding better performance from the government institutions in meeting their obligations and providing services. Corruption has remained high in various levels of the government with Uganda's ranking moving down from 126 to 130 out of 180 countries in the Transparency International's Corruption Perception Indices for 2008 and 2009.

While Uganda has signed some international standards and codes, many of them has not yet been ratified and domesticated by the government. These are relevant not only for benchmarking good practices for promoting good governance and sustainable development, but also enhancing the country's competitiveness through participation in key economic organizations. There is also a limited range of formal international standards related to the quality of goods and services that have been implemented in Uganda. There are 1,204 Ugandan standards which is an extremely low number compared to Kenya which has up to 8,000 standards. Out of the registered

25,000 firms, only 86 are certified by ISO. Only 203 firms are certified by The Uganda National Bureauof Standards(UNBS.)

There are many areas where citizens are pressing the government for better performance given that the poverty level was at 31 per cent in 2005/06 and inequity is evident in the relatively high Gini-coefficient data. Progress towards the MDG goals is not as fast as desirable, with poor outcomes for health, in particular. There are also gaps in social protection. As of March 2007, the pension scheme covered only 44,000 public servants while the (full name,pl) National Social Security Fund (NSSF) protected 1,282,994 workers between 1967 and 2003. The orphans and vulnerable children are estimated at 7.5 million of whom only 23 per cent are adequately cared for.

The management and administration of land in Uganda is still a major challenge and will require additional input to implement the Land Act as amended in 2009. The complex and multiple land tenure systems have severely limited the land use planning and utilization. At present, an approximately 10 per cent of the total land area is titled. In addition, there are problems of alteration of land titles to the extent that it has caused a number of land disputes.

The management of the environment is not effective and remains relegated to a few institutions that face significant challenges in enforcement. Over 7 per cent of the original wetland area had been reclaimed as of 2005. The forest cover declined from 85 per cent in 1980 to 24 per cent in 2000. Given this trend, the country will experience significant adverse environment and economic effects in the near future.

Constraint 2: Inadequate Financial Services

In the public sector, despite concern about inefficiencies in the use of public funds, the majority of Ministries, Departments and Agencies (MDAs) usually cite funding as a key constraint to delivery of public goods and services. The financing constraint in the public sector is manifested mainly in under-funding of priorities of public sector programmes and projects, thereby impending the quantity and quality of service delivery. In 2008, the Government domestic revenue, as a ratio of GDP was about 13 per cent, which is low compared to about 25 per cent and 17 per cent for Kenya and Tanzania respectively. While this ratio is projected to grow over the NDP period, the Government has to rely on external financing for much of the budget expenditures, especially in the category of capital development. In this situation, it will be constrained to some extent, in its ability to match expenditure to priorities in a timely way.

In the private sector, availability of and access to financial services is also one of the major constraints. Non-availability of financial services is manifested indicating lack of sufficient financial services infrastructure required to deliver appropriate financial services across the country. For instance, only 23 districts in Uganda have commercial banks. The situation is worse in most of the rural areas as most banks are situated in urban centres. The number of bank accounts in the country is about five million, representing a 16 per cent penetration. This low financial penetration limits access to safe and sound financial services and therefore, access to high quality and low-cost finance is a constraint. The capital market for long term funds is still under development. There are, currently, six local equities and five across border listings with the government and corporate bonds traded on stock exchange in Uganda, which compares poorly with that of Nairobi stock exchange in Kenya with 46 listed firms.

The nominal lending interest rates of banks range from 17 per cent to 23 per cent, which is high, except for very few areas where the rates of return on investments are higher, and such interest rates are bound to continue discouraging the investments. In the formal Marco finance sector, nominal lending interest rates are even higher, ranging from 24 per cent to 36 per cent per annum and the situation is even worse in the informal microfinance systems. Coupled with the low savings culture in the country, raising relatively cheap domestic capital for investment is expensive and will require boosting the mobilization of domestic resources for intermediation through promoting rural savings and further development of pension and insurance systems.

Constraint 3: Inadequate Quantity and Quality of Human Resource

The quality of human resource is important to organizational success. Despite the large and fast growing youthful labour force and the Government's efforts to provide education and training at various levels, the country continues to experience deficits in the supply of skilled human resources. This is evident in the limited availability of skilled labour as partly shown by wide wage differentials and the high number of vacant posts in technical areas. The lack of skilled human resources is associated with quality issues in the education system, including low school completion rates, limited capacity in the vocational and technical training institutions and the brain drain from the country. This is exacerbated by inadequate manpower planning in key areas of the economy.

The health sector, for example, continues to experience considerable workforce challenges arising from numbers, skills and motivational factors. The ratio of doctors to the patients in Uganda was 1:24,725 which is significantly lower compared to Kenya at 1:7,100, Cuba at 1:169 in 2002 and Malaysia at 1:1,430 in 2000. The ratio of nurses to the population in Uganda is 1:1,634 compared with 1:877 in Kenya, 1:740 in Malaysia in 2000 and 1:134 in Cuba in 2002. The low number of health professionals in Uganda adversely affects the delivery of health services. The situation is similar with regard to the low levels of qualified people in other professions.

Human resource development is further encumbered by low service delivery standards in health and education sectors. For national referral hospitals the ratio is 1:30,000,000 whereas the standard should be 1:10,000,000. At the level of Health Centre IV (HC IV), the ratio is 1:187,500 whereas the desired ratio should be 1:100,000. For Health Centre III, the ratio is 1:84,507 whereas the optimum should be 1:20,000. For the education sector, at primary level, the class room pupil ratio is 1:68 compared to the target of 1:53. For secondary school level, the student classroom ratio is 1:60 whereas the target is 1:35. In addition, the curriculum and methods of training employed in the national education system do not adequately prepare graduates to the standards that are globally competitive, thus rendering the country to continually depend on foreign experts in specialized fields.

Constraint 4: Inadequate Physical Infrastructure

The physical infrastructure plays a major role in the movement of financial goals and services as well as factor inputs from places of supply to places of demand. The inadequate physical infrastructure is a constraint to the production in many sectors of the economy. In the transport sector, the roads, which carries 96.4 per cent of the total cargo freight, only 4 per cent of it is paved. The rail network carries only 3.5 per cent of the freight cargo and only 26 per cent of it is functional. This contrasts with China and India where over 90 per cent of the cargo is transported by rail. The cost of cargo freight by rode is 3 times more than the cost of using the rail. There are no function wagon ferries and the country has only one entry-exit airport. Transport costs remain a significant trade barrier, equivalent to effective protection of over 20 per cent and imposes an implicit tax on exports of over 25 per cent (and up to 50 per cent on air freight). As a result, the cost of doing business in Uganda is high, undermining economic competitiveness.

Uganda has one of the lowest electricity consumption per capita in the world. The consumption per capita is 75kWh/capita compared to Malaysia about 3,668 kWh/capita, and Korea, about 8,502 kWh/capita. It compares poorly even with its peers like Kenya at 152 kWh/capita and Ghana at 253 kWh/capita. The low level of consumption per capita is due to inadequate power generation capacity to meet the energy requirements of the population. The access to electricity is low at 11 per cent in comparison to Malaysia, at 98 per cent and Kenya, at 15 per cent. The use of electricity is constrained by higher power tariffs and inadequate transmission and distribution network. The Tanzania electricity consumer pays USD 8 cents per kWh; Kenyan consumer pays 13 cents per kWh, while Uganda consumer pays 22 cents per kWh. Moreover, the supply of electricity in Uganda experiences disruptions due to load shedding and power outages, although this has improved in recent years as a result of short term but expensive interventions to mitigate the impact on the economy.

The communication infrastructure has improved significantly following liberalization of the sector that has attracted substantial private sector investments. However, the limited coverage cost of access and usage as well as limited diversity of communication media are still major constraints. For example, the internet infrastructure is still limited with most concentration in the capital, Kampala. The price of the internet in Uganda is still very high for everyday users at USD14 cents per minute compared to Malaysia at USD 0.155 cents per minute. The problem of high cost hence limited the use of the internet is related to inadequate bandwidth and other related infrastructure. The situation is expected to improve; following the connection of East Africa to the undersea cable and installation of the national ICT backbone in Uganda.

The consumptive use of water stands at 21 cubic meters per capita, which is far below the world average of 599 cubic meters per capita. This means that less water is used for irrigation and industry, estimated at 28 times less than the world average. Access to safe water stands at 63 per cent for the rural and 72 per cent for the urban population. This is low compared to Malaysia with 96 per cent and 100 per cent in the rural and urban areas respectively. The available data suggests that 30 per cent of the population has no access to safe water supply for domestic use.

Constraint 5: Gender Issues, Negative Attitudes, Mind-Set, Cultural Practices and Perceptions

Certain elements in Uganda's traditions, culture and religious norms are not supportive to modern approaches in the society and have, therefore, limited economic growth and structural transformation.

There is discrimination against women in Uganda through traditional rules and practices that explicitly exclude them or give preference to men. At the governance level, these rules and practices limit the political and economic participation of women. This in turn, leads to formulation and passing of policies and laws, which do not protect women's rights. At the community and household level, women are restricted from participating in important decisions such as resource use, family planning and access to services such as health and education. Women have been marginalized in access to ownership and control over land, education, business ownership, skills development access to financial resources; employment and inheritance rights. The culture of early marriages amongst girls increases the rate of early pregnancies and is partly responsible for the country's high maternal mortality rate and high fertility rate. At all levels, the culture of ignoring various forms of violence against women must also be reversed in order to unlock critical barriers towards women's empowerment.

In terms of culture, Uganda has lagged behind in development partly as a result of backward cultural practices, beliefs, attitudes and lack of national ethical values in political, social and economic spheres. These are manifested in the form of; poor time management, negative attitudes towards work reflected in low human productivity and low entrepreneurial spirit, high dependency levels which stood above 1.5 large family size (?) and rapidly growing population, female genital mutilation, property inheritance, which favours mainly the male gender, the lack of ethics integrity and patriotism; and a tolerance for corruption.

Negative attitudes and perceptions influence the use of appreciation of natural resources, adoption of science and technology and use of ICT. There is low utilization of contraceptives among the population and slow behaviour change resulting in increased HIV/AIDS prevalence rate. In addition, negative attitudes towards immunization still persist, contributing to lower coverage, which stands at about 46 per cent.

The poor business and entrepreneurial attitudes of the indigenous private sector is partly reflected in the country's high enterprise mortality rate, the highest in Global Entrepreneurship monitor reports of 2005, 2006 and 2007. The survey reports rated Uganda in the top countries, globally, in terms of enterprise creation. The high failure rate of not only fledging businesses but also those that accessed capital large export markets, combined with the low number of successful joint venture investments below local entrepreneurs and foreign investors, underscore the poor business attitudes of Uganda's SME's.

Constraint 6: Low Application of Science, Technology and Innovation

Global competitiveness is dependent on technological capability. This relates to the astuteness and competences of firms and productive enterprises to access, apply, influence, adopt, adapt and deploy technical knowledge to gain competitive advantage. The technical capabilities of a country (its national system of innovation), is known to affect the export competitiveness of firms within the country and in particular the sectoral structure of export performance. In addition, technological capabilities influence the existence of multinational enterprises (MNEs) in a particular country and affect the competitive advantage of those MNEs[19]. The low level of total factor productivity experienced in Uganda is partly due to limited application of STI in the production process and service delivery mechanisms. The constraint is manifested in terms of poor quality of products and services; a low technology component of exports as evidenced in the percentage of manufactured exports to primary exports, which was as low as 4.2 per cent in 2008/09; and low numbers of R&D personnel with one R&D personnel per 1000 labour force compared to OECD countries that range from 5 to 18 R&D personnel per 1,000 labour force. Technology Achievement Index (TAI)[20] is 0.24 compared to Malaysia at 0.34. The share of expenditure on R&D as a percentage GDP is 0.3 per cent compared to Malaysia at 0.63 (2005/06). In 2001, Uganda published only 91 articles in respected journals compared to about 10,000 by Israel. In the area of innovations, only 3 patent applications are submitted per year compared to Malaysia that has over 6000 patents per year. The ratio of science graduates to Arts graduates is 1:5 whereas it is 1:1.5 in Malaysia.

Beyond the limited application of STI for social transformation, Uganda's ambition to become a major player in the international market has been severely impeded by the country's lag in technological competitiveness. This is evident from the low technology component in exports as evidenced by the low per cent of manufactured exports to Primary products.

Constraint 7: Inadequate Supply and Limited Access to Critical Production Inputs

The value is created through a production and exchange process. Inputs are at the centre of the value chain creation. Examples of these inputs include raw materials such as fertilizers and water, human financial resources and physical capital. Some of these inputs are in serious short supply or are very expensive, thereby curtailing their widespread use. For example, fertilizer use in Uganda is at 1kg/hectare having increased from 0.37kg hectares in 2000 compared to 6kg/hectare in Tanzania, 16kg/hectare in Malawi and 31.3kg/hectare in Kenya. This is partly attributed to high cost, which has almost tripled from USD 252 per ton in January 2007 to USD 752 per ton in January 2008 (IFDC 2008). Use of fertilizers can boost the yields by 50 per cent. Currently irrigation is carried out on 14,418 hectares of land against an irrigation potential of 400,000 hectares. This represents 3.6 per cent of irrigation potential exploited, explaining the low consumptive use of water for production, which stands at 21 cubic meters per capita, far below the world average of 599 cubic meters per capita.

The cost of 50kg cement in Uganda is about USD 15 compared to USD 3 in Malaysia and about USD 10 in Kenya. Uganda currently has four steel mills, which use scrap as their basic raw material. National demand for steel products is estimated at 60,000 to 80,000 metric tonnes per annum while the current production level is estimated at only 7,000 metric tonnes per annum, which represents 9 per cent of the demand. This is very low compared to Kenya that produced over 220,000 tonnes in 2003. Lack of access to these critical production inputs severely affects the development.

Application of Science and Technology in Uganda

Achieving socio-economic transformation remains continuous improvement in the way the goods are produced and delivered and the services within the economy. This can be realized through accelerated use of applied technology, research and innovation, especially for the rural people where in Uganda, most of the population lives.

Currently, the research and development is mainly confined to institutions such as universities, colleges, vocational institutions and the government research institutions with limited applicability to production and delivery of services. It is also worth noting that adoptability of the little available applied Science and Technology is slow in both public and private sectors.

Globally, a country's efforts in this direction are measured against indicators such as researcher per workforce, R&D personnel per labour force, expenditure on STI as percentage of GDP, number of publication in scientific journals, number of scientific patents per year, exports with high technology constant, total manufactured exports and the composite index known as the overall technology achievement index.

Uganda has currently only one researcher per a thousand members of the workforce compared to over five researchers per a thousand in the developed world (UNCST 2005); ONE R&D personnel per a thousand of the labour force compared to five to eighteen R&D personnel per 1000 of the labour force in the OTCD countries. The ratio of GTS to S&T graduates is 5:1 compared to 1:1.5 in Malaysia.

The share of expenditure on R&D as a percentage of GDP is 0.3 per cent compared to the African Union target of 1 per cent. (repeated – check in the Constraint 6)

Only 91 scientific publications were produced in internationally recognized scientific journals in 2001, compared to 10,000 in Israel. However, world class discoveries have been made in HIV/AIDS prevention and vaccine trials, cassava mosaic eradication and development of clonal coffee.

According to a UMCST survey, only 40 patents have so far been granted for local inventions and usually less than 3 patent applications were submitted in any given year by 2004. In addition, out of the total recorded applications for industrial property rights, 75 per cent were owned by foreign enterprises suggesting low local involvement (WIDO 2005).

Science and Technology for Occupational Development

As already indicated earlier, abut 90 per cent of Uganda's population live in villages, though migration to urban areas is increasing at a very fast rate as shown in the Table 15.7.

Table 15.7: Trends in Urban population in Uganda 1980–2002

Census Year	*Urban Population (million)*	*Total Population (million)*	*Percentage Population in Urban areas*	*Growth (Per cent)*
1980	0.84	12.6	6.7	
1991	1.65	10.7	9.9	6.1
2002	3.00	24.2	12.3	5.1

The level of urbanization is still very low in most of the regions in Uganda ranging between 7 per cent and 10 per cent. Part of the migration is due to lack of occupation opportunities in the villages. This lack of opportunities to the village population could be improved by the application of science and technology to increase productivity.

To enhance science and technology for occupational village development, the government of Uganda is carrying out the following:

1. Enhance Educational Levels

Uganda has introduced a universal primary education where enrollment has jumped from 3 million pupils to 7 million pupils in the lat 3 years. This entity will feed into a universal technology education that will feed into tertiary institutions and colleges. While some of the students will go on to join Universities and colleges, the government is putting emphasis on science education. Further emphasis will be put on establishment of village polytechnics where students will be instructed into productive science and technology courses that will be a backbone for creation of technologies. Such attempts will enhance value addition to the numerous natural and agricultural resources that will be eventually occupied by the villages through provision of food and employment.

The overall aim of this is to create a skilled manpower base for the village population that will ensure prosperity for all by increasing incomes and wealth for Ugandans through value addition to local resources, human capital development and marketing of the products.

2. Establishment of ICT Infrastructure

In order to enable better communication between the rural and urban areas, the government has set up village telephone and internet services. With these facilities, villagers can have access to a lot of information such as market prices, etc., for their produce; technologies that can add value to the agriculture produce; and their children can access a lot of educational materials to supplement what they get in classrooms.

There are job opportunities as well because many people engage in providing services to the community who needs to make telephone calls, get market information *i.e.* by setting up kiosks in the villages; people get a livelihood that has changed their lives for the better. The cost of doing business has also gone down, *i.e.* getting a price for the coffee could take about a week in the past due to use of the postal system. As of now, the price of coffee can be obtained by just clicking on the Internet.

3. Technologies for Value Addition of the Agricultural Produce

3.1 Agricultural Resources

The economy of Uganda is based on agriculture. However, over the years all the agricultural produce was exported in the raw form without making any value addition due to the colonial set up. Now the government of Uganda through its programme for modernization of agriculture has set up value addition to agricultural produce.

Agriculture has for a long time been a core sector of Uganda's economy. By 2005, it employed 73 per cent of the labour force. However, in 2008/09, the sector accounted for 23.7 per cent of GDP. Agricultural exports accounted for 47 per cent of the total exports into 2007. The country's agro based industrial activities via increased investment in science and technology, agriculture can contribute to:

(a) Improved Seeds

Through Research and Development, new disease and drought resistant seeds have been introduced for use by the rural population. These improved seeds will enhance food production, increase income and enhance prosperity.

Examples of new seeds including new cassava planting materials: In recent years, cassava plants have been destroyed by the cassava mosaic disease. Others include new rice seeds for upland areas and high yielding bananas

(b) Introduction of irrigation

Most of Uganda's agricultural produce was dependant on rain. In order to mitigate the effects of the weather, new technologies for irrigated agriculture have been introduced to take advantage of Uganda's abundant water resources *i.e.* about 20 per cent areas of Uganda is covered by water.

(c) Food Processing

A lot of success has been achieved in adding value to most of the agricultural products. This includes:

- Tomatoes into tomato sauce
- Fish into fish fillet
- Fruits into fruit juices
- Bananas, peeled and frozen – earlier, the bananas were marketed long distances with their peels and the peelings clogged to Urban Sewerage systems.
- Potatoes into chips

(d) Postharvest Storage

New Science and technology research has established new and more cost effective methods for storage of agricultural produce. Before introduction of better post harvest storage, more than 50 per cent of the harvests were lost due to pests and vermin attacks. Now, the losses are below 40 per cent and people in the villages can store and prolong the shelf life of the produce. (This mitigates famine and guarantees them a year long income.)

4. Energy

Most of Uganda's rural areas have no access to power. 90 per cent of their energy source is wood whose extraction has a great effect on the environment. (Forest cover is about 20 per cent, decreased from about 70 per cent in the past 50 years). The government has now introduced technologies for harvesting of the large biomass, still available by introduction of biogas and other alternatives and renewable energy sources. Though still not affordable solar power is being encouraged and with more R&D efforts, the cost will eventually come down.

Although Uganda has recently discovered oil in the Albertine Graben, this oil will not provide an immediate relief. The oil will be used to offset the importation bill and the government has enacted a policy that will see some of the revenue obtained used to promote Science and Technology that will benefit the rural population.

5. Mineral Extraction

Uganda has considerable mineral resources. These include tin, wolfram, limestone, gold, copper, columbium-tantalite, kaolin, feldspar, salt, phosphates, among others. Most of these minerals are located in the rural areas. Their extraction has been on a small-scale and also, not environmental friendly. Simple and appropriate technologies to extract these minerals have been established to create a cleaner production. These include:

1. Non-use of mercury for gold extraction
2. Local production of crushers and ball mills that are energy efficient
3. Design and construction of fuel efficient kilns for lime production
4. Development of low cost fertilizers from phosphates, vermiculite, cow dung.
5. Production of simple ceramics based on kaolin, feldspar, ball clay, sand.
6. Production of dimension stone from granites and marbles using simple tools designed and made by local artisans from locally established foundries.

Conclusion and Recommendations

While a lot of Science and Technology is available world over, the rural population in Uganda is yet to benefit from it. This is due to various issues and constraints:

1. The Science and Technology needs to be re-packaged and made appropriate for use by the village population.

2. There is need to increase the literacy levels so that the village population can absorb and internalize the science and technology and be able to apply it.
3. Financial mechanisms need to be put in place to enable the population acquire and access the science and technology.
4. To enhance use of science and technology for production; markets need to be availed otherwise villagers will be discouraged if they produce and nobody buys their products.
5. Researchers need to be facilitated to carry out more research and development work geared to help the village population
6. There is a need for more interaction among Non Aligned Movement (NAM), scientists and technologists (as we are doing today) to enhance the diffusion of science and technology for occupational village development.

Chapter 16

Policy and Practical Measures Adopted by Ethiopia in Promoting Occupational Villages

Yared Awgichew

Agricultural Technology Transfer Expert,
Ministry of Science and Technology, Ethiopia
E-mail: yaredawgichew@yahoo.com

ABSTRACT

Eighty-three per cent of Ethiopians live in rural areas and derive their livelihoods from agriculture. Since the the Ethiopian People's Revolutionary Democratic Front (EPRDF) came into power in 1991, Ethiopia has achieved substantial reduction of poverty, mostly through the deliberate move in the country side. The Ethiopian government has adopted agricultural led development industrialization (ALDI) strategy, which continues to serve as a framework of rural investment planning in the area of infrastructure, social services, research and extension. The current Ethiopia's Plan for Accelerated and Sustainable Development to End Poverty (PASDEP, 2005-2010) has aimed at achieving the expansion of health, education, electricity, marketing, roads and other facilities to rural areas. As a result of PASDEP, the number of electrified towns and villages was increased from 800 to 5000 while 70 per cent of rural people were organized under co-operatives; 200 market information points in district level and 20 terminal centers in the remote areas were established; 25 Training and Vocational Education centers were established; 55,000 extension workers were trained; 18,000 Farmer Training Centers were established; 10 million of people gained training; average walking distance was reduced to 3.2 hours, 8 millions of telephone lines were accessed in 100 per cent within 5 km and increased ITC services. These infrastructure expansions at the exponential rate, relatively to the country way of coming, has resulted in knowledge based production and reduced reliance on subsistence agriculture as a main source of livelihoods. When we consider the power generation and credit service those are alone, but despite of many challenges the Electrification of rural areas can encourage the mechanization of traditional rural industries (weaving, carpentering, blacksmithing, milling, construction, etc), attract outside capital, extend the hours of business operation (increasing productivity), create better storage facilities, and change the way of rural people's live, especially by enabling them to use modern appliances (e.g. cooking and storage) and connecting them to the modern world (radio and TVs).

Keywords: *Ethiopia, Public policy measures, Poverty reduction strategies, Occupational villages.*

Introduction

Eighty-three per cent of Ethiopians live in rural areas and derive their livelihoods from agriculture. Considering this huge number, the Ethiopia government has adopted agricultural led development industrialization (ALDI) strategy, which continues to serve as a framework of rural investment planning in the areas of infrastructure, social services, research and extension. Hence the previous "Sustainable Development and Poverty Reduction Programme" (SDPRP) and the current programme namely "Plan for Accelerated Sustainable Development to End Poverty"(PASDEP) were aimed at achieving the expansion of electricity, marketing, roads, health, education, and other facilities in the rural areas.

Within the PASDEP framework, the growth of agriculture led development and the accelerated private sector involvement are taken as important pillar strategies to tackle the poverty, food insecurity challenges of rural areas and unemployment problems of the rural villages, towns and cities. The government has embarked on an aggressive programme to accelerate progress as rapidly as possible -including a big push on education to increase human resource capacity, expanding infrastructure to enhance competitive advantage of domestic firms that are export-oriented, opening the economy, building institutions, and decentralizing the Government. The ultimate goals of PASDEP are to ensure human development among the poor and among women in particular, as the financing and the capacity allows. The principles underlying the PASDEP are that all sectors broadly contribute for the reduction and ultimate eradication of poverty, by giving particular attention to the agriculture and rural development. There has been a change in the poverty level in the country since the implementation of the plan and promising progresses have been registering towards achieving the Millennium Development Goals (MDG). This period has also witnessed the beginnings of two peculiar changes in the economic landscape of Ethiopia: a shift to a higher growth path and more diversified sources of growth. In the past (up until 2002-03), the economic growth had been volatile due to trade term effects and the weather induced fluctuations in the agriculture sector. However, at the present time the economic growth was very much sustained and complemented by the strong performance in the construction, manufacturing, trade and tourism, banking and insurance, real estate, etc. The construction sector has been spurred by the much needed public sector investment in infrastructure (roads, rural infrastructure development including food security, telecommunication, electricity, irrigation etc.,). Microfinance opportunities have been successfully expanding in Ethiopia during the past fifteen years, including the remote villages, where the majority of people are engaged in small scale agriculture, which is little supported by the modern technology.

To understand the degree of effect of infrastructure on the productivity of citizens in rural Ethiopia, studies were conducted by so many researchers. Among them, a group of experts from the International Food Policy Research Institute was working in the west Ethiopia and in the capital, Addis Ababa to find out the effect of electricity on the productivity of the hand-loom clusters. Finally, they found that access to electricity matters to better performance of the firms because the workers or producers are now able to work for longer hours in lighting conditions. Therefore, it has been possible to facilitate the cluster growth through infrastructure investment.

This paper presents how the Government programmes initiated through various policy measures *viz.*, clearing the road for rural occupational villages' development, have achieved maximum efficiency in the rural Ethiopia with the following initiatives:

- Strengthening the rural-urban sectoral linkages;
- Supporting for strong off farm business progress ;
- Diversification of basic infrastructures;
- Development of human capital.

Policy Measures (Clearing the Road for Rural Occupational Villages' Development)

ALDI

The Ethiopian government has formulated a long-term economic development strategy known as –Agriculture Led Development Industrialization (ALDI), which is devised towards the renovation of the backward economic structure. It is a two-sided strategy, integrating a) the external sector (export -led part), and b) the internal sector, which explains the forward and the backward relationship between agriculture and industry.

In other words, agriculture will provide commodities for exports, domestic food supply and industrial input; and enlarge the market for domestic manufactures. The strategy for ALDI is the government's overarching policy response to Ethiopia's food security and challenges in agricultural productivity. The ALDI's distinctive features include:

- Commercialization of smallholder agriculture through product diversification;
- Shift to higher-valued crops;
- Promotion of high-value export crops;
- Support for the development of large-scale commercial agriculture;
- Effective integration of farmers with domestic and external markets; and
- Tailoring interventions to address the specific needs of the different agro-ecological zones of the country.

The ALDI strategy promotes the use of labor-intensive methods to increase output and productivity by applying chemical inputs, diversifying production, utilizing improved agricultural technologies etc. It also emphasizes the importance of distinguishing agro-ecological zones and tailoring strategies as well as interventions for optimal development outcomes. This distinction guides the differentiated interventions needed to promote cross-sectoral and integrated growth.

Accordingly, the rural development strategy divides the country into three main agro-ecological zones, which include regions with a) adequate rainfall, b) moisture stressed areas, and c) pastoral areas. In regions with adequate rainfall, the focus is on exerting all possible efforts to efficiently utilize available rainwater to bring about a maximum possible rate of agricultural development, and promoting irrigation in areas where it is feasible. In moisture stressed areas, major activities have been undertaken to enhance food security and are focused on increasing off-farm income opportunities, and voluntary resettlement to more productive areas. In terms of pastoral areas, special efforts are made to enhance specialization in livestock production and marketing through the provision of water supply for the community and their livestock as well. Major interventions in this area also include improving livestock quality, expansion of animal health services, water points, feed production and improvement of breeds and development of market infrastructure. While the ALDI strategy is based on the three main zones described above, it is realized that there is a complex diversity within each zone, and hence actual interventions are locally tailored accordingly.

In all regions, the emphasis will be ensured on every possible aspect, which is in place to facilitate the takeoff of commercial opportunities. The transformation of the smallholder farmer is to be achieved through area based specialization as well as diversification of agricultural commodities. In line with this, commercialization of smallholder agriculture will be the engine for rural growth in the coming decade or so.

The government will largely restrict itself to playing an intermediating role where there are gaps. The areas where the government would have a comparative advantage include setting the regulatory framework and standards, financing infrastructure, especially the a) roads *via* the Rural Roads and Transport Programme, b) the small-scale irrigation primarily focusing on water harvesting at the household level that help farmers build medium-scale irrigation on a cost-recovery basis, and c) construction of some multi-purpose dams that would support larger-scale irrigation. Finally, the government may help with the acquisition of technology (such as specialized seeds or genetic material) from outside the country. In addition, ALDI will also focus on areas such as sanitation; telecommunication; power; information and communication technology (ICT) and administration capacity of the country.

SDPRP

In 2002/03 Ethiopia prepared and launched a sustainable development and poverty reduction programme (SDPRP), following wide-ranging public consultations in 2001. The strategy focused on improving human and rural development, food security and capacity building through transformation of the agricultural sector towards the following four pillars.

- ☆ Agricultural Development Led Industrialization (ADLI);

- Justice system reform and civil service reform;
- Decentralization and empowerment; and
- Capacity building in public and private sector.

Major emphasis was given to ADLI as the agricultural sector is the source of the country's livelihood. In addition, it has the potential to generate primary surplus to fuel the growth of other sectors of the economy, specifically in the industrial sector.

The SDPRP Programme was fully implemented and an assessment report was prepared upon its completion to provide lessons learnt that could serve as a basis for the current Plan for Accelerated and Sustainable Development to End Poverty (PASDEP). Annual Progress Reports (APRs) were also prepared regularly to track the progresses achieved, challenges encountered and lessons learnt in implementing the Programme. The SDPRP was not a stand-alone strategy. Rather it was built on the existing Government policies, strategies and programmes, while exploring new options.

PASDEP

The Plan for Accelerated and Sustainable Development to End Poverty (PASDEP) is a five-year (2005/06-2009/10) strategic framework that was built on the strategic directions pursued under SDPRP but embodies a few strong new directions including a major focus on the growth with particular emphasis on commercialization of agriculture, development of private sector and the scaling up of resources to achieve the MDGs. The MDG required interventions span seven sectors (*i.e.*, agriculture and rural development; education; health; water and sanitation; road; urban development; and private sector and trade issues) as well as three cross cutting areas (*i.e.*, gender, HIV/AIDS and population). In effect, PASDEP is the Ethiopia's medium term plan for achieving the MDGs. The eight pillar strategies of the PASDEP are:

- Building all-inclusive implementation capacity;
- A massive push to accelerate growth;
- Creating the balance between economic development and population growth;
- Unleashing the potentials of Ethiopia's women;
- Strengthening the infrastructure backbone of the country;
- Strengthening human resource development;
- Managing risk and volatility;
- Creating employment opportunities.

In terms of sectoral coverage, the PASDEP is more comprehensive than the SDPRP further encompassing sectors that were not adequately articulated in the SDPRP such as tourism, small and medium-enterprise development and job creation, urban development and construction, etc. It has also articulated policy and institutional innovation in agriculture and rural development, rural-urban linkages, and pastoral development, and spatial dimensions of the growth strategy.

As part of a participatory preparation process, the PASDEP was discussed at various levels with various stakeholders (communities, CSOs, and donors). Within Government, the document was first discussed at the full federal Cabinet level followed by discussions with the pertinent regional officials, and finally revised and endorsed by the House of People's Representatives - Parliament of Ethiopia.

Both the MDGs and the SDPRP/PASDEP exercises are coordinated by the Ministry of Finance and Economic Development (MoFED). Accordingly, the PASDEP was conceived as the medium-term plan to attain the MDGs. It establishes concrete basis for human rights '-the right of every person in the country to health, education, shelter and security-' as pledged in the Universal Declaration of Human living in extreme poverty.

Ethiopia's Achievements through Various Development Plans

In this chapter we discussed the important practical measures that the Ethiopian government so far has implemented.

Credit Service

With ten years of history, the number of Ethiopian Microfinance Institutions (EMI) increased to twenty five in number with thousands of branches and sub branches through out the country. In some regions the penetration capacity reached in the lowest administration unit, a geographical area having an average of 2000 population. Some of the key strategies for the success of these thousands of microfinance institutions include: innovative adaptation of the group guarantee lending model; successfully customized to local Ethiopian realities; decentralization of operation; focus on using indigenous knowledge and resources in client screening and follow-up, appropriate strategies to deal with financing small scale (rain-fed) agriculture, which is often subject to seasonal changes. However, much remains to be done.

Association of Ethiopian Micro Finance Institutions (AEMFI) has established to provide microfinance services to the economically and socially disadvantaged Ethiopians. AEMFI helps increase the efficiency and effectiveness of existing MFIs while facilitating the establishment of new ones.

Table 16.1: The Incremental Change of Coverage in Credit Service

Year	*Number of Clients*
2003	755,073
2004	1,001,565
2005	1,277,939
2006	1,518,245
2007	1,711,539

Although, the above improvements have been observed, there are several challenges ahead. Some of them are given below:

- ☆ Donors are not keen about MFIs and reluctant to fund
- ☆ Less saving habits of people
- ☆ limited loan products
- ☆ Less willingness from commercial banks to lend money to MFIs without collateral

- ☆ Shortage of experienced and qualified human resources
- ☆ Lack of logistics in rural areas such as road, telephone, etc.

Electrification

With the hope that electrification would encourage mechanization of the traditional rural industries (weaving, carpentering, black smithing, milling, construction, etc.), attract outside capital, extend the hours of business operation (increasing productivity), create better storage facilities and change the way of rural people's lives (by enabling cooking and storage), and connecting them to use modern appliances (*e.g.* radio and televisions), the Ethiopian government has taken some aggressive measures for the last 5 years. Due to this approach, the country has achieved a high marginal change on the distribution of electricity in all corner of the country.

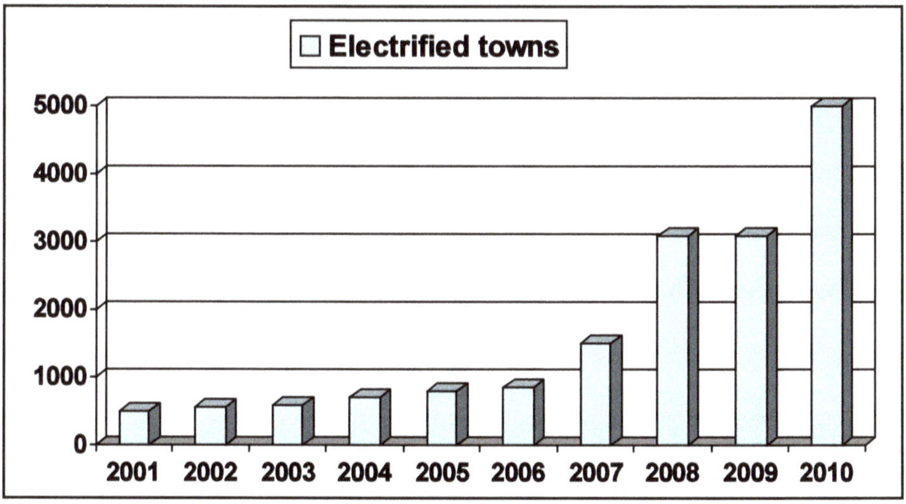

Figure 16.1: Number of Electrified Towns as of July 2010

In addition to such achievements, the Ethiopian Government works for another four (4) hydroelectric power stations that will be part of the national grids by the years 2014-2015 and electrified another 10,000 villages. These initiatives all are rooted from the Government programmes towards betterments of lives in the rural area and other economic activities.

Education

Understanding the influential power of education on the economy of the country, the Ethiopian government has taken aggressive measures on education at all three levels *viz.*, primary, secondary and tertiary education. During the past 10 years, the Ethiopian government has done its best to increase the number of students to 17,000,000 as of 2008 [update](1 through 12 grades in Ethiopia).

Among others let's see the progress on

Table 16.2: The Trend of School Increment in All Level (*Source:* NSA)

Schools	*2004/05*	*2005/06*	*2006/07*	*2007/08*	*2008/09*	*Average Annual Growth Rate*
Kindergarten	1,497	1,794	2,313	2,740	2,893	17 per cent
Primary	16,513	19,412	20,660	23,354	25,212	11 per cent
Secondary	706	835	952	1,087	1,197	14 per cent
TVET	199	264	388	458	458	23 per cent
Higher Education	23	40	55	61	72	33 per cent

Cooperatives

In the last 15 years, as a result of the actions of the Ethiopian government, from none to 8000 cooperatives were established. They are different in type and nature and comprise 3.7 million of members with an 18.4 million of family members. The total capital of cooperatives amounts to more than 61.4 million USD. These cooperatives mainly provide services to its members such as input supply, credits, marketing, supply of consumer goods, savings, flour mill services, tractor services, water pumps etc.

Telecommunication

In the area of telecommunications, the government exerts maximum effort to diversify telecom service in to the rural Ethiopia. According to statistics, the coverage of the fixed, mobile and internet access has grown in many folds. The number of mobile subscribers had increased up to around 400,000 in 2004 and at present, it is around 7,000,000. Also there is a huge progress in the fixed and wireless telephone expansion. The government of Ethiopia has established some district net programmes and school net programmes, which are appreciated by the other countries.

Roads

According to the Ethiopian Road Authority, in 1997, the total road network was only 26,550 km, of which over half was in a poor condition. Only 22 per cent of the road network at that time was in a good and serviceable condition. Twelve years ago the Road Sector Development Programme (RSDP) set out a clear strategic direction for the restoration and expansion of Ethiopia's road network. The construction and rehabilitation of the road network was coupled with actions that would see to improved management and road maintenance activities. At the same time, steps were to be taken to strengthen the capacity of domestic road construction industry.

Today, some twelve years later, Ethiopia takes some pride in her accomplishments and achievements under the banner of the RSDP. The total road network has grown by some 20,262 km to 46,812 km (Figure 16.2) of which 54 per cent is now in a good and serviceable condition.

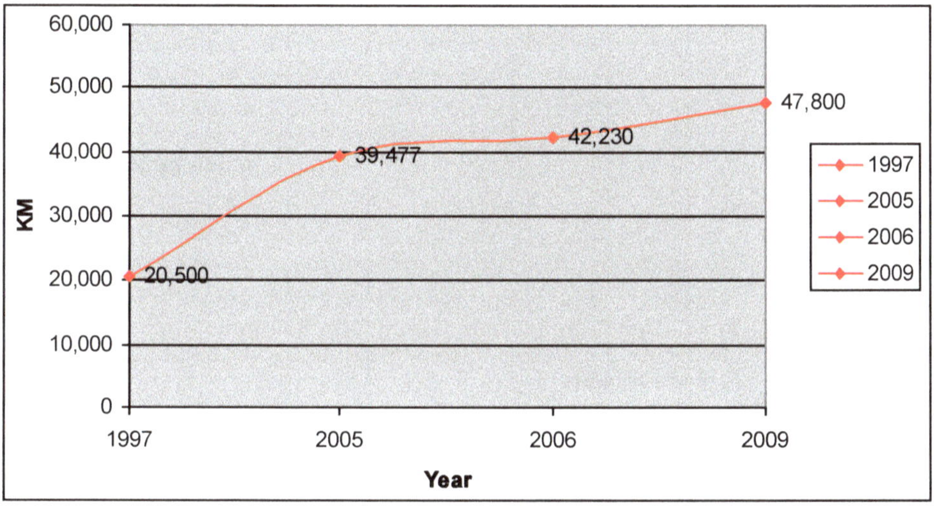

Figure 16.2: The Government Aggressive Action to Diversify the Road Network

Case Study on the Effect of Infrastructure in the Efficiency of Handloom Clusters

A case study presented below, which was conducted by Getachew Ayele *et al.* (2010) explains the success stories of the policy measures taken by the government of Ethiopia to promote occupational villages.

The basic hypothesis proposed by the researchers in this study is that the infrastructure improvements facilitate cluster productivity. The study had considered different parameters *viz.*, urban versus rural clusters, and within rural clusters electrified versus no electrified clusters while demonstrating different productivities based on the specific environment available. To fill in the knowledge gap, the investigators undertook an in-depth case study of handloom clusters in Ethiopia, with particular interest in the working network of rural clusters.

Background

According to the 2003 Cottage/Handicraft Manufacturing Industries Survey of the Central Statistics Agency, the textiles industry has the second highest number of establishments in the cottage and handicraft manufacturing (221,847), representing 23 per cent of the total number of cottage and handicraft enterprises, with almost 55 per cent of these located in rural areas. Twenty per cent of the rural employments are in the cottage and handicraft manufacturing industries. The study covered both the urban and rural clusters. The survey sites included three cluster areas within Addis Ababa (Shiro-Meda, Adisu-Gebeya, Kechene-Medhaniyalem), classified as "urban" and referred to as "Addis Ababa" and six sites in the Gamo zone, 500 kilometers south of Addis Ababa in the Southern Nations, Nationalities, and Peoples Region (SNNPR). The Gamo zone is largely rural, and the survey sites were centered in the Chencha district. The Chencha district includes 45 rural and 5 "urban" *kebeles*, or

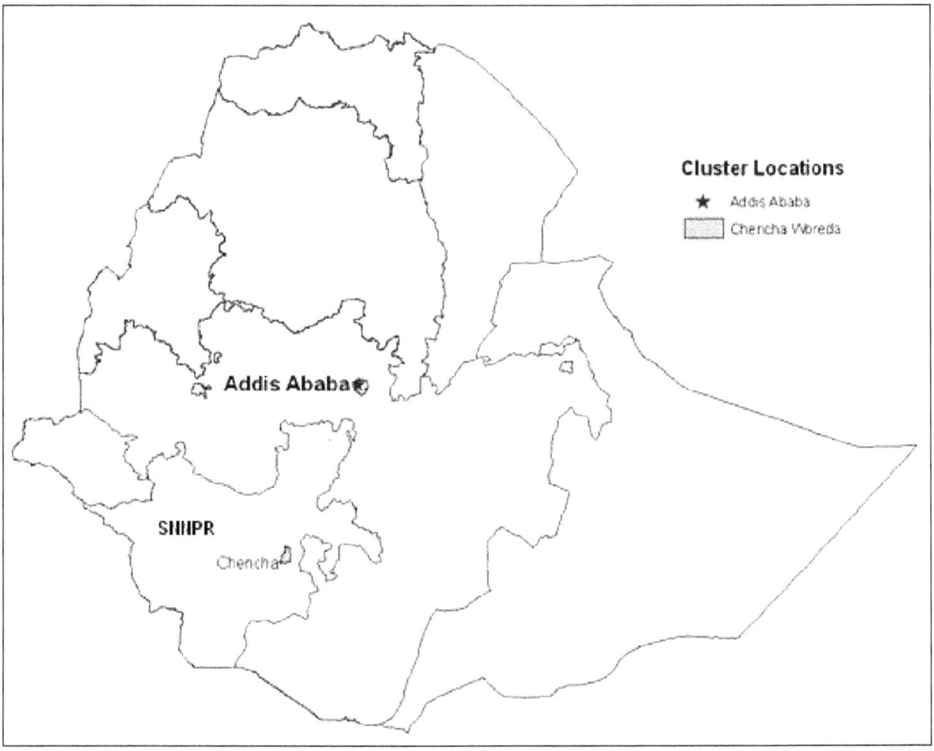

Figure 16.3: Map of Ethiopia

neighborhoods, with a total about 12,045 household heads. Of these kebeles, only five are electrified.

Table 16.3: Composition of the Sample

Chencha	*'Addis Ababa'*		*'Chencha' Electrified*		*'Chencha' Not Electrified*	
Total number of observations	195		145		146	
Types of producer enterprise	109	55.9%	72	49.7%	114	78.1%
Household						
Rented workspace	74	38.0%	72	49.7%	32	21.9%
Workshop	12	6.2%	1	0.7%	0	0.0%
Technology of production (# of producers)						
Traditional/wooden	147	75.4%	145	100.0%	146	100.0%
Improved wooden	0	0.0%	0	0.0%	0	0.0%
Improved metal	43	22.1%	0	0.0%	0	0.0%
Modern/semiautomatic	5	2.6%	0	0.0%		00.0%

Table 16.4: Productivity Measures, 2008

Chencha	*'Addis Ababa'*	*'Chencha' Electrified*	*'Chencha' Not Electrified*
Hours worked per day (average)	10.3	10.73	7.21
Average number of workers per enterprise	1.83	1.31	1.67
Average annual revenue per worker (in Ethiopian birr/worker)	14,859.38	8,460.75	6,021.49
Average annual value-addedper worker (in Ethiopian birr/worker)	4,427.22	2,543.74	1,796.28

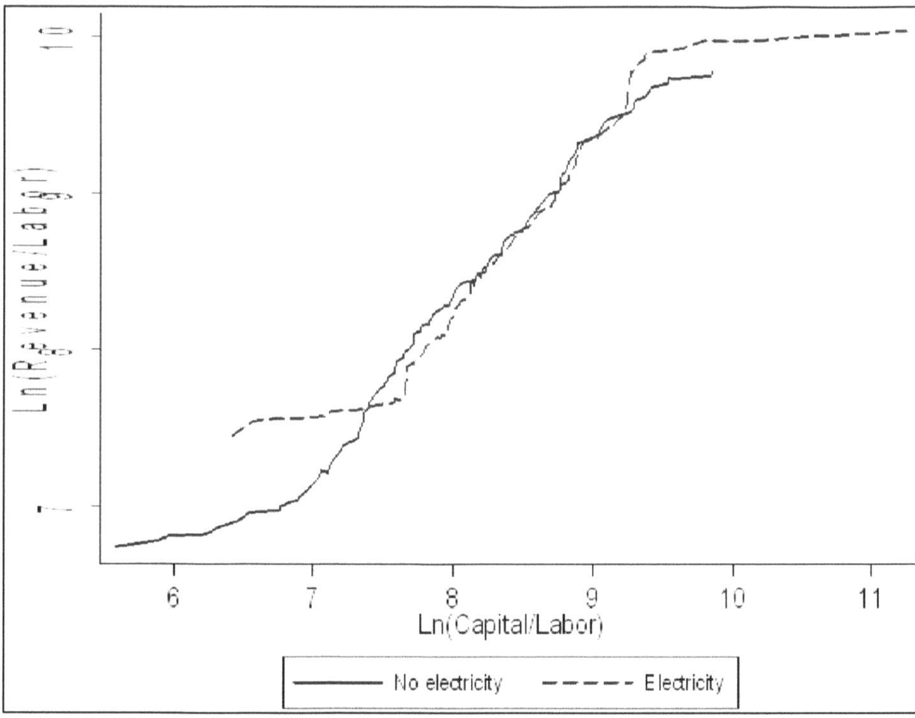

Figure 16.4: Chencha Revenue/Labor on Capital/Labor

Based on the data collected, the following observations were made.

- The average monthly production for the most important product was 25.31 units [2] in the urban clusters, compared with 15.55 units in rural electrified and 14.83 units in rural nonelectrified clusters.
- The results indicate that in the electrified rural clusters, on average, it takes 1.19 days to produce one 'netella', whereas in the non electrified rural clusters, it takes an average of 1.32 days.
- The average sales price of one unit of the most important product is significantly higher for urban clusters than for the rural ones, and again for electrified clusters as opposed to non-electrified ones.

Figure 16.5: Ethiopian Ladies Wearing Handloom Weaving Clothes

Despite of the high degree of adaptability inherent in the organizational structure of clusters, improvements in infrastructure can further boost labor productivity. Clusters with access to electricity can work longer hours and increasing labor productivity. Electricity enables and helps many poor producers who could not afford to participate in the marketplace to share workspace with access to light at rather low cost. Both average revenues and value-added per worker are higher for clusters with access to infrastructure. Even though after the controlling of other factors, it is evident that access to electricity greatly contributes to higher labor productivity for those with limited financial asset. It was further observed that infrastructures such as proper roads, technical training etc., could increase the options available to producers and traders.

Conclusion

As a developing country, Ethiopia has no other way than practicing the labor intensive economic activities, which may later fuel the other sectors. Based on the findings of the hand-loom weavers in the *'Chencha'* district and the Ethiopian government initiative to diversify the major infrastructures into the countryside, it is possible to understand two things *viz.*, the unavailability of infrastructure affect the economy of Ethiopia a lot and the government way of doing things, both on policy and actual intervention, is the right track to break what the case study show us, Forced idleness.

Acknowledgement

The Author is thankful to His Excellency Ato Juneydin Sado, Hon. Minister of Science and Technology of Ethiopia, for granting logistical and financial assistance for participation at the workshop. The cooperation extended by Ato Teshome S/ Mariam, Director, Technology Transfer Directorate in preparation of this paper is gratefully acknowledged along with the services of all staff of NAM S&T Center and the Ministry of Science and Technology of Vietnam.

Abbreviations

ALDI: Agricultural Led Development Industrialization
SDPRP: Sustainable Development and Poverty Reduction Program
PASDEP: Plan for Accelerated Sustainable Development to End Poverty
MDG: Millennium Development Goal
HIV/AIDS. Human Immune Deficiency Virus/Acquired Immune Deficiency Syndorome
CSOs. Community Based Organizations
MoFED. Ministry of Finance and Economic Development
MFIs. Microfinance Institutions
AEMFI. Association of Ethiopian Micro Finance Institutions
TV. Television
TVET. Technical Vocational Education Training
NSA: National Statistics Authority

References

1. A plan for accelerated development to end poverty. 2005-2010. Federal Democratic Republic of Ethiopia, 2005. September 2006.Addis Ababa: Ministry of Finance and Economic Development Federal Democratic Republic of Ethiopia
2. Ayele, G., L. Mormam, K. Wamisho, and X. Zhang, 2010. Infrastructure and Cluster Development: a Case Study of Handloom Weavers in Ethiopia.
3. Central Statistics Agency. 2003a. Main report for the 2002 cottage/handicraft manufacturing industries. Addis Ababa: Central Statistical Authority of the Federal Democratic Republic of Ethiopia.
4. Central Statistics Agency. 2003b. *Report on cottage/handicraft manufacturing industries survey: November 2002.* Addis Ababa: Central Statistical Authority of the Federal Democratic Republic of Ethiopia.
5. Educational statistics annual abstract. 2008-2009. march 2010. Addis Ababa. Ministry of education federal democratic republic of Ethiopia
6. Maquanent, G. 2009, Rural Industrialization in Ethiopia: Time for Action, Ottawa, Canada.
7. Microfinance associations the case of the association of microfinance associations of Ethiopia. 2002. GTZ.
8. PASDEP annual progressive report. 2006/2007. December 2007. Addis Ababa. Ministry of Finance and Economic Development Federal Democratic Republic of Ethiopia.

Chapter 17

Policy Approaches of Cambodia in Development Programmes

Hoeung Kimsay

Science and Technology Office, Department of Industrial Techniques, Ministry of Industry, Mines and Energy (MIME), Phnom Penh, Cambodia
E-mail: kimsay_hks@yahoo.com

ABSTRACT

Cambodia is an agricultural country with more than 81.5 per cent of Cambodians living in rural areas. More than 91 per cent of the rural household depends on agriculture as their primary source of income. In preparing the plan for the period 2006-2010 the Royal Government of Cambodia (RGC) decided to move away from the traditional comprehensive planning approach to one that focuses on strategic goals and actions. The author here discusses in detail the National Strategic Development Plan (NSDP). Points out that in line with full government ownership and leadership of the entire development process, RGC declared that the NSDP as the single, overarching, guiding and reference document for pursuing prioritized goals, targets and actions. The RGC Rectangular Strategy and the National Poverty Reduction Strategy builds upon a policy stance of economic growth and poverty reduction. The reform agenda of the RGC essentially focuses on deepening of economic reforms and macroeconomic stability, including fiscal and monetary reforms. The purpose of the NSDP is to operationalise the "Rectangular Strategy (RS) for Growth, Equity, and Efficiency" that was adopted by the Government in July 2004. Both the RS and the NSDP have, at their core, a comprehensive good governance reform programme. States that the Cambodian Millennium Development Goals (CMDGs) are nine in all, including all the 8 UN Millennium Development Goals (MDG). Currently, the Royal Government is committed to achieve the human development oriented CMDGs by 2015 through a Rectangular Strategy following the Triangle Strategy of the Royal Government. Further the percentage of Cambodians below the poverty line fell from an estimated 47 per cent in 1994 to 35 per cent in 2004. The National Green Growth Roadmap is also conceived while envisioning a combination of objectives. Concludes that in relation to all the above mentioned issues, Cambodia requires many development programmes to enhance performance of all sectors but lacks a lot of legal frameworks, specialized human resources, equipments and other resources at the present time. Besides, the country needs external assistance as development partners especially for financial and technical support in order to conduct such programmes and implementing the said action plans of the government.

Keywords: *Cambodia, Development programme, Plans and policy, Industrial sector, Green growth, poverty, Development strategy.*

Introduction

Cambodia is an agricultural country with more than 81.5 per cent of Cambodians living in rural areas and agriculture is their primary livelihood including farming, fishing, raising live-stock, harvesting forest and non-forest products. According to 2004 data, the country's population is poor, with a higher poverty rate in less accessible rural areas (45. 6 per cent) than in the urban and more accessible rural areas (28 per cent). More than 91 per cent of the rural household depends on agriculture as their primary source of income.

Cambodia has achieved a remarkable progress over the past decades in a number of areas that are important to human development. Peace and stability have been restored and maintained. Economic growth has risen to double-digit rates, with an average of 11.2 per cent in the last several years (MoP &UNDP, 2007). Access to education and health services has improved, while poverty has steadily declined, with a moderate progress towards a number of other Millennium Development Goals (CMDGs) of Cambodia. These are major achievements in the light of challenges that had to be overcome. Nevertheless, poverty and near-poverty remain high and other human development indicators continue to reflect difficult living conditions especially in rural areas, where the vast majority of Cambodians living and working, largely at a subsistence level. Most of the high growth over the past decade had been urban-based and narrowly focused on surging garment exports that could be vulnerable, the record levels of tourism and a boom in the construction industry. Similarly, the base of major beneficiaries of development over the past decade also has increasingly narrowed.

The Government's Rectangular Strategy and the National Strategic Development Plan, (NSDP) 2006-2010 provide a broad roadmap towards improved governance, increased public investments in rural areas, and support to the development of a job-creating private sector. However, in preparing the plan for the period 2006-2010, the Royal Government of Cambodia (RGC) decided to move away from the traditional, comprehensive planning approach to one that focuses on strategic goals and actions. In line with full government ownership and leadership of the entire development process, RGC declared that the NSDP as the single, overarching, guiding and reference document for pursuing prioritized goals, targets and actions. However, the task of implementation of same would be critical. Given the still-weak institutional capacities, implementation efforts will be best prioritized and targeted at areas offering the highest returns for improving human well-being, achieving shared prosperity and securing longer-term peace and stability. Agriculture, infrastructure, spatial development, employment and development of human resources are the priority sectors identified in the strategy.

Economic Conditions

Cambodia's population has sharply rebounded in the past decade, growing from an estimated 9.3 million in 1993 to 13.4 million in 2008 (MoP, 2008). The primary sector (agriculture, hunting, forestry and fishing) had accounted for 77.5 per cent in 1998 and 74.2 per cent in 2004. The secondary sector (Mining, and quarrying, electricity, gas and water supply, and construction) had accounted for 4.3 per cent in 1998 and 7.0 per cent in 2004. The tertiary sector (trade, hotel and restaurant, transport, business, administration service activities, etc.,) had accounted for 18.2 per cent in 1998 and 18.8 per cent in 2004.

By 2007, the total employment had increased to 7.8 million persons. The agriculture sector had accounted for approximately 60.4 per cent of employment, industry sector for 14.0 per cent, and services sector for 25.6 per cent. An estimated 60,000 new job-seekers join the labor force each year and most employed persons were multiple job-holders. The employment rate was estimated about 99.3 per cent and unemployment rate is about 0.7 per cent in 2007 (MoP, 2008).

Currently, the economic base is shifting away from agriculture, fisheries and forestry, with the sector accounting for around 31.4 per cent of Gross Domestic Product (GDP). The industry sector has continued to grow strongly and now accounts for around 27 per cent of GDP, with the textiles, garments and footwear industries accounting for 55 per cent of the industry sector. The services sector is the largest sector in the economy and it accounts for 36 per cent of GDP and is largely driven by the tourism industry. Most of the economic growth since 1995 has been due to the growth in the textiles, garments and footwear industries. The GDP growth registered a double digit growth since 2004 to 2007 but was expected to decline in 2008 to 7 per cent because of inflation. In 2007, GDP per capita was 594 USD. While consumer price inflation is estimated to have been stabled at around 5 per cent from 2005 to 2007 (NSDP 2005), there has been a significant increase in consumer prices over 2008, due mainly to significantly higher increases in the consumer prices of food, international oil prices and other factors.

Nevertheless, by the end of 2007, despite the fact of high pressure of oil and food prices, the performance of Cambodia's economy was generally satisfactory, with the GDP growth of around 6.8 per cent in 2008. When looking at the quarterly data, it is noted that the economic activities contract slightly in the second quarter due to the seasonality and then pick up during the third and fourth quarters. However in 2008, the economy declined notably in the second quarter with negative growth as high as 8.2 per cent of GDP (affected by high prices of energy and food, compounded with the uncertainties during the pre-election period in the country). Then it recovered up to 7.3 per cent of GDP in the third quarter partly owing to: (*i*) introduced policies to mitigate adverse effect of soaring food and petroleum product prices in order to curb inflation; (*ii*) political stability following the general election at the end of July 2008;

and (*iii*) increase of agricultural activities during the harvest season. Besides, the fiscal and monetary policies have been prudent to manage inflation and ensure macro-economic stability. In 2008, due to cautious and carefully interventions from the government, the inflation rate was brought under the control. It reached its peek of 25.72 per cent during the second quarter of 2008 because of the high oil price, the soaring food price plus appreciation of the US dollar. However, it was maintained around 13.46 per cent by the end of the year and on the average the inflation rate was at about 19.73 per cent. The inflation is expected to reduce further to a single digit in the coming years.

Furthermore, in following up on the Millennium Declaration, the country has launched several initiatives to meet its global and national commitment to fight with extreme poverty, including several institutional reforms. The RGC Rectangular Strategy and the National Poverty Reduction Strategy builds upon a policy stance of economic growth and poverty reduction. The reform agenda of the RGC essentially focuses on deepening of economic reforms and macroeconomic stability, including fiscal and monetary reforms, trade and investment promotion, administrative reforms, military demobilization, and improved fishery and forestry management.

Development Efforts

National Strategic Development Plan (NSDP) 2006-2010

The purpose of the NSDP is to operationalise the "Rectangular Strategy (RS) for Growth, Equity, and Efficiency" that was adopted by the Government in July 2004. Both the RS and the NSDP have, at their core, a comprehensive good governance reform programme.The issues of Governance issues have been identified by the Government and donors alike as a major obstacle to poverty reduction, economic growth, and enhanced aid effectiveness in Cambodia. But the consolidated donor initiative will strengthen leverage in terms of ensuring adherence to the reform agenda set down in the NSDP. In addition, it should be noted that the fiduciary risks present in Cambodia are best tackled by the different donors acting together, both in order to strengthen leverage and to share information and expertise. It is also anticipated that the World Bank together with the other participating donors will put in place special fiduciary safeguard measures for the RGC considers the NSDP as the single, overarching development strategy for pursuing prioritized goals and actions for the period 2006-2010. The NSDP has been framed as the implementation of the Rectangular Strategy, linking the vision in the Rectangular Strategy to concrete goals, targets and strategies. It synthesized goals and targets contained in the Second Five-Year Socio-Economic Development Plan for 2001-2005, the National Poverty Reduction Strategy for 2003-2005 and the Cambodian Millennium Development Goals. The NSDP highlights most essential strategies, targets and actions, but it leaves more details to be spelled out in the sectoral and sub-national plans, which fed into the first annual review of the NSDP scheduled for March 2007. The Government ministries and agencies, donors and civil society organizations were involved in the formulation of the NSDP.

Rectangular Strategy (RS) of the Royal Government of Cambodia

The Rectangular Strategy (RS) for growth, employment, equity and efficiency is a holistic and integrated document proclaimed by RGC during the early part of its third mandate in 2004. It is still relevant in the fourth mandate in which the paragraph (40) states that "in order to ensure sustainable economic growth, which is *sine qua non* for employment creation and income generation, the Royal Government will continue to foster the diversification and strengthen the competitiveness of the Cambodian economy by assuring a highly conducive climate for both public and private (domestic and foreign) investments for the development of the following important sectors: (1) Agriculture, (2) Water and Irrigation System (3) Transport Infrastructure (4) Electricity (5) Human Resource Development (6) Labor-Intensive Industry and Food Processing Industry for Exports (7) Tourism (8) Exploitation of Oil, Natural Gas and other Minerals (9) Information and Communication Technology, and (10) Trade" (RGC, 2008).

At the forefront of the political agenda of the Royal Government in the third legislature of the National Assembly are the:(a) promotion of economic growth; (b) generation of employment for all Cambodian workers; (c) implementation of the Governance Action Plan and, thereby, ensuring equity and social justice; and (d) enhancing efficiency and effectiveness of the Royal Government in implementing the reform programs in all sectors to reduce poverty and achieve sustainable development. In this regard, Samdach Akka Moha Sena Padei Techo *Hun Sen*, the Prime Minister of Cambodia, set out the Rectangular Strategy as a supporting tool to implement the political platform of the Royal Government in the third legislature of the National Assembly and continue to the second phase of the rectangular strategy to attain poverty reduction, development, progress, prosperity, national harmony and happiness of the Cambodian people. The Rectangular Strategy is an integrated structure of interlocking rectangles (Figure 17.1).

Cambodian Millennium Development Goals (CMDGs)

The Cambodian Millennium Development Goals (CMDGs) consist of 9 goals, 25 overall targets, and 106 specific targets. CMDGs are nine in all, including all the 8 UN Millennium Development Goals (MDG). Each CMDG has been disaggregated into several sub-CMDGs (MoP, 2003) with clear targets as follows:

- ☆ Eradicate extreme poverty and hunger: it is intended to halve, between 1993 and 2015, the proportion of people whose income is less than the national poverty line;
- ☆ Achieve universal nine year basic education: the target has set to ensure all children complete primary schooling by 2010 and nine-year basic schooling by 2015;
- ☆ Promote gender equality and women's empowerment: the overall target is to reduce significantly gender disparities in upper secondary education and tertiary education;

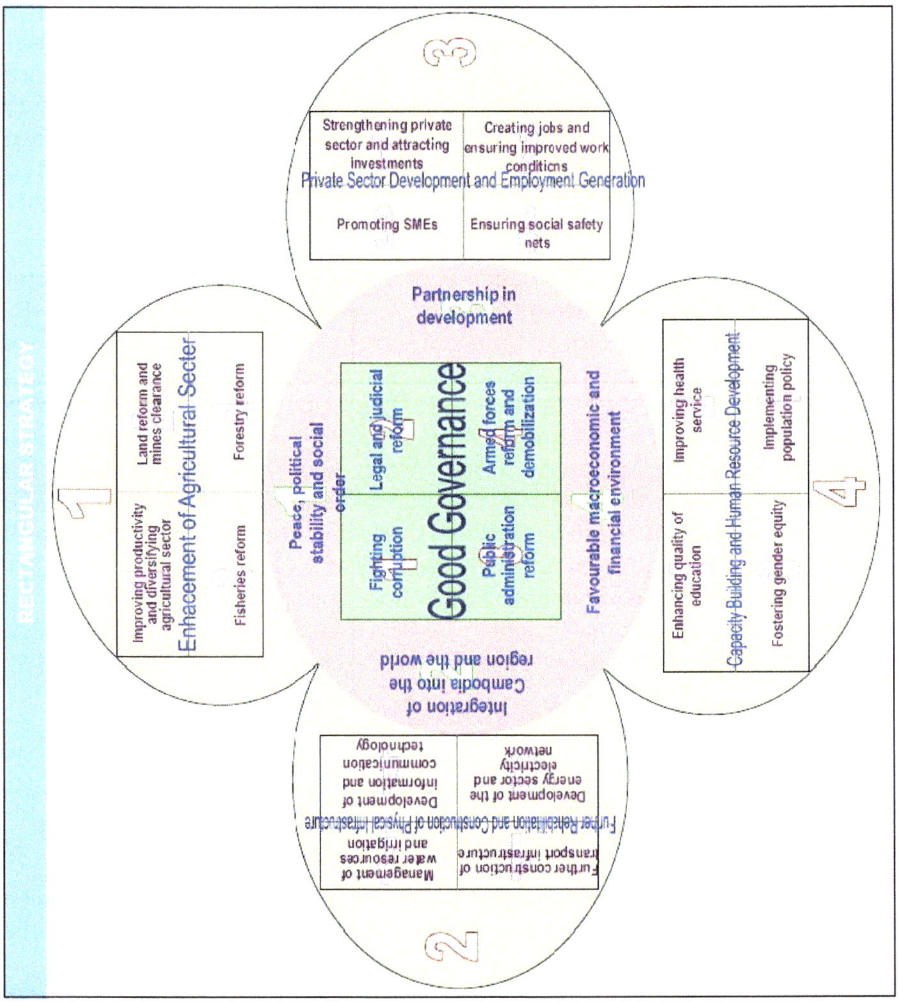
RECTANGULAR STRATEGY
Good Governance
Fighting corruption
Legal and judicial reform
Public administration reform
Armed forces reform and demobilization
Peace, political stability and social order
Partnership in development
Favourable macroeconomic and financial environment
Integration of Cambodia into the region and the world
Enhacement of Agricultural Sector
Improving productivity and diversifying agricultural sector
Land reform and mines clearance
Fisheries reform
Forestry reform
Private Sector Development and Employment Generation
Strengthening private sector and attracting investments
Creating jobs and ensuring improved work conditions
Promoting SMEs
Ensuring social safety nets
Capacity Building and Human Resource Development
Enhancing quality of education
Improving health service
Fostering gender equity
Implementing population policy
Further Rehabilitation and Construction of Physical Infrastructure
Management of water resources and irrigation
Further construction of transport infrastructure
Development of information and communication technology
Development of the energy sector and electricity network
1
2
3
4

Figure 17.1

- Reduce child mortality: the government has set the target to reduce the under-five mortality rate;
- Improve maternal health: reduce the maternal mortality rate;
- Combat HIV/AIDS, malaria and other diseases: it is intended to decrease the spread of HIV/AIDS;
- Ensure environmental sustainability: integrate the principles of sustainable development into country policies and programmes and reverse the loss of environmental resources;
- Forge a global partnership for development: develop further an open, rule-based, predictable, non-discriminatory trading and financial system; and
- De-mining, UXO and Victim Assistance: move towards zero impact from landmines and UXOs by 2012

CMDG Achievements

The poverty levels in Cambodia have registered a significant decline since 1993, in those parts, which were surveyed then. Such decline has occurred mainly through widespread economic growth. The latest survey figures highlight the areas of still lingering poverty that need focused attention and targeted action to reach the Cambodia Millennium Development Goals (CMDGs) in reducing poverty. Now it is time that resources begin to be properly directed and effectively used to maximize benefits for the disadvantaged and the deprived to lift them into the mainstream. Consolidating and building on the achievements of the past, the Royal Government is committed to achieve the human development oriented CMDGs by 2015 through a Rectangular Strategy following the Triangle Strategy of the Royal Government for well-orchestrated attention to, as well as investment and growth in, priority sectors. Social and economic growth should be equitable, and opportunities and benefits affordable and accessible to all, both geographically and between the rich and the poor. We shall use scarce resources for the maximum benefit of all Cambodians, particularly the poor, and not leave investment in the public sector to the choices of others.

The Royal Government of Cambodia gratefully acknowledges that its progress since 1993 has been made possible largely by generous financial and technical support from our external development partners. We are very thankful to them. However, much of the external resources in the past had been used directly by the providers on programmes and activities of their choice, outside government purview or scrutiny. A large share has also been spent on intangible items like technical assistance. In the future however, Cambodia will take full charge of the entire process and ensure that all resources, government and external development partners, are clearly directed to priorities and sectors chosen by the government. It will further pursue to ensure that the highest proportion of external funds is used as resource transfers for development.

The National Strategic Development Plan (2006-2010), was formulated using the comprehensive Rectangular Strategy of the Royal Government of Cambodia and it synthesizes various policy documents (Cambodia Millennium Development Goals,

National Poverty Reduction Strategy, National Population Policy, etc.,) and through extensive consultations that were also held among all stakeholders. It provides the framework and compass for growth, employment, equity, and efficiency to reach CMDGs and well-focused and directed future equitable development, pro-poor and pro-rural. It will take the country on an assured growth path, once sustainable. It lays out the vision, goals, strategies and priority actions for the next five years. It makes balanced, realistic and feasible allocations to various sectors on a priority basis. Its implementation will be closely and regularly monitored to make annual adjustments.

Cambodia also recognizes the crucial necessity to install and strengthen proper institutions of governance, processes and procedures through well thought-out reforms to ensure sustainable development. Step-by-step, steady progress is seen at a rapid pace not witnessed elsewhere in the world. Actual implementation of reforms and practice of laws and procedures would over time imbue the entire system with best practices and cement and solidify various institutions essential in any democracy, as has happened elsewhere in the world, albeit over a much longer period. The NSDP is a blue print for further progress on building the edifice of a new Cambodia, and is the single, all-encompassing RGC document for the next five years.

Moreover, in order to reach the CMDGs, it has shown the commitment of the RGC in protecting and conserving the natural resources and environment and also improving the livelihood and access to proper sanitation for the people.

National Poverty Reduction Strategy

In June 2006, the Government adopted the National Strategic Development Plan (NSDP) 2006–2010, which contains the Government's strategies to reduce poverty rapidly and to achieve other CMDGs. The results of a survey show that poverty has declined steadily since the end of the civil war: the percentage of Cambodians below the poverty line fell from an estimated 47 per cent in 1994 to about 35 per cent in 2004 (NSDP 2005). However, it is reported that while poverty incidence has improved, the gap between the rich and the poor has increased. Cambodia has witnessed in the past 12 years, a robust and steady macro-economic growth. GDP at constant prices grew at an average of 7.0 per cent per year, led by rapid growth mainly in the garments manufacturing, tourism and construction. The per capita GDP registered a similar growth of 4.7 per cent despite increase in total population (NSDP 2005). It is reported that the inflation was kept at a rate less than 5 per cent. The riel-US$ exchange rate was fairly stable, declining gradually. Government revenues and expenditure, both in actual terms and as percentage of GDP, improved at a strong pace. In the Financial Sector, reforms are progressing and various measures have been putting in place to regulate and foster the sector to grow in an orderly and assured manner.

Agriculture plays a very important role in improving incomes, particularly the poor, who live in rural areas. The total production in crops, particularly rice, grew steadily resulting in exportable surpluses in recent years. The average paddy yield per hectare has also grown, but is still below those in neighboring countries. Several measures are underway to improve and diversify agricultural production. In regard to forestry, many old concessions were terminated for non-performance and new concessions were granted with strict conditions to follow the plans according to Environmental and Social Impact Assessments (ESIAs).

In 2008, the newly re-elected Royal Government of Cambodia adopted a Rectangular Strategy II to guide its development efforts aimed at growth, employment, equity and efficiency. Cambodia also recognizes that achievement of poverty reduction and CMDGs critically depend on steady and synergistic progress in several socio-economic spheres such as: robust and equitable economic growth; strong check on inflation; significant increase in agriculture production, productivity and value added efforts for products; protection and enhancement of environment; strengthening and improvement of infrastructure especially connecting remote rural areas; robust industrial growth; creation of employment and incomes; reforms in public administration and judiciary, indeed in all aspects of governance; fast growth in private sector investments; unhindered growth in Trade; growth in services sector including tourism; and above all, directed assistance to the most vulnerable. The Cambodia's goals are to ensure to bring overall poverty levels rapidly down to below 25 per cent in the country by the year 2010 (NSDP, 2005).

Cambodia's Population Policy

According to the preliminary results of the General Population Census of Cambodia 2008, it has a population of 13.4 million, of which 51.5 per cent were females, 48.5 per cent males, and growing at an estimated rate of 1.54 per cent per annum (MoP, 2008). The national average population density was 75 persons per square kilometer. Around 80.5 per cent of the total population lives in rural areas and 19.5 per cent in urban areas.

Cambodia has set a Population Policy to induce changes in population trends so as to bring the size, composition and distribution of population in line with the needs of sustainable development and spatial planning needs for poverty alleviation and improvement in the quality of life of all Cambodians. The Ministry of Health has considered Maternal and Child health as priority issue in the New Health Strategic Plan 2008-2015; birth spacing is one of priority area for reducing maternal, new born and child morbidity and mortality that improves reproductive health. Furthermore, the Second Socio-Economic Development Plan recognizes that high fertility has detrimental effects on the health and nutrition of mothers and their infants.

Besides, the Government reaffirms its respect and support for the right for all couples and individuals to have the basic right to decide freely and responsibly on the number and spacing of their children, and to have access to the information, education, services and means to do so (RGC, 2003). It is stated that the population issues facing the country, the policy sets forth an overall goal and several objectives. Its overall goal is achieve sustainable development, reduce poverty and improve the quality of life of all Cambodians through changes in the size, composition and distribution of the population.

The National Green Growth Roadmap

Vision: Prosperity and Sustainable Livelihoods

The National Green Growth Roadmap is conceived while envisioning a combination of objectives for economic growth and development, human well-being, and environmental quality, which will promote the simultaneous attainment of prosperity and sustainable livelihoods for the betterment of lives of the Cambodian people. Just like the economic growth is a precondition for human development, environmental sustainability is integral to both economic growth and to the well-being of the Cambodian nation as a whole.

In light of this, the National Green Growth Roadmap aims at laying out a number of possibilities for *win-win-win* situations among economy, environment and society. It argues that rapid economic growth, environmental sustainability and human well-being can be achieved without any trade-offs. It is based largely on ESCAP's holistic Green Growth Approach, the Roadmap argues that the economy, the environment and society are not mutually exclusive but rather inter-dependent, deeply intertwined and integral components of each other.

The National Green Growth Roadmap will focus on addressing the seven "A"s:

- ☆ Access to clean water and sanitation;
- ☆ Access to energy;
- ☆ Access to information and knowledge;
- ☆ Access to means for better mobility;
- ☆ Access to finance and investments;
- ☆ Access to food security (agriculture) and non-chemical products; and
- ☆ Access to green jobs

These seven needs for access will be addressed through green economic growth projects and programmes, low carbon solutions, including eco-efficient/resource efficient innovations, which can create income generating opportunities and new "green jobs" for prosperity. While an instant paradigm shift towards Green Growth is not to be expected, an increased focus on access and on the implementation of Green Growth policies, will gradually lead the country towards an economic growth model that is more conducive for human development and for environmental sustainability than the current business-as-usual approach.

Short-Term Interventions: Next Five Years 2009-2014

1. Create a National Ministerial Green Growth Council;
2. Create a national public awareness and consultation process;
3. Develop national strategy for fostering Green industries: based on 3Rs;
4. Develop stimulus measures for promotion of Sustainable Agriculture;
5. Develop a scheme for innovative investments:
   a. Payment for Environmental Services;
   b. Green Tax and Budget Reforms;
   c. Swap of debts schemes;
   d. Green capital investment – disclosure.

Initiate the Green Growth pilot projects to be completed in the medium term section (see next section).

Medium-Term Interventions: Next Ten Years 2010-2020

Eco-Village

Eco-villages are urban or rural communities of people, who strive to integrate a supportive social environment with a low-impact way of life. To achieve this, they integrate various aspects of ecological design, Parma culture, ecological building, green production, alternative energy, community building practices, and much more.

Consequently, Cambodia has undergone a chronic destruction of natural resources with various kinds of activities. Recognizing that the environment plays a critical role in providing services to sustain the peoples' livelihoods and well-being, the Ministry of Environment has set strategies and action plans to respond to overall development plans and to reach the CMDGs. Phnom Penh Municipality should continue to implement and further strengthen its 10-year development strategy to promote clean water supply, build the physical infrastructure such as sewage and drainage systems; flood prevention bank; to replace the existing dumping site to a new one, bigger and better facility; finally, strictly enforce the traffic law, promoting public transports and support the concept of side street walk, promote the tree planting and construct and repair resorts or recreations. Dealing with Eco-Village development, the country needs to promote environmental education among local community to strengthen the environmental management. It is also necessary to build the local community capacity and raise awareness on the impact of waste and other chemical substances on health and the environment and identify both proactive and reactive measures to ensure better living conditions of the local communities.

Furthermore, MIME should promote the SMEs development to ensure the sustainable livelihoods of the local community and need to continue to build capacity of SMEs through training and awareness raising to enhance the productivity, improve quality of production with setting reasonable price to be competitive in both local and foreign markets. MIME also needs to continue the project on cleaner production and careful management of chemical products, improve energy efficiency and climate-related issues. It is reported that the tourism sector has emerged as a major driver of economic growth with having 2,125,465 international visitors in 2008, is expected to increase in the next several years (MoT, 2008). Indirect support will be provided to ecotourism, so the benefits of tourism are extended to the rural poor. The development of eco-village concept could attract more tourists to visit and stay. Thus, the tourism areas must be environmentally friendly like a slogan "Clean City, Clean Resort and Good Service + Green Village". It is also a great opportunity to expand employments and improve livelihood conditions of local communities.

The land tenure in the rural can benefit rural development through making guidelines for land use, increase ownership and security, and increase efficiency of land markets. The government can promote eco-city and/or eco-village concepts to create diversified income sources thereby further building on the positive momentum created by land tenure. Furthermore, the country has set up two main empirical fields: a) Labour field and b) Technical-Vocational Education and Training (TVET) field to ensure the appropriate eco-village development. Provincial governments can implement resource use taxes (green taxes) that can be recycled either to help the financing of TVET for green skills, such as sustainable agriculture practices or to grant government backed loans for environmentally sound technologies to the rural population. Alternatively, the government can use revenues from resource-use taxes to establish an index-based micro insurance scheme granting farmers an option for managing weather risks. These schemes can be designed to favour organic agriculture to create incentives for green business growth in this sector and enhance environmentally sustainable food security. The requirements also need to further promote farmers to use non-chemical products or organic products to ensure the sustainable agriculture practice and protect human health. Furthermore, the country continues to promote planting trees along the roads and streets, in the villages, in the pagodas and at schools and other in the public areas as well. In summary, supporting such measures in rural areas will have great potential to restore the environment, whilst also improving the livelihood of the rural population. It is also necessary to further develop the rural water supply and sanitation sector to provide water that meets the appropriate standards for domestic consumption for all inhabitants of the rural areas; to identify ways and means of safe disposal of human urine, excreta and domestic wastewater and to promote a clean environment to minimize the transmission of water and excreta-related diseases.

Sustainability of Water Resources

The water plays an important role in achieving the main goal of the RGC to reduce poverty; to develop economy and food security; and to conserve the environment. In other words, water is a basic resource to support all sectors *viz.*, agriculture, industry, energy, domestic use, navigation, tourism, fisheries, and maintaining eco-system. Currently, about 90 per cent of the urban population enjoys access to water, but it is limited to only 44 per cent among the rural population. The lack of access to proper sanitation and clean water has increased the incidence of water and vector borne diseases in the country. The expansion of centralized modes of water infrastructure is very costly and there are too many ministries currently focusing on this issue, creating confusion about responsibilities among them. The end-of pipe solutions are not cost-effective, as Cambodia currently has only insufficient and low-level technology for wastewater treatment.

In response, the RGC has started developing a strategic action plan to further rehabilitate and repair rural roads while ensuring more access to clean water and sanitation to the rural community. It should continue to implement the policy on the services of rural water supply and sanitation without any harmful effects on the environment and public health and to ensure that every person has access to safe water supply and sanitation services and lives in a hygienic environment. Furthermore, the MoRD has developed a policy on development of indigenous people, and the formulation of a rural development policy and strategy conducted by the Republic of Korea through the Korean International Cooperation Agency (KOICA) also emerged in 2009. Moreover, MIME has been assigned to prepare the National

Policy on Water Supply and Sanitation, through close collaboration with ministries and the line agencies concerned. The said National Policy is intended to promote the people's quality of daily living and welfare to ensure sufficient water supply and sanitation services with low costs as well as using a sustainable, and environmentally friendly sanitation system, are clearly identified nation wide. Meanwhile, MIME should continue to develop and effectively implement this policy to ensure and sustain the access of clean water to the public with sufficient quantity, high quality and affordable price and the drinking water standard should be continued.

To further improve the situation, there is a need for effective enforcement and implementation of the water resource management law and a policy for sustainable irrigation that places greater focus on integrated water management. It is necessary to continue its existing policy on the major functions of water for: (*i*) agriculture; (*ii*) energy; (*iii*) industry, small manufacturing enterprises and services; (*iv*) domestic use; and (*v*) navigation and tourism and this policy is not merely to cover on the fresh water resources, but also to cover coastal and marine waters including the marine water protection and conservation. MoWRM should continue to promote the people to understand the benefits of water for daily life and socio-economic development, and to participate in effective, sustainable maintenance, management, conservation, and use of water resources. Moreover, there is a need for sub-decrees on: (*i*) farmer water user community; (*ii*) river basin management; (*iii*) water licensing; and (*iv*) water quality. The above policy measures can be addressed by the government through implementation of a life-line tariff and progressive taxation to create incentives for the industrial sector to improve their efficiency of delivering water resources. By implementing water pricing through metering consumers, the government can set a percentage for recycling to finance the expansion of water services to excluded households. In addition, there is a need for continuing to improve and fortify the national drinking water standard and a necessity for integrating water harvesting technology into building codes to fully harness the potential for rainwater harvesting, to increase resilience to climate change and improve the quality of life of citizens.

Sustainability of Agriculture, Energy and Forests

Sustainable agriculture is an effective strategy for improving food security and reducing poverty. It promotes the production of ample food without depleting the earth's resources or polluting its environment. Sustainable agriculture offers opportunities to achieve economic development, save and create jobs, reduce poverty, and even cut down Green House Gas (GHG) emissions. It also helps ensuring food safety of consumers, providing food security to the poor and smallholder farmers, offering trade opportunities for developing countries, and restoring and improving ecosystems. Meanwhile, the sustainable agriculture, energy and forests are critical to improve the lives of small farmers and other rural poor by enabling them to increase food production and income, access to energy while ensuring environmental sustainability and maintaining strong social cooperation. The cooperation and collaboration among line ministries are of importance to reach this goal. In reality, the use of fuel wood for cooking and lighting has both environmental and health impacts. In addition, the use of large quantities of wood in the brick manufacturing process; deforestation in infrastructure development and the extension activities of agriculture also cause stress on existing forests. However, it is necessary to replant trees along the road after they are cut for building the roads. The planted trees play vital roles such as maintaining the forest cover, stabilizing ecosystems, preventing soil erosion, and to cleaning the poisonous environment from carbon emission while serving as rainfall catchments.

In this context, the government representatives have suggested pilot/demonstration projects to promote the use of biogas for energy production. A concerted strengthening of the existing forestry laws for outlining the Protected Environment Areas, as well as increasing the use of renewable energy technologies etc., have the potential to minimize illegal logging. In addition, the introduction of an auctioning mechanism to control logging, coupled with performance guarantee mechanisms, concession fees or excise taxes at sawmills can also lessen the strain on forests. In other words, it is essential to continue to effectively managing forest concessions and forest communities, while promoting the reforestation, agro-industry, agro-forests, and intensive rice cultivation.

Apart from that, the country has adopted the agriculture development plan with the overall goal of poverty reduction and economic growth through agriculture sector development including sound chemical management and non-chemical use options. The Sectoral Plan is intended to ensure food security, increase income, create employment and improve nutrition status of all people by improving the productivity, diversification and commercialization of agriculture with environmentally sound protection and food safety measures. To build on this plan, the increased use of renewable energy in rural areas combined with an increased protection of existing forests can have positive effects on human health and poverty reduction. The RGC can aid and facilitate the mainstreaming of renewable energy options by reducing import tariffs on ESTs. The government backed loans can be provided to improve the sustainability of the agriculture sector by supporting people in agriculture and agro-forestry, both in terms of switching to organic agriculture and providing off-grid renewable energy options.

However, making energy affordable, improving rural access to electricity, and meeting the rising demand for power for a growing economy pose considerable challenges. It may be required to shift from traditional public sector lending operations for electricity to (*i*) effective implementation of ongoing public sector projects, (*ii*) support for private sector operations, and (*iii*) support for analytical, policy and technical work on oil and gas and on renewable energy (ADB, 2007). Renewable energy technologies will be developed to meet the energy demand in rural and remote areas, especially in the off-grid areas. The government of Cambodia is placing great emphasis on the rural electrification, on-grid where possible, and off-grid by renewable energy, to raise the living standards of the rural population. The renewable energy strategy plan was adopted consisting of hydropower, solar power, biomass and biogas, wind power, etc., to facilitate the electricity access (MIME, 2006).

Sustainability of Waste Management

To achieve a sustainable waste management, the most pertinent challenges must be addressed. Such challenges include: (*i*) toxic pollution from laundry factories and from the garment sector; (*ii*) unduly disposed waste from vehicle servicing stations (oils, lubricants, coolants); (*iii*) household waste due to lack of awareness; (*iv*) medical waste (*i.e.* needles and chemicals); and (*v*) the lack of proper monitoring of imported second-hand products in the automobile sector (vehicles, tires, etc.), and electronic wastes (E-Waste).

The suggested management methods in line with the Green Growth objectives include: (*i*) environmental awareness campaigns at the community level; (*ii*) identifying and monitoring of proper dumping sites; (*iii*) effective enforcement; (*iv*) education of youth; (*v*) control of the importation of second-hand goods; and (*vi*) promotion of local level self-governance and leadership for better coordination of these efforts. Also the application of 3Rs principle is very important to ensure the effective waste management. The requirements also need to improve and effective enforcement of the sub-decree on solid waste management applying not only to manage simple solid wastes and/or garbage properly, but also to manage hospital wastes, industrial wastes and other hazardous wastes such as waste of transportation, storage, disposal, treatment, etc. The implementation of policy and strategy on the transformation of waste to energy should be in place. Fiscal policy tools such as taxes on environmentally harmful substances and products *e.g.*, plastic bags, chemicals, hazardous waste, and nitrogen based fertilizers, can successfully limit the use of these unsustainable products while at the same time, create an environment favorable for green products and services. In the identification of proper dumping sites, the Cambodian government can also implement quantity-based pricing for domestic and municipal solid waste such as tipping fees to foster greater recycling practices.

Tourism has emerged as a major driver of economic growth with 2,125,465 international visitors in 2008, expected to increase in the next several years (MoT, 2008) and in the first semester of year 2009, received the international visitors 1,086,518. Actually, MoT plans in short to develop tourism projects along the GMS (there are 4 projects) and Core Environmental Program (there are 4 initiative projects) Indirect support will be provided to eco-tourism so the benefits of tourism are extended to the rural poor. However, Cambodia meets a significant challenge in terms of tourism development, lacking of adequate infrastructure for tourism zones, including proper waste management, sanitation, and lack of standards (standard price) for goods and souvenirs and the provision of electricity to some areas that enable the tourists to visit only in the day time. Security also represents another concern for further tourist destinations. In response, MoT has developed a strategic action plan to improve the tourism sector in order to accommodate increasing number of visitors to Cambodia. The action plan can highlight options for creating environmentally sustainable solutions to the challenges mentioned above, creating incentives for eco-tourism development, green-certification schemes for hotels, guesthouses and eco-lodges. Criteria could include energy-efficiency, composting of waste, use of renewables such as biogas, organic rice growing, and other initiatives.

Sustainability of Transportation

There is a need to cooperate and collaborate among line ministries towards work relevant to the current challenges *viz.*, (*i*) increase of vehicles on the streets leading to traffic congestion; (*ii*) lack of affordable and convenient public transport options; (*iii*) violations of traffic rules; (*iv*) little or no control over the number of vehicles (Phnom Penh has 2 million citizens and 1 million cars); (*v*) high single occupancy of vehicles; and (*vi*) lack of proper facilities for pedestrians in general (sidewalks, traffic lights, etc.). The methods identified to cope with such challenges include: further strengthen and effectively enforce the existing traffic law, particularly pertaining to proper helmets for motorcycles, drink and drive campaigns, use of seatbelts, and limited speed in the urban areas. Phnom Penh Municipality is strictly enforcing the traffic law, while promoting public transport and supporting the concept of side street walk. It is necessary to control the vehicle engines every month or year to ensure the quality and function of the engines and to ensure the environmentally sustainable transportation. The private-owned public transport should also be in place especially in the cities. It is necessary to promote vehicles that use gas or vegetable oil for transportation to minimize the carbon emission. Sustainability of transportation in the forest areas is also important especially to facilitate transport of forest products and by-products

Lack of Legal Framework

Cambodia lacks regulations and trained personnel for effectively monitoring the sustainable development activities. The existing legislation related to sustainable development needs to be reviewed and strengthened in order to reach the green development approaches. There are many laws and regulations developed, but no specific law on green growth and the dissemination of these laws to the people is still limited. There exists ineffective enforcement of laws and regulations. Furthermore, there is still limited understanding of the green growth from the stakeholders. The Ministry of Environment has a "core" set of responsibilities that must be implemented as part of the governmental mandate in order to introduce the concept of green growth.

Conclusion

In relation to all the above mentioned issues, Cambodia requires many development programmes to enhance performance of all sectors but lacks a lot of legal frameworks, specialized human resources, equipments and other resources at the present time. Besides, the country needs external assistance as development partners especially for financial and technical support in order to conduct such programmes and implementing the said action plans of the government.

Reference

http://www.cambodiajobs.biz/2011/11/national-development-economist-at.html.

www.ingramcontent.com/pod-product-compliance
Ingram Content Group UK Ltd.
Pitfield, Milton Keynes, MK11 3LW, UK
UKHW022215051026
15624UKWH00001B/167

9 789351 241713